AF508611

Policies and Strategies in Sexual and Reproductive Health

Policies and Strategies in Sexual and Reproductive Health

Editors

Juan Miguel Martínez Galiano
Miguel Delgado-Rodríguez

MDPI • Basel • Beijing • Wuhan • Barcelona • Belgrade • Manchester • Tokyo • Cluj • Tianjin

Editors
Juan Miguel Martínez Galiano
University of Jaen
Spain

Miguel Delgado-Rodríguez
University of Jaen
Spain

Editorial Office
MDPI
St. Alban-Anlage 66
4052 Basel, Switzerland

This is a reprint of articles from the Special Issue published online in the open access journal *International Journal of Environmental Research and Public Health* (ISSN 1660-4601) (available at: https://www.mdpi.com/journal/ijerph/special_issues/Policies_Strategies_Sexual_Reproductive_Health).

For citation purposes, cite each article independently as indicated on the article page online and as indicated below:

LastName, A.A.; LastName, B.B.; LastName, C.C. Article Title. *Journal Name* **Year**, *Volume Number*, Page Range.

ISBN 978-3-0365-0816-0 (Hbk)
ISBN 978-3-0365-0817-7 (PDF)

Contents

About the Editors

Juan Miguel Martínez Galiano Ph.D.) is currently a midwife at a public hospital and Professor at the University of Jaén (Spain). He holds a degree in nursing and midwifery, master's degree in Research and Advances in Preventive Medicine and Public Health, and Ph.D. degree, received in 2012 from the University of Granada for research on the impact of prenatal education programs on mother and newborn (cum laude). Galiano's research expertise is in sexual and reproductive health. He is a member of the Consortium for Biomedical Research in Epidemiology and Public Health (CIBERESP) and the group for Research, Development and Innovation in Epidemiology, Preventive Medicine and Surgery (CTS-435). He is also the principal investigator of numerous research projects with competitive financing.

Miguel Delgado-Rodríguez (Ph.D.) is Professor of Preventive Medicine and Public Health at the University of Jaén and a medical specialist in preventive medicine and public health. His Master of Public Health in Epidemiology was completed at UCLA. He has supervised more than 50 doctoral theses and published more than 400 articles and 200 books and book chapters. He has participated in more than 30 national and international projects with competitive financing. He has served as Scientific Director of the CIBERSP and coordinator of Public Health of the Health Research Fund (FIS) of the Carlos III Health Institute.

International Journal of
Environmental Research and Public Health

Editorial

The Relegated Goal of Health Institutions: Sexual and Reproductive Health

Juan Miguel Martínez-Galiano [1,2,*] **and Miguel Delgado-Rodríguez** [2,3]

1   Department of Nursing, University of Jaen, 23071 Jaén, Spain
2   Consortium for Biomedical Research in Epidemiology and Public Health (CIBERESP), 28029 Madrid, Spain; mdelgado@ujaen.es
3   Department of Health Sciences, University of Jaen, 23071 Jaén, Spain
*   Correspondence: jgaliano@ujaen.es

Academic Editor: Paul B. Tchounwou
Received: 5 February 2021; Accepted: 9 February 2021; Published: 11 February 2021

Sexual and reproductive health does not always receive the attention it deserves and frequently is not supported with the necessary resources to guarantee its maintenance. The World Health Organization (WHO) is aware of its importance and in its structure has specific programs focused on this health aspect.

Although conceptually sexual and reproductive health is seen as a united whole, this does not occur in some parts of the world. Sexual health, according to the WHO, is a state of physical, mental, and social well-being concerning sexuality. It requires a positive and respectful approach to sexuality and sexual relations, as well as the possibility of having pleasant and safe sexual experiences that are free from all coercion, discrimination, and violence. Reproductive health focuses on all aspects of the reproductive system, its functions, and its processes. [1]

The WHO aims to give a joint approach to sexual and reproductive health with which it intends to support the development of research and learning policies in this field. Even the United Nations General Assembly has established the specific goal of guaranteeing universal access to sexual and reproductive health services by 2030.

One of the biggest challenges in sexual and reproductive health is sexually transmitted infections (STI). For the development of well-oriented health programs on this issue, it is necessary to know the attitude of health professionals towards patients with STIs. One of the most relevant STIs, due to the absence of curative treatments and vaccines, is human immunodeficiency virus (HIV); the results of a study carried out in Spain with nursing students show that they have a positive attitude towards HIV patients, and this may improve their future healthcare practice in tackling this condition [2]. The identification of groups vulnerable to STIs has a special relevance for prevention programs; Perez-Morente et al. have found that immigrant women in periods of crisis are more susceptible to acquiring an STI [3,4]. Screening and diagnostic methods, improved treatments, as well as the availability of more effective vaccines also deserve special attention. Alderete has identified a new *Trichomonas vaginalis* epitope chain protein and the epitope chain proteins of infectious agents that may have potential applicability for a possible vaccine against this parasite [5]. Health education plays a special role in the implementation of STI prevention programs, for which it is necessary to identify risky practices. In this issue, Santa-Barbara et al. show in their results that the practice of oral sex, the practice of anal sex, and the non-use of condoms are clear risk factors. [6] Valid instruments are available, as shown by Sánchez-Mendoza et al., to determine the self-efficacy of condom use in adolescents, [7] one of the social strata with the highest risk of acquiring STIs. This favours the design of adequate strategies designed to better promote the use of condoms as a method of protection against STIs.

IJERPH **2021**, *18*, 1767; doi:10.3390/ijerph18041767

Family planning is another basic component of sexual and reproductive health. Contraceptive methods allow people to enjoy sexual acts without the worry of unwanted pregnancy and help in family planning. Adolescents are the most susceptible group to unwanted pregnancies, and this occurs in people residing in higher-risk environments or more disadvantaged places, as shown by the results of Ávila-Burgos et al. in Mexico. [8] Accessibility to contraceptive methods is critical for providing the correct medication and giving adequate instructions to avoid unwanted pregnancies. This is displayed in this issue by Langer et al. in a study carried out in pharmacies in Germany. [9]. In the case of having to perform an abortion due to an unwanted pregnancy or for any other reason, it must be carried out in safe conditions for the woman, and therefore it is necessary to implement training and awareness programs [10]. Besides this, it is a priority to have adequate legislation to guarantee the safety of women. Legal restrictions on abortion can increase maternal and infant morbidity and mortality. This is what Pabayo et al. conclude in a study performed in the United States in which infant mortality in states with restrictive abortion laws are compared with that of states with more permissive laws [11].

Women are the main recipients of health programs on sexual and reproductive health. They may suffer from chronic diseases, such as endometriosis, which causes chronic fatigue, a worse quality of life, greater anxiety, and worse sleep, among other consequences, as Mundo-López et al. [12] show. On the other hand, clinical and educational practices during pregnancy and childbirth influence the health status of women. For example, Cano-Ibáñez et al. [13] show that correct nutrition and promoting a dietary pattern based on the Mediterranean diet improve the outcome of childbirth. Another example is provided by Infante-Torres et al. in a systematic review with meta-analysis [14] in which they show that care without prolonged expulsive periods, since an excessive duration of this labour stage leads to greater neonatal morbidity, improves newborns' health. Postpartum post-traumatic stress disorder is an increasingly attention-grabbing disorder. For this reason, predictive models are useful, such as the one developed by Hernandez-Martinez et al. that helps to predict the risk of this disease [15]. Prenatal care is essential, but it is not practised in the same way in all areas or by all women. Thus, for example, in China women belonging to ethnic minorities and living in rural areas initiate pregnancy control later [16]. This necessitates the development of special programs for these situations. Not only must the woman be cared for, but her satisfaction with the healthcare process must also be taken into account. Instruments are available to gain knowledge of this aspect concerning attendance at delivery, such as the satisfaction scale with the assistance received, as validated by Pozo-Cano et al. [17].

Violence must be taken into account in reproductive health. The WHO in 2014 made a declaration for the prevention and eradication of disrespect and mistreatment during childbirth care in health centres; however, the figures of obstetric violence a few years later remain high [18]. Gender-based violence is a serious public health problem; it is exercised in various ways, with pregnancy being a clear risk factor. The umbrella review carried out by Roman-Galvez et al. [19] makes the magnitude of this problem visible and shows the high prevalence figures around the world. Another form of violence against women is female genital mutilation. Turkmani et al. present a conceptual model, based on cultural values and the physical and emotional needs of women, as a framework to guide maternity services to better address this problem [20]. The same authors conclude that empowering women and increasing their awareness of their health care rights can help involve women as active partners in the design and delivery of health information.

To progress in research and therefore in its application to the protection of sexual and reproductive health, it is necessary to take into account the social and health professionals, scientific societies, and institutions involved. García-Martin et al. stress the importance of cooperation between health professionals, academic and health institutions, and the community as an essential point to improve research quality and make significant advances in the field of women's health [21].

Every health program needs to address popular beliefs that do not have any scientific proof. For instance, some Chinese women maintain a preference for male children, which can cause a greater number of pregnancies with their associated risks [22].

In summary, it is necessary to develop policies and strategies tackling sexual and reproductive health in collaboration with the different agents involved to address its different components, such as family planning, STIs, women's health, pregnancy, and violence, among others. These policies must be translated into primary prevention, screening, early treatment, and rehabilitation programs. For the development of these programs, it is necessary to identify the most vulnerable populations, as well as their risk factors. They should also be based on scientific evidence and apply validated tools. Research in sexual and reproductive health should be promoted so that its results are the basis for the design of appropriate policies and strategies in sexual and reproductive health.

Author Contributions: Both coauthors contributed equally to this work. Both authors have read and agreed to the published version of the manuscript.

Funding: This research received no external funding.

Conflicts of Interest: The authors declare no conflict of interest.

References

1. OMS. La Salud Sexual Y Su Relación Con La Salud Reproductiva: Un Enfoque Operativo. Available online: https://apps.who.int/iris/bitstream/handle/10665/274656/9789243512884-spa.pdf (accessed on 30 January 2021).

2. Álvarez-Serrano, M.A.; Martínez-García, E.; Martín-Salvador, A.; Gázquez-López, M.; Pozo-Cano, M.D.; Caparrós-González, R.A.; Pérez-Morente, M.Á. Spanish Nursing Students' Attitudes toward People Living with HIV/AIDS: A Cross-Sectional Survey. *Int. J. Environ. Res. Public Health* **2020**, *17*, 8672. [CrossRef]

3. Pérez-Morente, M.Á.; Gázquez-López, M.; Álvarez-Serrano, M.A.; Martínez-García, E.; Femia-Marzo, P.; Pozo-Cano, M.D.; Martín-Salvador, A. Sexually Transmitted Infections and Associated Factors in Southeast Spain: A Retrospective Study from 2000 to 2014. *Int. J. Environ. Res. Public Health* **2020**, *17*, 7449. [CrossRef] [PubMed]

4. Pérez-Morente, M.Á.; Martín-Salvador, A.; Gázquez-López, M.; Femia-Marzo, P.; Pozo-Cano, M.D.; Hueso-Montoro, C.; Martínez-García, E. Economic Crisis and Sexually Transmitted Infections: A Comparison Between Native and Immigrant Populations in a Specialised Centre in Granada, Spain. *Int. J. Environ. Res. Public Health* **2020**, *17*, 2480. [CrossRef]

5. Alderete, J.F. Advancing Prevention of STIs by Developing Specific Serodiagnostic Targets: Trichomonas vginalis as a Model. *Int. J. Environ. Res. Public Health* **2020**, *17*, 5783. [CrossRef] [PubMed]

6. Santa-Bárbara, R.C.; Hueso-Montoro, C.; Martín-Salvador, A.; Álvarez-Serrano, M.A.; Gázquez-López, M.; Pérez-Morente, M.Á. Association between Sexual Habits and Sexually Transmitted Infections at a Specialised Centre in Granada (Spain). *Int. J. Environ. Res. Public Health* **2020**, *17*, 6881. [CrossRef]

7. Sanchez-Mendoza, V.; Soriano-Ayala, E.; Vallejo-Medina, P. Psychometric Properties of the Condom Use Self-Efficacy Scale among Young Colombians. *Int. J. Environ. Res. Public Health* **2020**, *17*, 3762. [CrossRef]

8. Avila-Burgos, L.; Montañez-Hernández, J.C.; Cahuana-Hurtado, L.; Villalobos, A.; Hernández-Peña, P.; Heredia-Pi, I. Government Expenditure on Maternal Health and Family Planning Services for Adolescents in Mexico, 2003–2015. *Int. J. Environ. Res. Public Health* **2020**, *17*, 3097. [CrossRef]

9. Langer, B.; Grimm, S.; Lungfiel, G.; Mandlmeier, F.; Wenig, V. The Quality of Counselling for Oral Emergency Contraceptive Pills—A Simulated Patient Study in German Community Pharmacies. *Int. J. Environ. Res. Public Health* **2020**, *17*, 6720. [CrossRef] [PubMed]

10. Sanitya, R.; Marshall, A.I.; Saengruang, N.; Julchoo, S.; Sinam, P.; Suphanchaimat, R.; Phaiyarom, M.; Tangcharoensathien, V.; Boonthai, N.; Chaturachinda, K. Healthcare Providers' Knowledge and Attitude Towards

Abortions in Thailand: A Pre-Post Evaluation of Trainings on Safe Abortion. *Int. J. Environ. Res. Public Health* **2020**, *17*, 3198. [CrossRef] [PubMed]

11. Pabayo, R.; Ehntholt, A.; Cook, D.M.; Reynolds, M.; Muennig, P.; Liu, S.Y. Laws Restricting Access to Abortion Services and Infant Mortality Risk in the United States. *Int. J. Environ. Res. Public Health* **2020**, *17*, 3773. [CrossRef]

12. Mundo-López, A.; Ocón-Hernández, O.; San-Sebastián, A.P.; Galiano-Castillo, N.; Rodríguez-Pérez, O.; Arroyo-Luque, M.S.; Arroyo-Morales, M.; Cantarero-Villanueva, I.; Fernández-Lao, C.; Artacho-Cordón, F. Contribution of Chronic Fatigue to Psychosocial Status and Quality of Life in Spanish Women Diagnosed with Endometriosis. *Int. J. Environ. Res. Public Health* **2020**, *17*, 3831. [CrossRef]

13. Cano-Ibáñez, N.; Martínez-Galiano, J.M.; Luque-Fernández, M.A.; Martín-Peláez, S.; Bueno-Cavanillas, A.; Delgado-Rodríguez, M. Maternal Dietary Patterns during Pregnancy and Their Association with Gestational Weight Gain and Nutrient Adequacy. *Int. J. Environ. Res. Public Health* **2020**, *17*, 7908. [CrossRef]

14. Infante-Torres, N.; Molina-Alarcón, M.; Arias-Arias, A.; Rodríguez-Almagro, J.; Hernández-Martínez, A. Relationship Between Prolonged Second Stage of Labor and Short-Term Neonatal Morbidity: A Systematic Review and Meta-Analysis. *Int. J. Environ. Res. Public Health* **2020**, *17*, 7762. [CrossRef]

15. Hernández-Martínez, A.; Martínez-Vazquez, S.; Rodríguez-Almagro, J.; Delgado-Rodríguez, M.; Martínez-Galiano, J.M. Elaboration and Validation of Two Predictive Models of Postpartum Traumatic Stress Disorder Risk Formed by Variables Related to the Birth Process: A Retrospective Cohort Study. *Int. J. Environ. Res. Public Health* **2021**, *18*, 92. [CrossRef] [PubMed]

16. Yan, C.; Tadadej, C.; Chamroonsawasdi, K.; Chansatitporn, N.; Sung, J.F. Ethnic Disparities in Utilization of Maternal and Child Health Services in Rural Southwest China. *Int. J. Environ. Res. Public Health* **2020**, *17*, 8610. [CrossRef]

17. Pozo-Cano, M.D.; Martín-Salvador, A.; Pérez-Morente, M.Á.; Martínez-García, E.; Luna del Castillo, J.d.D.; Gázquez-López, M.; Fernández-Castillo, R.; García-García, I. Validation of the Women's Views of Birth Labor Satisfaction Questionnaire (WOMBLSQ4) in the Spanish Population. *Int. J. Environ. Res. Public Health* **2020**, *17*, 5582. [CrossRef] [PubMed]

18. Martínez-Galiano, J.M.; Martinez-Vazquez, S.; Rodríguez-Almagro, J.; Hernández-Martinez, A. The magnitude of the problem of obstetric violence and its associated factors: A cross-sectional study. *Women Birth* **2020**. [CrossRef]

19. Román-Gálvez, R.M.; Martín-Peláez, S.; Martínez-Galiano, J.M.; Khan, K.S.; Bueno-Cavanillas, A. Prevalence of Intimate Partner Violence in Pregnancy: An Umbrella Review. *Int. J. Environ. Res. Public Health* **2021**, *18*, 707. [CrossRef]

20. Turkmani, S.; Homer, C.S.E.; Dawson, A.J. Understanding the Experiences and Needs of Migrant Women Affected by Female Genital Mutilation Using Maternity Services in Australia. *Int. J. Environ. Res. Public Health* **2020**, *17*, 1491. [CrossRef] [PubMed]

21. García-Martín, M.; Amezcua-Prieto, C.; H Al Wattar, B.; Jørgensen, J.S.; Bueno-Cavanillas, A.; Khan, K.S. Patient and Public Involvement in Sexual and Reproductive Health: Time to Properly Integrate Citizen's Input into Science. *Int. J. Environ. Res. Public Health* **2020**, *17*, 8048. [CrossRef]

22. Wang, X.; Nie, W.; Liu, P. Son Preference and the Reproductive Behavior of Rural-Urban Migrant Women of Childbearing Age in China: Empirical Evidence from a Cross-Sectional Data. *Int. J. Environ. Res. Public Health* **2020**, *17*, 3221. [CrossRef] [PubMed]

Publisher's Note: MDPI stays neutral with regard to jurisdictional claims in published maps and institutional affiliations.

 International Journal of
Environmental Research and Public Health

Article

Elaboration and Validation of Two Predictive Models of Postpartum Traumatic Stress Disorder Risk Formed by Variables Related to the Birth Process: A Retrospective Cohort Study

Antonio Hernández-Martínez [1], Sergio Martínez-Vazquez [2], Julián Rodríguez-Almagro [1,*], Miguel Delgado-Rodríguez [3,4] and Juan Miguel Martínez-Galiano [2,4]

[1]   Department of Nursing, Physiotherapy and Occupational Therapy, University of Castilla-La Mancha, 13600 Ciudad Real, Spain; Antonio.HMartinez@uclm.es
[2]   Nursing Department, University of Jaen, 23071 Jaen, Spain; sergio_sk8_9@hotmail.com (S.M.-V.); jgaliano@ujaen.es (J.M.M.-G.)
[3]   Division of Preventive Medicine and Public Health, University of Jaén, 23071 Jaén, Spain; mdelgado@ujaen.es
[4]   Consortium for Biomedical Research in Epidemiology and Public Health (CIBERESP), 28029 Madrid, Spain
*   Correspondence: julianj.rodriguez@uclm.es; Tel.: +34-676-683-843

Received: 31 October 2020; Accepted: 21 December 2020; Published: 24 December 2020

Abstract: This study aimed to develop and validate two predictive models of postpartum post-traumatic stress disorder (PTSD) risk using a retrospective cohort study of women who gave birth between 2018 and 2019 in Spain. The predictive models were developed using a referral cohort of 1752 women (2/3) and were validated on a cohort of 875 women (1/3). The predictive factors in model A were delivery type, skin-to-skin contact, admission of newborn to care unit, presence of a severe tear, type of infant feeding at discharge, postpartum hospital readmission. The area under curve (AUC) of the receiver operating characteristic (ROC) in the referral cohort was 0.70 (95% CI: 0.67–0.74), while in the validation cohort, it was 0.69 (95% CI: 0.63–0.75). The predictive factors in model B were delivery type, admission of newborn to care unit, type of infant feeding at discharge, postpartum hospital readmission, partner support, and the perception of adequate respect from health professionals. The predictive capacity of model B in both the referral cohort and the validation cohort was superior to model A with an AUC-ROC of 0.82 (95% CI: 0.79–0.85) and 0.83 (95% CI: 0.78–0.87), respectively. A predictive model (model B) formed by clinical variables and the perception of partner support and appropriate treatment by health professionals had a good predictive capacity in both the referral and validation cohorts. This model is preferred over the model (model A) that was formed exclusively by clinical variables.

Keywords: post-traumatic stress disorder; predictive model; validation

---

1. Introduction

Post-traumatic stress disorder (PTSD) has been described as "the complex somatic, cognitive, affective, and behavioral effects of psychological trauma" [1]. PTSD affects the newborn, the mother–child relationship, and the mother's health and quality of life [2–4]. PTSD prevalence can vary considerably, and in a systematic review including 28 studies the average prevalence was 4.0% in the general perinatal population and 18.5% in women at risk [5].

Several factors have been associated with the development of postpartum PTSD [6]. However, research into the obstetric factors involved in PTSD development is scarce. Most studies focus on the influence of birth type [7–15] and newborn variables such as prematurity [16,17], low weight [16], low Apgar scores [16,18], type of newborn feeding [7], and newborn hospitalization [9]. Several studies have recently observed a relationship between certain obstetric practices and the risk of PTSD [2,15].

One of the biggest challenges in health is creating tools for predicting the risk of particular health problems. The purpose of these tools is the early identification of those most susceptible to developing the problem or anticipating the appearance of the first symptoms. Several published prediction models exist for PTSD after childbirth, using different variables and populations [9,18–27]. Although several models exist, only one studied the prediction capacity with ROC curves (19), and none were validated in populations other than those used to create the model. No prediction tool has been made based on the factors related to either the birth process or obstetric practices, and that can be easily used as a screening tool by health professionals. One of the difficulties in creating these prediction models is the choice of initial predictive factors, which may be purely objective or also subjective. We consider that predictors of an objective clinical nature, such as the clinical data of problems and interventions during pregnancy and childbirth, are easily obtainable but may be insufficient to predict complex phenomena such as PTSD. Conversely, the use of variables of a more subjective nature such as anxiety, emotions, depression, among others, although used in various predictive models [19,20,22–24,26], is more complex to use. Many of these subjective variables are assessed through scales and questionnaires, some of them long and complex, limiting their application in clinical practice.

Therefore, the objective of this study was to develop and validate two postpartum PTSD risk prediction models, one based exclusively on objective clinical predictive factors and another including subjective factors such as perceptions of partner support and treatment received by professionals.

2. Materials and Methods

An observational study was conducted using a retrospective cohort of women who gave birth between 2018 and 2019. The Research Ethics Committee of the province of Jaen approved this study with reference number TD-VCDEPP-2019/1417-N-19. The main tool employed to collect the relevant data for this study was medical records. The women were required to read an information sheet about the study and its objectives and check a box in which they showed their consent to participate in it; that is, they signed an ad hoc digital informed consent. The STROBE statement has been followed in the reporting of this study. [28]

2.1. Design and Participants

This analytical and observational study used a retrospective cohort of women who gave birth between 2018 and 2019. A total of 1752 women were included in the referral cohort for the predictive models, which were subsequently validated with a cohort of 875 women. The Clinical Research Ethics Committees of Universidad de Jaen gave ethical approval prior to the start of the study. All of the participants received written information on the study, including the fact that participation was completely voluntary and anonymous.

We used the maximum modeling principle to estimate the sample size. We needed 10 events (women at risk of PTSD) per each incorporated variable [29]. If we consider that our initial model may contain 15 variables, we would need 150 women at risk of PTSD. Taking into account that the prevalence of PTSD in other Spanish studies is reported as 10.6% [2], we would need a minimum of 1415 women for the derivation cohort and at least half as many for validation, about 708 women. Nevertheless, the researchers' team opted to recruit the maximum number of women.

2.2. Data Collection and Information Sources

The main tool employed to collect the relevant data for this study was medical records. The primary outcome variable—risk of PTSD—and the following objective independent variables were collected from the medical records.

Independent variables were:

1. Maternal: maternal age, education level, nationality, attendance at maternal education classes, and the use of a birth plan.
2. Obstetric: previous cesarean section (CS), number of deliveries, induction of labor, type of labor, use of regional analgesia, use of general anesthesia, use of natural analgesic methods, episiotomy, and perineal tear.
3. Fetal: prematurity, twin pregnancy, breastfeeding in the first hour, skin-to-skin contact, admission of the newborn to care unit, type of feeding at hospital discharge.
4. Subjective variables evaluated with a Likert-type scale (scores 1–5): degree of support from the partner during pregnancy, delivery, and postpartum; respectful treatment by professionals during pregnancy, delivery, and postpartum. The different categories used for each variable are detailed in Table 1.

Table 1. Bivariate analysis of potential predictive factors of post-traumatic stress disorder (PTSD) risk.

Predictor	PTSD (Score PPQ) Derivation Cohort		*p*-Value
	<19 Points *n* (%)	≥19 Points *n* (%)	
Maternal age			0.199
35 years	638 (84.6)	116 (15.4)	
>35 years	886 (86.8)	132 (13.2)	
Education level			0.401
Primary school	22 (75.9)	7 (24.1)	
Secondary school	77 (87.5)	11 (12.5)	
High school	323 (85.0)	57 (15.0)	
University	1082 (86.2)	173 (13.8)	
Nationality			0.986
Spanish	1443 (85.8)	238 (14.2)	
Other	61 (85.9)	10 (14.1)	
Parity			<0.001
Primiparous	1004 (83.0)	205 (17.0)	
Multiparous	499 (92.1)	43 (7.9)	
Live newborn			0.056
No	8 (66.7)	4 (33.3)	
Yes	1496 (86.0)	244 (14.0)	
Twin pregnancy			0.319
No	1471 (86.0)	240 (14.0)	
Yes	33 (80.5)	8 (19.5)	
Previous cesarean section			<0.001
No	1102 (89.7)	126 (10.3)	
Yes	402 (76.7)	122 (23.3)	
Place of birth			0.099
Public hospital	1210 (86.2)	194 (13.8)	
Private hospital	262 (83.2)	53 (16.8)	
Midwife-led hospital	6 (85.7)	1 (14.3)	
Home	26 (100.0)	0 (0.0)	
Labor induction			0.007
No	913 (87.7)	128 (12.3)	
Yes	591 (83.1)	120 (16.9)	
Regional analgesia			0.001
No	442 (90.4)	47 (9.6)	
Yes	1062 (84.1)	201 (15.9)	
General anesthesia			<0.001
No	1461 (86.5)	228 (13.5)	
Yes	43 (68.3)	20 (31.7)	
Natural analgesia			0.139
No	1214 (85.3)	210 (14.7)	
Yes	290 (88.4)	38 (11.6)	

Table 1. *Cont.*

Table 1. *Cont.*

Predictor	PTSD (Score PPQ) Derivation Cohort		*p*-Value
	<19 Points *n* (%)	≥19 Points *n* (%)	
Type of birth			**<0.001**
Normal vaginal delivery	912 (91.6)	84 (8.4)	
Instrumental	273 (84.0)	52 (16.0)	
Elective CS	109 (84.5)	20 (15.5)	
Emergency CS	210 (69.5)	92 (30.5)	
Episiotomy			0.925
No	1069 (85.8)	177 (14.2)	
Yes	435 (86.0)	71 (14.0)	
Perineal tear			**<0.001**
No	940 (83.9)	180 (16.1)	
Mild	512 (90.9)	51 (9.1)	
Severe (III–IV)	52 (75.4)	17 (24.6)	
Prematurity			0.023
No	1141 (86.4)	223 (13.6)	
Yes	93 (78.8)	25 (21.2)	
Maternal antenatal classes			0.123
No	295 (87.5)	42 (12.5)	
Yes (less than 5 classes)	208 (81.9)	46 (18.1)	
Yes (more than 5 classes)	1001 (86.2)	160 (13.8)	
Breastfeeding 1 h after childbirth			**<0.001**
No	338 (76.0)	107 (24.0)	
Yes	1166 (89.2)	141 (10.8)	
Skin-to-skin contact			**<0.001**
No	302 (73.1)	111 (26.9)	
Yes	1202 (89.8)	137 (10.2)	
Birth plan			**<0.001**
No	803 (87.8)	112 (12.2)	
Yes, but not respected	164 (65.3)	87 (34.7)	
Yes, and was respected	537 (91.6)	49 (8.4)	
Admission of the newborn to care unit			**<0.001**
No	1324 (87.6)	187 (12.4)	
Yes	180 (74.7)	61 (25.3)	
Hospital length of stay			**<0.001**
1 day	122 (91.0)	12 (9.0)	
2 day	779 (90.2)	85 (9.8)	
3 day	365 (82.4)	78 (17.6)	
4 days or more	238 (76.5)	73 (23.5)	
Infant feeding on discharge			**<0.001**
Maternal	1226 (88.2)	164 (11.8)	
Mixed	233 (78.5)	64 (21.5)	
Artificial	45 (69.2)	20 (30.8)	
Postpartum surgical intervention			0.001
No	1449 (86.5)	227 (13.5)	
Yes	55 (72.4)	21 (27.6)	
Hospital readmission			**<0.001**
No	1474 (86.4)	232 (13.6)	
Yes	30 (65.2)	16 (34.8)	
	Mean (SD)	Mean (SD)	
Perception of adequate treatment by health professionals during pregnancy, childbirth and the puerperium. Likert scale 1–5	3.4 (0.93)	2.88 (1.28)	**<0.001 ***
Perception of support by the couple during pregnancy, childbirth and the puerperium. Likert scale 1–5	2.99 (0.97)	1.67 (1.22)	**<0.001 ***

Bold: statistically significant differences. * Student–Fisher *t*-test. PPQ: Perinatal Post-Traumatic Stress Disorder Questionnaire.

The primary outcome variable, risk of PTSD, was determined using the modified Perinatal Post-Traumatic Stress Disorder Questionnaire (PPQ) [30] (Appendix A: Spanish version). The PPQ is a 14-item measure assessing post-traumatic symptoms related to the childbirth experience, including intrusiveness or re-experiencing, avoidance behaviors, and hyperarousal or numbing of responsiveness. The PPQ also contains one item about feelings of guilt. Response options were modified from the original dichotomous scale to a five-level Likert scale (scored 0 to 4). The total possible score on the modified PPQ ranged from 0 to 56. In the current study, internal consistency was higher than in previous investigations using the dichotomous scaling, with an $\alpha = 0.90$ [30].

2.3. Statistical Analysis

A descriptive statistical analysis using absolute and relative frequencies for qualitative variables and means and standard deviation (SD) for quantitative variables was performed.

The analysis of potential predictive factors, which have been previously identified in the literature as risk factors of delayed onset breastfeeding, was carried out in a bivariate analysis using the chi-square and Student's *t*-test to estimate qualitative and quantitative variables, respectively. Of these variables, and following Lemeshow's statistical criteria, associations with p-values < 0.25 were selected for inclusion in the multivariate binary logistic regression model [31,32] (Table 1). These analyses were performed in the derivation cohort.

Then, two models were created (Table 2): model A based on exclusively clinical criteria and model B based on clinical criteria plus maternal perceptions of the degree of partner support and the treatment received by healthcare professionals. These models were constructed using backward elimination (RV in SPSS) with the derivation of cohort women's data. To assess the prediction qualitatively, we used Swets's criteria, which uses the following category values: 0.5–0.6 (bad), 0.6–0.7 (poor), 0.7–0.8 (satisfactory), 0.8–0.9 (good), and 0.9–1.0 (excellent) [33]. In addition, the Nagerlkerkes R-square and the calibration were determined using the Hosmer–Lemeshow test p-value of both models.

Table 2. Predictive models of PTSD risk during the postpartum.

Model Properties	Model A			Model B		
Number of Events in Derivation Cohort	248 (14.2%)					
Number of Events in Validation Cohort	95 (10.9%)					
Nagerlkerkes R-Square	0.127			0.310		
Hosmer–Lemeshow Test p-Value	0.133			0.732		
Risk Factor	**Coef * Beta Value**	**OR (95% CI)**	**p-Value**	**Coef * Beta Value**	**OR (95% CI)**	**p-Value**
Type of birth						
Normal vaginal delivery		1 (Ref)			1 (Ref)	
Instrumental	0.484	**1.62 (1.10–2.41)**	0.016	0.344	1.22 (0.81–1.86)	0.344
Elective CS	0.341	1.41 (0.77–2.57)	0.267	0.200	1.22 (0.68–2.18)	0.499
Emergency CS	1.121	**3.07 (1.96–4.80)**	<0.001	0.827	**2.29 (1.56–3.35)**	<0.001
Initiate skin-to-skin contact	−0.428	**0.65 (0.45–0.96)**	0.028			
Admission of the newborn to care unit						
No		1 (Ref)			1 (Ref)	
Yes	0.452	**1.57 (1.26–2.87)**	0.015	0.503	**1.65 (1.12–2.44)**	0.012
Perineal tear						
No		1 (Ref)				
Type I–II	−0.020	0.98 (0.67–1.44)	0.919			
Type III–IV	0.795	**2.21 (1.17–4.19)**	0.015			
Infant feeding on discharge						
Maternal		1 (Ref)			1 (Ref)	
Mixed	0.404	**1.50 (1.06–2.12)**	0.022	0.122	1.13 (0.77–1.65)	0.530
Artificial	0.740	**2.10 (1.16–3.79)**	0.014	0.803	**2.23 (1.13–4.04)**	0.021
Hospital readmission	0.934	**2.55 (1.30–5.00)**	0.007	1.160	**3.19 (1.43–7.11)**	0.005
Partner's perception of support (Likert scale 1–5)				−0.234	**0.79 (0.69–0.91)**	0.001
Perception of respect by professionals (Likert scale 1–5)				−0.863	**0.42 (0.37–0.48)**	<0.001
Constant	−2.177			0.545		
AUC-ROC derivation cohort		0.70 (0.67–0.74)	<0.001		0.82 (0.79–0.85)	<0.001
AUC-ROC validation cohort		0.69 (0.63–0.75)	<0.001		0.83 (0.78–0.87)	<0.001

OR: odds ratio; Bold: statistically significant differences.

The derivation and validation cohorts were compared after using chi-square and Student's *t*-test for qualitative and quantitative variables, respectively (Table 3). Finally, the AUC-ROC in the validation cohort was estimated for the predictive model that we created (Table 2). In this case, the probabilities used proceed from applying the predictive model created with the derivation cohort using the data of the women in the validation cohort. SPSS 20.0. (SPSS Inc., Chicago, IL, USA) was used for all statistical analyses.

Table 3. Comparison of characteristics between the derivation and validation cohort.

Characteristics	Derivation Cohort N = 1752 n (%)	Validation Cohort N = 875 n (%)	p-Value *
PPQ			**0.018**
<19	1504 (85.8)	780 (89.1)	
≥19	248 (14.2)	95 (10.9)	
Maternal age			0.930
≤35 years	754 (43.0)	375 (42.9)	
>35 years	998 (57.0)	500 (57.1)	
Education level			0.478
Primary school	29 (1.7)	9 (28.6)	
Secondary school	88 (5.0)	37 (4.2)	
High school	380 (21.7)	193 (22.1)	
University	1255 (71.6)	636 (72.7)	
Nationality			0.351
Spanish	1681 (95.9)	846 (96.7)	
Other	71 (4.1)	29 (3.3)	
Parity			0.092
Primiparous	1209 (69.0)	575 (65.8)	
Multiparous	542 (31.0)	299 (34.2)	
Live newborn			0.130
No	12 (0.7)	2 (0.2)	
Yes	1740 (86.0)	873 (99.8)	
Twin pregnancy			0.333
No	1711 (97.7)	849 (97.0)	
Yes	41 (2.3)	26 (3.0)	
Previous cesarean section			0.167
No	1228 (70.1)	636 (72.7)	
Yes	524 (29.9)	239 (27.3)	
Place of birth			0.526
Public hospital	1404 (80.1)	697 (79.7)	
Private hospital	315 (18.0)	155 (17.7)	
Midwife-led hospital	7 (0.4)	3 (0.3)	
Home	26 (1.5)	20 (2.3)	
Labor induction			0.213
No	1041 (59.4)	542 (61.9)	
Yes	711 (40.6)	333 (38.1)	
Regional analgesia			0.413
No	489 (27.9)	231 (26.4)	
Yes	1263 (72.1)	644 (73.6)	
General anesthesia			0.404
No	1689 (96.4)	849 (97.0)	
Yes	63 (3.6)	26 (3.0)	
Natural analgesia			0.768
No	1424 (81.3)	707 (80.8)	
Yes	328 (18.7)	168 (19.2)	
Type of birth			0.152
Normal vaginal delivery	996 (56.8)	536 (61.3)	
Instrumental	325 (18.6)	146 (16.7)	
Elective CS	129 (7.4)	64 (7.3)	
Emergency CS	302 (17.2)	129 (314.7)	

Table 3. *Cont.*

Characteristics	Derivation Cohort N = 1752 n (%)	Validation Cohort N = 875 n (%)	p-Value *
Episiotomy			0.965
No	1246 (71.1)	623 (71.2)	
Yes	506 (28.9)	252 (28.8)	
Perineal tear			0.157
No	1120 (63.9)	529 (60.5)	
Mild	563 (32.1)	314 (35.9)	
Severe (III–IV)	69 (3.9)	32 (3.7)	
Prematurity			0.821
No	1634 (93.3)	814 (93.0)	
Yes	118 (6.7)	61 (7.0)	
Maternal antenatal classes			0.133
No	337 (19.2)	185 (21.1)	
Yes (less than 5 classes)	254 (14.5)	104 (11.9)	
Yes (more than 5 classes)	1161 (66.3)	586 (67.0)	
Breastfeeding 1 h after childbirth			0.556
No	445 (25.4)	213 (24.3)	
Yes	1307 (74.6)	662 (75.7)	
Skin-to-skin contact			0.422
No	413 (23.6)	194 (22.2)	
Yes	1339 (76.4)	681 (77.8)	
Birth plan			0.739
No	915 (52.2)	459 (52.5)	
Yes, but not respected	251 (14.3)	116 (13.3)	
Yes, and was respected	586 (33.4)	300 (34.3)	
Admission of the newborn to care unit			0.886
No	1511 (86.2)	753 (86.1)	
Yes	241 (13.8)	122 (13.9)	
Hospital length of stay			0.987
1 day	134 (7.6)	69 (7.9)	
2 day	864 (49.3)	434 (49.6)	
3 day	443 (25.3)	216 (24.7)	
4 days or more	311 (17.8)	156 (17.8)	
Infant feeding on discharge			
Maternal	1390 (79.3)	681 (77.8)	0.563
Mixed	297 (17.0)	163 (18.6)	
Artificial	65 (3.7)	31 (3.5)	
Postpartum surgical intervention			
No	1676 (95.7)	844 (96.5)	0.331
Yes	76 (4.3)	31 (3.5)	
Hospital readmission			
No	1706 (97.4)	128 (12.3)	0.480
Yes	46 (2.6)	19 (2.2)	
	Mean (SD)	**Mean (SD)**	
Perception of adequate treatment by health professionals during pregnancy, childbirth, and the puerperium. Likert scale 1–5	3.33 (1.00)	3.28 (1.04)	0.334
Perception of support by the couple during pregnancy, childbirth, and the puerperium. Likert scale 1–5	2.80 (1.11)	2.83 (1.26)	0.493

* Student–Fisher *t*-test.

3. Results

Characteristics of Participants

The derivation cohort consisted of 1752 women and the validation cohort 875 women, with a prevalence of PTSD risk of 14.2% (248) and 10.9% (95), respectively. First, we built predictive models using the derivation cohort. The variables associated with the risk of PTSD (screening criterion *p*-value < 0.25) selected for the multivariate analysis were: maternal age, parity, live birth, place of delivery, induced delivery, use of natural methods for pain, regional analgesia, general anesthesia, type of delivery, perineal tear, skin-to-skin contact, breastfeeding the first hour of life, admission of the newborn to care unit, hospital stay, breastfeeding at discharge, postpartum surgical intervention, postpartum readmission, degree of partner support during pregnancy, delivery and postpartum, and degree of respect received from professionals during pregnancy, delivery and postpartum. (Table 1).

Then, two predictive models were created. Model A was based exclusively on clinical variables, and model B consisted of clinical variables plus subjective variables on support received from their partner and treatment received from healthcare professionals (see Table 2). The variables to be included in the final predictive models were selected automatically by the SPSS program, through backward step instruction.

When performing the multivariate analysis, model A included the following variables: type of delivery, skin-to-skin contact, admission of the newborn to care unit, perineal tear, type of infant feeding at discharge, and postpartum hospital readmission. The predictive capacity (AUC-ROC) in the referral cohort was 0.70 (95% CI: 0.67–0.74) (Figure 1), while in the validation cohort it was 0.69 (95% CI: 0.63–0.75) (Figure 2), which is considered as satisfactory in Swets's criteria.

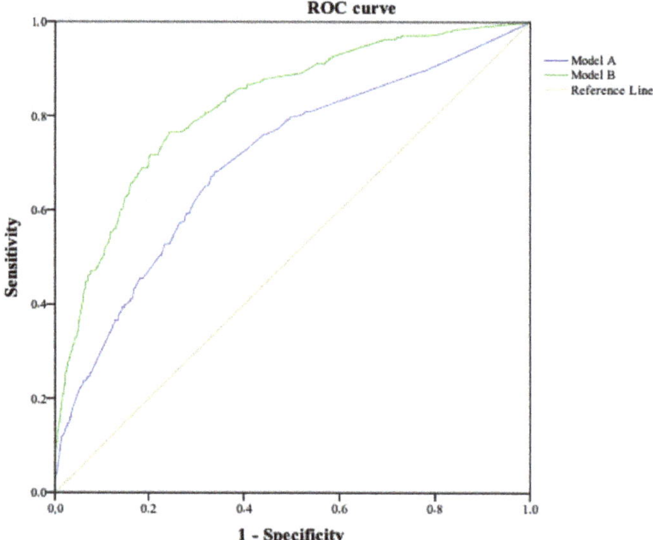

Figure 1. Predictive capacity of model A and model B in the derivation cohort. Area under the ROC curve to determine the predictive ability of the model in the validation cohort, representing the sensitivity on the *y* axis and 1- specificity in the *x* axis.

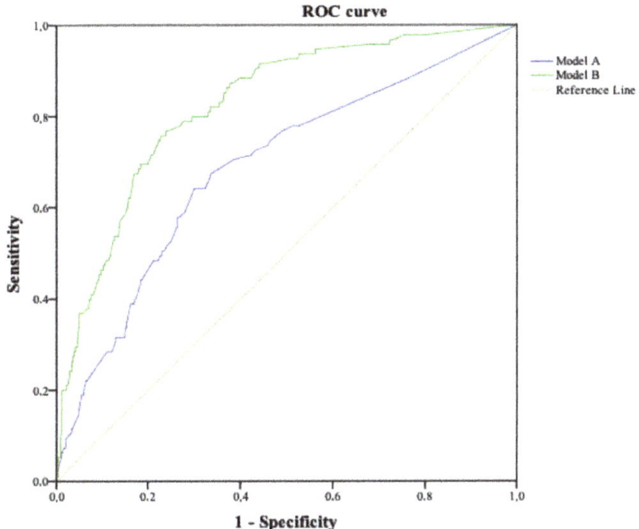

Figure 2. Predictive capacity of model A and model B in the validation cohort. Area under the ROC curve to determine the predictive ability of the model in the validation cohort, representing the sensitivity on the *y* axis and 1-specificity on the *x* axis.

The predictive factors in the final model B were: type of delivery, admission of the newborn to care unit, type of infant feeding at discharge, postpartum hospital readmission, support received by the partner, and the perception of respect from healthcare professionals. The predictive capacity (AUC-ROC) in the derivation cohort was 0.82 (95% CI: 0.79–0.85) (Figure 1), while in the validation cohort it was 0.83 (95% CI: 0.78–0.87) (Figure 2). This predictive capacity is considered good per Swets's criteria. Finally, we examined comparability issues in both cohorts, and found no statistically significant differences with any variable except for the risk of PTSD (p = 0.018), which was 14.2% (248) in the referral cohort and 10.9% (95) in the validation cohort (see Table 3).

4. Discussion

This study presents the main results of the development of two postpartum PTSD risk prediction models. Model A, constructed using only clinical variables, presented a satisfactory predictive capacity (AUC-ROC = 0.70), while model B, constructed with clinical variables and subjective patient perceptions, presented a good predictive capacity (AUC-ROC = 0.82). The predictive variables common to both models were: type of delivery, admission of the newborn to care unit, type of infant feeding at discharge, and postpartum hospital readmission. However, model A also included skin-to-skin contact and the presence of a severe tear as exclusive variables. In contrast, model B included variables relating to the support received from the partner and the perception of respect from healthcare professionals during childbirth.

Currently, there are several studies published with PTSD risk prediction models [9,19–27]. However, only the study by Van Heumen et al. studied the prediction capacity with ROC curves [19], presenting an AUC-ROC of 0.795, lower than our best model. Moreover, none have been validated in populations other than those used to create the model, which is an important limitation. The sample sizes were also all smaller than ours, and only one exceeded 1000 subjects [19], and only three studies exceeded 500 subjects [9,19,20]. Some of these studies have been carried out on very specific population groups, such as the study by López et al. [21], who used a sample of women who had a cesarean delivery, excluding most women who give birth vaginally. Other models have included other scales and assessments based on questionnaires of anxiety, emotions, depression, among others, as predictive factors [19,20,22–24,26]. The use of multiple questionnaires and scales could hinder their application due to the complexity in obtaining this information and does not allow universal use because these scales and questionnaires were designed and validated to be used in specific populations. In terms of obstetric predictors, only three authors have included this type of clinical variable. Concerning these models, we agree on the inclusion of the variable "type of delivery" as a predictive factor of PTSD risk [9,18,23]; specifically, instrumental delivery [9] and emergency cesarean section [9,18] as they present forms of childbirth with the greatest risks. Although multiple studies associate the type of delivery with the risk of PTSD [7–15], and the presence of perineal tears [13,18,34], no predictive models have been developed that include these.

Regarding neonatal variables as factors that influence the risk of PTSD, various variables related to the newborn were identified, including the newborn's hospital admission [9]. Along similar lines, the risk was also related to the lack of skin-to-skin contact and formula feeding. Although these variables have not been included in other models, they have been related to an increased risk of PTSD in other studies [2,7,15,35]. Although these three variables are related to each other, the authors believe that they also have a partially independent effect. In the first place, not all hospitalized children stop skin-to-skin contact, as in many cases, admission occurs several hours after birth. Second, many women whose children are hospitalized continue to breastfeed despite the great obstacle it poses. The predictive model of Fairtbrother et al. [18] also includes low birth Apgar scores as a factor. In our sample, this variable was not assessed.

Another variable included in our model B was the perception of respectful treatment by healthcare professionals toward women. This aspect is closely related to the concept of obstetric violence and has not been evaluated in other predictive models, despite the existence of publications that identify a relationship between the treatment received by healthcare professionals during childbirth care and the presence of PTSD [36,37]. This aspect takes on particular relevance as the World Health Organization [38] and the United Nations [39,40] report an upward trend in women who perceive inadequate treatment during childbirth care.

Finally, the support provided by the partner plays a relevant role in the risk of PTSD, in such a way that women who perceived that their partners supported them during pregnancy, childbirth and the postpartum period had a lower risk of PTSD, coinciding with the model of Czarnocka and Slade based on a study carried out with 264 women [27].

Strengths and Limits

One of the potential limitations of this study was that the observed prevalence of PTSD risk was high compared to other studies. In a systematic review with a meta-analysis carried out by Yildiz et al., average rates of 4.0% were found overall (95% CI: 2.77–5.71), and 18.5% (95% CI: 10.6–30.38) in women at risk [5]. The higher prevalence of our sample can be attributed to the use of a screening tool (PPQ) as we did not diagnose PTSD; instead, the risk of presenting PTSD was estimated.

Another limitation of the study was that it was carried out in a population residing in Spain, and even though the validation results were good, they need validation in other countries and cultural contexts. Regarding strengths, in addition to satisfactory predictive capacity, these models have other positive characteristics, such as including only five variables (parsimony principle), using variables that are usually recorded in medical records, and having a justified relationship with the risk of PTSD. We should also highlight that the model was validated in a population different than the one used to create the models, and they also had different prevalences for PTSD risk. These differences are interesting for the extrapolation of results; this validation cohort could almost be considered external validation. Additionally, creating two predictive models may be useful for clinicians because it will expand application possibilities. For example, in situations where verbal contact with patients is not possible, model A, based on clinical variables recorded in medical records, could be used. While when verbal contact with the patient is possible, model B would be the instrument of choice. In particular, this tool is especially useful for professionals who have initial contact with women after childbirth. These professionals can use this tool as screening to identify patients who require further evaluation by more specialized professionals in the field of postpartum PTSD, such as psychologists and psychiatrists.

5. Conclusions

In short, two predictive models formed by clinical variables and perceptions of support from their partner and the care received from health professionals presented adequate predictive capacities to predict the risk of postpartum PTSD both in the referral cohort and in the validation cohort. The model of choice includes the woman's perceptions of support received from her partner and the relationship with healthcare professionals. These models can help identify women at increased risk for postpartum PTSD, increasing the early detection of this increasingly prevalent problem. On the other hand, they can also be useful in primary prevention if health policies are applied that reduce risk factors such as cesarean delivery and inadequate treatment by health professionals and encourage other factors such as skin-to-skin contact and breastfeeding.

Author Contributions: Conceptualization, S.M.-V. and J.R.-A.; methodology, M.D.-R. and J.M.M.-G.; formal analysis, A.H.-M.; writing—original draft preparation, S.M.-V. and J.M.M.-G.; writing—review and editing, S.M.-V., M.D.-R. and J.M.M.-G.; supervision, J.R.-A., J.M.M.-G. and A.H.-M.; project administration, A.H.-M. All authors have read and agreed to the published version of the manuscript.

Funding: This research received no external funding.

Appendix A

Table A1. Modified perinatal post-traumatic-stress questionnaire symptoms (modified PPQ).

1.	Did you have bad dreams of giving birth or of your baby's hospital stay?
2.	Did you have upsetting memories of giving birth or of your baby's hospital stay?
3.	Did you have any sudden feelings as though your baby's birth was happening again?
4.	Did you try to avoid thinking about childbirth or your baby's hospital stay?
5.	Did you avoid doing things that might bring up feelings you had about childbirth or your baby's hospital stay (e.g., not watching a TV show about babies)?
6.	Were you unable to remember parts of your baby's hospital stay?
7.	Did you lose interest in doing things you usually do (e.g., did you lose interest in your work or family)?
8.	Did you feel alone and removed from other people (e.g., did you feel like no one understood you)?
9.	Did it become more difficult for you to feel tenderness or love with others?
10.	Did you have unusual difficulty falling asleep or staying asleep?
11.	Were you more irritable or angry with others than usual?
12.	Did you have greater difficulties concentrating than before you gave birth?
13.	Did you feel more jumpy (e.g., did you feel more sensitive to noise, or more easily startled)?
14.	Did you feel more guilt about the childbirth than you felt you should have felt?

* Notes: the response and weight of scores for each question: (0) nothing; (1) once or twice; (2) sometimes; (3) often, but less than 1 month; (4) often for more than a month. The clinical range for high-risk mothers is established at 19 or more points.

References

1. Van der Kolk, B.A.; Pelcovitz, D.; Roth, S.; Mandel, F.S.; McFarlane, A.; Herman, J.L. Dissociation, somatization, and affect dysregulation: The complexity of adaptation of trauma. *Am. J. Psychiatry* **1996**, *153*, 83–93. [CrossRef] [PubMed]
2. Hernandez-Martinez, A.; Rodriguez-Almagro, J.; Molina-Alarcon, M.; Infante-Torres, N.; Donate Manzanares, M.; Martinez-Galiano, J.M. Postpartum post-traumatic stress disorder: Associated perinatal factors and quality of life. *J. Affect. Disord.* **2019**, *249*, 143–150. [CrossRef] [PubMed]
3. Cook, N.; Ayers, S.; Horsch, A. Maternal posttraumatic stress disorder during the perinatal period and child outcomes: A systematic review. *J. Affect. Disord.* **2018**, *225*, 18–31. [CrossRef]
4. Dikmen-Yildiz, P.; Ayers, S.; Phillips, L. Factors associated with post-traumatic stress symptoms (PTSS) 4-6 weeks and 6 months after birth: A longitudinal population-based study. *J. Affect. Disord.* **2017**, *221*, 238–245. [CrossRef] [PubMed]
5. Yildiz, P.D.; Ayers, S.; Phillips, L. The prevalence of posttraumatic stress disorder in pregnancy and after birth: A systematic review and meta-analysis. *J. Affect. Disord.* **2017**, *208*, 634–645. [CrossRef]
6. Geller, P.A.; Stasko, E.C. Effect of Previous Posttraumatic Stress in the Perinatal Period. *J. Obstet. Gynecol. Neonatal. Nurs.* **2017**, *46*, 912–922. [CrossRef]
7. Halperin, O.; Sarid, O.; Cwikel, J. The influence of childbirth experiences on women's postpartum traumatic stress symptoms: A comparison between Israeli Jewish and Arab women. *Midwifery* **2015**, *6*, 625–632. [CrossRef]
8. Söderquist, J.; Wijma, K.; Wijma, B. Traumatic stress after childbirth: The role of obstetric variables. *J. Psychosom. Obstet. Gynecol.* **2002**, *23*, 1. [CrossRef]
9. Adewuya, A.O.; Ologun, Y.A.; Ibigbami, O.S. Post-traumatic stress disorder after childbirth in Nigerian women: Prevalence and risk factors. *BJOG* **2006**, *113*, 284–288. [CrossRef]
10. Susan, A.; Harris, R.; Sawyer, A.; Parfitt, Y.; Ford, E. Posttraumatic stress disorder after childbirth: Analysis of symptom presentation and sampling. *J. Affect. Disord.* **2009**, *119*, 200–204. [CrossRef]
11. Andersen, L.B.; Melvaer, L.B.; Videbech, P.; Lamont, R.F.; Joergensen, J.S. Risk factors for developing post-traumatic stress disorder following childbirth: A systematic review. *Acta Obstet. Gynecol. Scand.* **2012**, *91*, 1261–1272. [CrossRef] [PubMed]
12. Ayers, S.; Bond, R.; Bertullies, S.; Wijma, K. The aetiology of post-traumatic stress following childbirth: A meta-analysis and theoretical framework. *Psychol. Med.* **2016**, *46*, 1121–1134. [CrossRef] [PubMed]
13. Cohen, M.M.; Ansara, D.; Schei, B.; Stuckless, N.; Stewart, D.E. Posttraumatic stress disorder after pregnancy, labor, and delivery. *J. Womens. Health* **2004**, *13*, 315–324. [CrossRef] [PubMed]
14. Srkalović Imširagić, A.; Begić, D.; Šimičević, L.; Bajić, Ž. Prediction of posttraumatic stress disorder symptomatology after childbirth—A Croatian longitudinal study. *Women Birth.* **2017**, *30*, e17–e23. [CrossRef]
15. Hernández-Martínez, A.; Rodríguez-Almagro, J.; Molina-Alarcón, M.; Infante-Torres, N.; Rubio-Álvarez, A.; Martínez-Galiano, J.M. Perinatal factors related to post-traumatic stress disorder symptoms 1–5 years following birth. *Women Birth.* **2020**, *33*, e129–e135. [CrossRef]
16. DeMier, R.L.; Hynan, M.T.; Harris, H.B.; Manniello, R.L. Perinatal stressors as predictors of symptoms of posttraumatic stress in mothers of infants at high risk. *J. Perinatol.* **1996**, *16*, 276–280. [PubMed]
17. Elklit, A.; Hartvig, T.; Christiansen, M. Stress disorder in parents of premature neonates—Secondary publication. *Ugeskr. Laeger.* **2008**, *170*, 3643–3645. [PubMed]
18. Fairbrother, N.; Woody, S.R. Fear of childbirth and obstetrical events as predictors of postnatal symptoms of depression and post-traumatic stress disorder. *J. Psychosom. Obstet. Gynecol.* **2007**, *28*, 239–242. [CrossRef]
19. Van Heumen, M.A.; Hollander, M.H.; van Pampus, M.G.; van Dillen, J.; Stramrood, C.A.I. Psychosocial Predictors of Postpartum Posttraumatic Stress Disorder in Women with a Traumatic Childbirth Experience. *Front. Psychiatry* **2018**, *9*, 348. [CrossRef]
20. O'Donovan, A.; Alcorn, K.L.; Patrick, J.C.; Creedy, D.K.; Dawe, S.; Devilly, G.J. Predicting posttraumatic stress disorder after childbirth. *Midwifery* **2014**, *30*, 935–941. [CrossRef]

21. Lopez, U.; Meyer, M.; Loures, V.; Iselin-Chaves, I.; Epiney, M.; Kern, C. Post-traumatic stress disorder in parturients delivering by caesarean section and the implication of anaesthesia: A prospective cohort study. *Health Qual. Life Outcomes.* **2017**, *15*, 118. [CrossRef] [PubMed]

22. Shlomi Polachek, I.; Dulitzky, M.; Margolis-Dorfman, L.; Simchen, M.J. A simple model for prediction postpartum PTSD in high-risk pregnancies. *Arch. Women's Ment. Health* **2016**, *19*, 483–490. [CrossRef] [PubMed]

23. Haagen, J.F.G.; Moerbeek, M.; Olde, E.; Van Der Hart, O.; Kleber, R.J. PTSD after childbirth: A predictive ethological model for symptom development. *J. Affect. Disord.* **2015**, *185*, 135–143. [CrossRef] [PubMed]

24. Vossbeck-Elsebusch, A.N.; Freisfeld, C.; Ehring, T. Predictors of posttraumatic stress symptoms following childbirth. *BMC Psychiatry* **2014**, *14*, 200. [CrossRef] [PubMed]

25. Ford, E.; Ayers, S. Support during birth interacts with prior trauma and birth intervention to predict postnatal post-traumatic stress symptoms. *Psychol. Health* **2011**, *26*, 1553–1570. [CrossRef] [PubMed]

26. Boudou, M.; Séjourné, N.; Chabrol, H. Douleur de l'accouchement, dissociation et détresse périnatales comme variables prédictives de symptômes de stress post-traumatique en post-partum. *Gynecol. Obstet. Fertil.* **2007**, *35*, 1136–1142. [CrossRef] [PubMed]

27. Czarnocka, J.; Slade, P. Prevalence and predictors of post-traumatic stress symptoms following childbirth. *Br. J. Clin. Psychol.* **2000**, *39*, 35–51. [CrossRef] [PubMed]

28. Vandenbroucke, J.P.; von Elm, E.; Altman, D.G.; Gøtzsche, P.C.; Mulrow, C.D.; Pocock, S.J.; Poole, C.; Schlesselman, J.J.; Egger, M. STROBE Initiative. Strengthening the Reporting of Observational Studies in Epidemiology (STROBE): Explanation and elaboration. *PLoS Med.* **2007**, *4*, e297. [CrossRef] [PubMed]

29. Peduzzi, P.; Concato, J.; Kemper, E.; Holford, T.R.; Feinstem, A.R. A simulation study of the number of events per variable in logistic regression analysis. *J. Clin. Epidemiol.* **1996**, *49*, 1373–1379. [CrossRef]

30. Callahan, J.L.; Borja, S.E.; Hynan, M.T. Modification of the Perinatal PTSD Questionnaire to enhance clinical utility. *J. Perinatol.* **2006**, *9*, 533–539. [CrossRef] [PubMed]

31. Hosmer, D.W., Jr.; Lemeshow, S.; Sturdivant, R.X. *Model-Building Strategies and Methods for Logistic Regression*; John Wiley & Sons, Ltd.: Hoboken, NJ, USA, 2013; pp. 89–151.

32. Mickey, R.M.; Greenland, S. The impact of confounder selection criteria on effect estimation. *Am. J. Epidemiol.* **1989**, *129*, 125–137. [CrossRef] [PubMed]

33. Swets, J.A. Measuring the accuracy of diagnostic systems. *Sci. Sci.* **1988**, *240*, 1285–1293. [CrossRef] [PubMed]

34. Ford, E.; Ayers, S.; Bradley, R. Exploration of a cognitive model to predict post-traumatic stress symptoms following childbirth. *J. Anxiety Disord.* **2010**, *24*, 353–359. [CrossRef] [PubMed]

35. Abdollahpour, S.; Khosravi, A.; Bolbolhaghighi, N. The effect of the magical hour on post-traumatic stress disorder (PTSD) in traumatic childbirth: A clinical trial. *J. Reprod. Infant. Psychol.* **2016**, *34*, 403–412. [CrossRef]

36. Silveira, M.F.; Mesenburg, M.A.; Bertoldi, A.D.; De Mola, C.L.; Bassani, D.G.; Domingues, M.R. The association between disrespect and abuse of women during childbirth and postpartum depression: Findings from the 2015 Pelotas birth cohort study. *J. Affect. Disord.* **2019**, *256*, 441–447. [CrossRef]

37. De Schepper, S.; Vercauteren, T.; Tersago, J.; Jacquemyn, Y.; Raes, F.; Franck, E. Post-Traumatic Stress Disorder after childbirth and the influence of maternity team care during labour and birth: A cohort study. *Midwifery* **2016**, *32*, 87–92. [CrossRef]

38. World Health Organization. *Global Status Report on Violence Prevention*; World Health Organization: Geneva, Switzerland, 2014.

39. United Nations. A Human Rights-Based Approach to Mistreatment and Violence against Women in Reproductive Health Services with a Focus on Childbirth and Obstetric Violence. 2019. Available online: https://digitallibrary. un.org/record/3823698 (accessed on 23 December 2020).
40. Martínez-Galiano, J.M.; Martinez-Vazquez, S.; Rodríguez-Almagro, J.; Hernández-Martinez, A. The magnitude of the problem of obstetric violence and its associated factors: A cross-sectional study. *Women Birth* **2020**. [CrossRef]

Publisher's Note: MDPI stays neutral with regard to jurisdictional claims in published maps and institutional affiliations.

International Journal of
Environmental Research and Public Health

MDPI

Article

Spanish Nursing Students' Attitudes toward People Living with HIV/AIDS: A Cross-Sectional Survey

María Adelaida Álvarez-Serrano [1], Encarnación Martínez-García [2,3], Adelina Martín-Salvador [4,*], María Gázquez-López [1,*], María Dolores Pozo-Cano [2], Rafael A. Caparrós-González [2,5] and María Ángeles Pérez-Morente [6]

[1]   Department of Nursing, Faculty of Health Sciences, University of Granada, 51001 Ceuta, Spain; adealvarez@ugr.es
[2]   Department of Nursing, Faculty of Health Sciences, University of Granada, 18016 Granada, Spain; emartinez@ugr.es (E.M.-G.); pozocano@ugr.es (M.D.P.-C.); rcg477@ugr.es (R.A.C.-G.)
[3]   Guadix High Resolution Hospital, 18500 Guadix, Granada, Spain
[4]   Department of Nursing, Faculty of Health Sciences, University of Granada, 52005 Melilla, Spain
[5]   Mind, Brain and Behavior Research Center (CIMCYC), Faculty of Psychology, University of Granada, 18016 Granada, Spain
[6]   Department of Nursing, Faculty of Health Sciences, University of Jaén, 23071 Jaén, Spain; mmorente@ujaen.es
*   Correspondence: ademartin@ugr.es (A.M.-S.); mgazquez@ugr.es (M.G.-L.)

Received: 25 August 2020; Accepted: 19 November 2020; Published: 22 November 2020

Abstract: Human immunodeficiency virus (HIV) infection is still a public health issue. Human immunodeficiency virus/acquired immune deficiency syndrome (HIV/AIDS) creates, in society, stigmatizing attitudes, fear, and discrimination against infected people; even health professionals do not feel trained enough to adequately take care of these patients, which affects the quality of care provided to such patients. The purpose of this study was to explore nursing students' attitudes and other related factors toward people with HIV/AIDS, as well as their evolution in subsequent academic years. A cross-sectional study was performed with students in four academic years from four Spanish health sciences institutions (n = 384). Data were collected voluntarily and on an anonymous basis, utilizing the "Nursing students' attitudes toward AIDS" (EASE) validated scale. The students' attitudes toward people with HIV/AIDS were relatively positive, with a total mean EASE value of 85.25 ± 9.80. Statistically significant differences were observed according to the academic year (p = 0.041), in 4 out of 21 items of the scale and among students with no religious beliefs. By adjusting every variable, only the weak association with religion was maintained (p = 0.045).

Keywords: attitudes; HIV/AIDS; students; nursing

---

1. Introduction

Human immunodeficiency virus (HIV) infection is still an international public health problem [1,2]. In 2018, an estimated 37.9 million people worldwide were infected, of which 24.5 million people were treated with antiretroviral medications by mid-2019 [3]. In this sense, the World Health Organization, within the Sustainable Development Goals for the year 2030, included as a global goal to reduce HIV infection to zero contagions, zero discrimination, and zero deaths caused by the virus [2,4].

Today, particularly in developed countries, people with HIV live longer and experience the disease as a chronic condition, the most common comorbidities being cardiovascular diseases, kidney diseases, psychiatric disorders, and neoplasias, mainly associated with aging [3,5,6]. Currently, having HIV/AIDS still has a significant impact on the economic, political, legal, and social levels [1,2].

Regarding the latter, it has been documented that HIV/AIDS generates, among people, stigmatizing attitudes, fear, and discrimination against infected persons [7].

Furthermore, with the passing of time, as the disease settles in society, knowledge and education about HIV/AIDS have been reduced [8]. In the field of health, several authors state that even healthcare workers who are more exposed to work accidents with biological materials [9], such as nursing students and professionals, do not feel prepared enough to treat such patients adequately [6,8,10,11]. Variables such as ideology, age, stress, and negative opinions on homosexuality affect healthcare professionals' attitudes toward people with HIV/AIDS [1]. In addition, the lack of knowledge and the existence of erroneous concepts as regards transmission of HIV not only foster stigmatizing attitudes and fear of contagion, but they also impact clinical practices related to HIV/AIDS and the quality of the care received by seropositive patients [6,10,12–14]. In relation to these attitudes, in a study carried out by Valencia et al., it was reflected that women with HIV felt that they were mistreated by health professionals and that their rights to confidentiality and privacy were violated [15], which undoubtedly undermines legislation and ethics of care [12,13]

With respect to the assessments of HIV/AIDS patients performed by nursing students, there is research that affirms that these negative evaluations are related to stigmas related to promiscuity, drug addiction, and homosexuality [8,16,17]. In order to change such perceptions, many research contributions have shown that an increase in knowledge in this area reduces fear and anxiety [6,11].

Although it is reasonable to think that nursing students' negative attitudes toward people with HIV/AIDS will diminish as their academic level increases, there is a discrepancy in scientific literature regarding such a hypothesis [1,18,19]. Therefore, the objective of this study is to explore nursing students' attitudes and other related factors toward HIV/AIDS and their evolution in subsequent academic years.

2. Materials and Methods

2.1. Data Collection

An online, cross-sectional survey was conducted anonymously by self-administration. The participants were nursing students from four Spanish health sciences, institutions with a four-year-long study program of 240 credits. Although these four institutions are located in different geographical areas, student´s sociocultural levels of similar. The collection of the questionnaires was administered simultaneously in the four institutions during one month of the 2019–2020 academic year. It was sent through the teaching support platform, and participation was voluntary outside of class hours through the teaching support platform. The members of the research group invited all students enrolled across the four academic years to participate. After the first dissemination of the questionnaire, a reminder was sent to those students who had not yet answered the survey.

2.2. Measurement Tool

For data collection, the "Nursing students' attitudes toward AIDS" (EASE) one-dimensional scale was employed, the Spanish version of which was initially created and validated by Tomás-Sábado (Appendix A) [20]. This tool showed good internal validity in the identification of nursing students' attitudes toward HIV/AIDS in Spain (Cronbach's alpha coefficient = 0.779). This scale comprises 21 items with Likert-type responses ranging from 1 = totally agree to 5 = totally disagree. The score for each item depends on its directionality. Items 3, 5, 7, 8, 11, 14, 15, and 21 are scored from 5 to 1, 5 being totally agree and 1 being totally disagree, and items 1, 2, 4, 6, 9, 10, 12, 13, and 16–20 are scored from 1 to 5, 1 being totally agree and 5 being totally disagree. The scale's maximum score is 105, which indicates the most positive attitudes, and the minimum score is 21, which indicates the most negative and preconceived attitudes. For each participant, the scale's global value was calculated by adding each item's score (with a total of 21 scores). The original scale was used in most of the studies analyzed [1,21,22]. Additionally, the following sociodemographic variables were considered:

age, sex (male or female), academic year, marital status (in a relationship or not in a relationship), sexual orientation (heterosexual or non-heterosexual), and religion (with religious beliefs or with no religious beliefs).

2.3. Data Analysis

Prior to data analysis, the internal consistency of the scale was assessed by determining Cronbach's alpha, with a satisfactory result of 0.776. Next, a descriptive analysis of the entire sample was carried out by means of absolute and relative frequencies and the calculation of measures of central tendency (i.e., median and standard deviation) according to the quantitative or qualitative nature of the study's variables. The scale's normality of values was explored, and as a result of not being possible to verify their parametricity, for the inferential statistics, a nonparametric analysis was adopted. Differences among the scale's values according to the sociodemographic variables and academic year categories were analyzed through the Mann–Whitney U test, and for variables with more than two factors, the Kruskal–Wallis H test was employed. In cases of significant differences among academic years, post-hoc multiple comparisons were carried out through Bonferroni correction. The effect size was calculated using Kerby's formula [23]. To establish a correlation between age and the final score of the EASE scale, Spearman's Rho was used. Finally, the influence of the sociodemographic variables and academic year on the EASE scale's final score was explored by using a multiple linear regression model through the "introduce" method, in which the academic year, re-categorized into three dummy variables, was included, taking the first year as reference. Data were treated with the Statistical Package for the Social Sciences (SPSS) program, version 25, (IBM, New York, NY, USA, for Windows).

2.4. Ethical Considerations

This study complies with the good clinical practice regulations, as stated in the European Directive 2001/20/CE and Law 14/2007, of 3 July, on biomedical research. Treatment of personal data in health research is governed by Organic Law 3/2018, of 5 December, on Data Protection and Guarantee of Digital Rights. Every participant checked a box indicating consent to participate in the study.

3. Results

The sample included 384 students, with the response rate being 18%. Twenty-nine percent of the sample ($n = 115$) were in their first year of study. With regards to sex, 84% were women, and the total mean age was 23 ± 6.7 years. In terms of academic year, the mean age was significantly higher for those in their fourth year of study ($p < 0.001$). With respect to students' sexual orientation, 337 (87.8%) were heterosexual, a higher proportion of which was observed in the third year of study ($p = 0.012$). Regarding relationship status, 207 (53.9%) stated they were not in a relationship, the fourth year of study being the year with a higher number of students in a relationship ($p = 0.008$). Of the total number of students, 265 (69%) considered themselves as having religious beliefs (Table 1).

The total EASE score in the analyzed sample was 85.25 ± 9.80. Table 2 shows such values according to sociodemographic variables and academic year. Among the second-year students and among those with no religious beliefs, the highest statistically significant scores were observed ($p = 0.041$, and $p = 0.006$, with an effect size of 0.17, respectively) (Table 2). Significant causes according to academic year were identified in the second and first year of study ($p = 0.050$ with an effect size of -0.22) (results not shown).

The scale items that reached a higher mean score were P1 ("AIDS does not affect heterosexual couples") with 4.62 ± 1.01 and P11 ("being an AIDS carrier should not be a barrier to accessing education and employment") with 4.68 ± 0.87. Such items show the highest percentages of "disagree" and "agree" answers, with 83.1% and 83.9%, respectively (results not shown).

Table 1. Sociodemographic variables of the sample according to academic year.

Variables	1st Year (*n* = 115)	2nd Year (*n* = 104)	3rd Year (*n* = 86)	4th Year (*n* = 79)	TOTAL (*n* = 384)	*p*
	M (SD)	M (SD)	M (SD)	M (SD)	M (SD)	
Age (years)	21.37 (6.34)	22.79 (6.46)	23.71 (6.22)	25.46 (7.57)	23.12 (6.70)	<0.001 **
	n (%)	*n* (%)	*n* (%)	*n* (%)	*n* (%)	
Sex						
Male	15 (13)	19 (18.3)	19 (22.1)	9 (11.4)	62 (16.1)	ns
Female	100 (87)	85 (81.7)	67 (77.9)	70 (88.6)	322 (83.9)	
Sexual orientation						
Heterosexual	106 (92.2)	94 (90.4)	76 (88.4)	61 (77.2)	337 (87.8)	0.012 *
Non-heterosexual	9 (7.8)	10 (9.6)	10 (11.6)	18 (22.8)	47 (12.2)	
Relationship status						
Not in a relationship	67 (58.3)	60 (57.7)	51 (59.3)	29 (36.7)	207 (53.9)	0.008 **
In a relationship	48 (41.7)	44 (42.3)	35 (40.7)	50 (63.3)	177 (46.1)	
Religion						
With religious beliefs	74 (64.3)	71 (68.3)	64 (74.4)	56 (70.9)	265 (69)	ns
With no religious beliefs	41 (35.7)	33 (31.7)	22 (25.6)	23 (29.1)	119 (31)	

M = mean; SD = standard deviation; ns = *p* > 0.05; * = *p* ≤ 0.05; ** = *p* ≤ 0.01.

Table 2. Descriptive statistics of the EASE scale according to the sociodemographic variables and academic year.

Variables	M (SD)	Mann–Whitney *U* Test	*p*	Kerby´s
Sex		8836.5	ns	
Male	84.52 (8.59)			
Female	85.57 (9.68)			
Sexual orientation		9212	ns	
Heterosexual	85.08 (9.45)			
Non-heterosexual	87.7 (9.75)			
Relationship status		19,259.5	ns	
Not in a relationship	85.82 (9.58)			
In a relationship	84.91 (9.45)			
Religion		13,023.5	0.006 **	0.17
With religious beliefs	84.67 (9.14)			
With no religious beliefs	87.03 (10.14)			
Kruskal–Wallis *H* Test				
Academic year		8.27	0.041 *	
Bonferroni Correction				
First	85.15 (8.33)	7283	0.50 *	−0.22
Second	86.39 (10.86)			
Third	85.58 (8.48)			
Fourth	85.72 (10.23)			
Total	85.25 (9.80)			

M = mean; SD = standard deviation; ns = *p* > 0.05; * = *p* ≤ 0.05; ** = *p* ≤ 0.01.

With respect to the questions asked in the scale, there were significant differences among academic years in 4 out of the 21 items included, namely in item 2 ("fetuses infected with AIDS should be aborted"), 3 ("there is no risk arising from AIDS carriers' use of restaurants and pubs"), 4 ("seropositive women should not be allowed to get pregnant"), and 16 ("in hospitals, AIDS carriers should not share a room with non-infected persons") (Table 3). The causes for the significance among academic years in item 2 were observed between the first and second years (p = 0.041), between the first and third years (p = 0.0001), and between the first and fourth years (p = 0.021); in item 3, between the first and second years (p = 0.022); in item 4, between the first and third years (p = 0.018); and in item 16, between the first and second years (p = 0.014) (results not shown). In all cases, the first-year score was lower than that obtained in the subsequent years.

Table 3. Descriptive statistics of the EASE scale items, according to academic year.

	1st Year (n = 115)	2nd Year (n = 104)	3rd Year (n = 86)	4th Year (n = 79)	Kruskal–Wallis H Test	p
	M (SD)	M (SD)	M (SD)	M (SD)		
P1	4.61 (0.99)	4.61 (0.99)	4.62 (1.04)	4.68 (0.96)	0.76	ns
P2	4.11 (0.83)	4.34 (0.95)	4.58 (0.77)	4.46 (0.73)	21.48	<0.001 **
P3	3.87 (1.33)	4.34 (1.09)	4.05 (1.24)	4.03 (1.36)	8.69	0.034 *
P4	4.07 (0.95)	4.31 (0.87)	4.43 (0.83)	4.38 (0.85)	10.85	0.013 *
P5	4.37 (1.06)	4.14 (1.35)	4.06 (1.27)	4.28 (1.09)	3.04	ns
P6	4.57 (0.64)	4.62 (0.77)	4.47 (0.84)	4.52 (0.83)	2.68	ns
P7	4.43 (1.04)	4.19 (1.22)	4.01 (1.27)	4.25 (1.17)	6.9	ns
P8	3.57 (1.27)	3.67 (1.30)	3.6 (1.37)	3.75 (1.25)	1.03	ns
P9	4.29 (0.91)	4.42 (0.9)	4.45 (0.76)	4.25 (1.01)	3.09	ns
P10	2.97 (1.22)	3.19 (1.34)	3.2 (1.26)	3.35 (1.26)	5.21	ns
P11	4.71 (0.69)	4.67 (0.98)	4.7 (0.84)	4.7 (0.82)	1.61	ns
P12	3.8 (1.05)	4 (1.08)	3.91 (1.04)	4.01 (1.15)	3.95	ns
P13	3.62 (1.11)	3.78 (1.12)	4.03 (1.02)	3.78 (1.17)	6.99	ns
P14	2.03 (1.1)	2.12 (1.14)	2.21 (1)	2.15 (1.05)	2.33	ns
P15	4.5 (0.86)	4.43 (1.03)	4.44 (1.08)	4.43 (0.96)	0.73	ns
P16	4 (0.98)	4.36 (0.94)	4.19 (1.00)	4.1 (1.03)	9.57	0.023 *
P17	4.38 (0.94)	4.58 (0.87)	4.47 (0.93)	4.38 (0.97)	4.44	ns
P18	4.57 (0.65)	4.66 (0.71)	4.49 (1.02)	4.44 (0.98)	3.39	ns
P19	4.23 (0.95)	4.38 (0.89)	4.4 (0.77)	4.47 (0.88)	4.39	ns
P20	4 (1.30)	3.94 (1.34)	3.86 (1.31)	3.71 (1.40)	2.75	ns
P21	3.45 (1.28)	3.64 (1.32)	3.43 (1.39)	3.59 (1.32)	1.96	ns

M = mean; SD = standard deviation; ns = p > 0.05; * = p ≤ 0.05; ** = p ≤ 0.01. Definitions of P1–21 can be found in Appendix A.

In the multiple linear regression model, it was observed that by adjusting all variables, only religion was associated with the scale values, with those stating no religious beliefs showing more favorable attitudes toward HIV/AIDS ((95% CI: 0.05–4.26); p = 0.045) (Table 4).

Table 4. Multiple linear regression model for the EASE scale.

Variables	ß	SD	p	95% CI	
Age	−0.029	0.08	0.704	−0.18	0.12
Sex					
Male	Reference				
Female	1.63	1.36	0.228	−1.03	4.30
Relationship status					
In a relationship	Reference				
Not in a relationship	0.54	1.03	0.595	−1.48	2.58
Sexual orientation					
Heterosexual	Reference				
Non-heterosexual	2.12	1.56	0.171	−0.93	5.18
Religion					
With religious beliefs	Reference				
With no religious beliefs	2.15	1.07	0.045b *	0.05	4.26
Academic year					
First	Reference				
Second	2.42	1.29	0.060	−0.11	4.95
Third	1.78	1.37	0.194	−0.91	4.47
Fourth	1.60	1.45	0.269	−4.46	1.25

ß = Beta coefficient; * = p ≤ 0.05; CI: confidence interval.

4. Discussion

The attitudes toward people with HIV/AIDS of the students that participated in the study were relatively positive or favorable, since they indicated a mean EASE score above 85 points, as recommended by Tomás-Sábado and Aradilla-Herrero [1]. This result coincides with that reported in several similarly designed studies [1,21,24], reinforcing the idea that nursing, as a profession, through its humanistic approach, generates among future professionals positive attitudes toward an infection such as HIV or a disease such as AIDS [25].

According to the academic level, some differences were found in the EASE scores, the first academic year's score being lower than those obtained in subsequent years. Such a result coincides with that published by Leyva-Moral et al., where the lowest percentage of positive attitudes was observed among first-year students [25]. Nevertheless, differences in scores were found between the second- and first-years, but not in subsequent years with respect to the first one, which contradicts previous studies regarding an increase in favorable attitudes as the academic level increases [1,18].

Some authors have stated that when facing problems that are especially sensitive at a social level and that are ideologically charged, such as HIV and AIDS, an increase in knowledge alone is not enough to change people's beliefs or attitudes [22,24–27]. Furthermore, contact with actual patients occurring in subsequent years during clinical practice, on the contrary, could increase fear of contagion among students, thus obstructing an advance toward higher levels of disease tolerance, which is an issue that requires further investigation [22].

The median age of the analyzed sample was in accordance with those published in similar studies [1,9,21,22,24,25,28], corresponding to young adults, although this variable was not associated with attitudes toward HIV/AIDS. However, the observed tendency, that as age increased, attitudes became more negative, coincides with the results of other studies that confirmed such an association [25,29–31].

Despite the fact that the analysis was reconducted to assess the EASE scale's internal consistency by obtaining a Cronbach's alpha satisfactory result, some items showed a change opposite to the one expected. That is, it was expected that in cases of positive assertions the level of agreement would increase and in cases of negative assertions, the level of disagreement would increase. However, in terms of item 14, "persons infected with AIDS should be considered victims of the social system," 66.4% of students disagreed or totally disagreed, and for item 10, "seropositive persons should be identified as such," the opposite occurred, i.e., 30% of students agreed or totally agreed. Serrano-Gallardo and Giménez-Maroto already identified such contradictions due to a certain degree of ambiguity in the wording of these items [9]. Although some of the items on the EASE scale could be considered outdated at present due to clinical and cultural changes, we believe that the scale allows general and non-specific attitudes toward HIV/AIDS among nursing students to be explored [25].

Having no religious beliefs was the only analyzed variable that showed an association with better attitudes toward HIV/AIDS, regardless of the influence of the other variables. This result coincides with the findings of other studies, where nursing professionals who practiced a religion, or students who believed religion had a significant role in their lives or firmly thought that the disease etiology was due to divine retribution, showed more stigmatizing attitudes toward their HIV/AIDS patients [20,32–35].

Limitations

This study has some limitations. Although the number of students who agreed to answer the questionnaire was higher than the figures published in other studies [1,9,21,22,24,25,28], we believe it could be improved, since students who are more aware of the subject being analyzed could be overrepresented. It is also possible that given the sensitive nature of the subject addressed, when asking for personal opinions, there could have been a social desirability bias that was resolved by guaranteeing the anonymous nature of all participants. Finally, and as we have commented previously, the clinical and cultural changes that have occurred in the last decade with respect to HIV/AIDS could mean that some items on the scale are out of date, although this possible limitation is minimized by establishing a general analysis of attitudes toward HIV/AIDS.

Despite the limitations referred to above, this study contributes current information on the persistence of false opinions regarding HIV/AIDS in a varied sample of nursing students, since four health sciences institutions were analyzed, which were located in different geographic areas but shared the same curriculum. Results may be useful to rethink the type of training our students receive and about the need, therefore, to establish educational content that fosters closer humanization of care and the reduction of HIV/AIDS-associated stigma.

5. Conclusions

Nursing students' attitudes toward HIV/AIDS are relatively positive and get better following the first academic year. When considering all variables of the study, the influence of the academic level waned and the association with religion was maintained. Thus, students with no religious beliefs showed more positive attitudes toward this problem. An improvement in the level of knowledge and experiences among nursing students does not seem to be enough to cause a change of attitude, which may reflect the strong ideological charge that still leads to this health issue today.

It is therefore necessary to carry out a critical analysis of teaching strategies in current nursing programs in order to achieve not only better and more in-depth knowledge of the disease and its modes of transmission, but also a change of attitude, free of negative preconceptions toward HIV/AIDS. This approach would result in a better quality and humanization of the care provided to people with HIV/AIDS. It would also be pertinent in future research to reformulate some of the items on the EASE scale, adapting them to the current reality of people with HIV/AIDS.

Author Contributions: Conceptualization, M.G.-L.; methodology, M.A.Á.-S., E.M.-G., and A.M.-S.; investigation, M.A.Á.-S., E.M.-G., A.M.-S., M.G.-L., M.D.P.-C., R.A.C.-G., and M.Á.P.-M.; data curation, M.A.Á.-S. and E.M.-G.; visualization, A.M.-S.; writing—original draft preparation, M.A.Á.-S., E.M.-G., A.M.-S., M.G.-L., M.D.P.-C., R.A.C.-G., and M.Á.P.-M.; writing—review and editing, M.A.Á.-S., E.M.-G., A.M.-S., M.G.-L., M.D.P.-C., R.A.C.-G., and M.Á.P.-M.; supervision, E.M.-G. and M.Á.P.-M. All authors read and agreed to the published version of the manuscript.

Funding: This research received no external funding.

Acknowledgments: To all students who participated in this study.

Conflicts of Interest: The authors declare no conflict of interest.

Appendix A

Scale of attitudes toward AIDS for nursing.

1. AIDS does not affect heterosexual couples.
2. Fetuses infected with AIDS should be aborted.
3. There is no risk arising from AIDS carriers' use of restaurants and pubs.
4. Seropositive women should not be allowed to get pregnant.
5. AIDS is a problem that affects us all.
6. Continued care of an AIDS patient is synonymous with contagion.
7. AIDS carriers are entitled to doctor–patient confidentiality.
8. In daily activities, there is no risk of transmission of HIV.
9. Persons infected with AIDS should be separated from other sick persons.
10. Seropositive persons should be identified as such.
11. Being an AIDS carrier should not be a barrier to accessing education and employment.
12. Specific hospitals for AIDS patients and carriers should be created.
13. AIDS is the greatest affliction of our time.
14. Persons infected with AIDS should be considered victims of the social system.

15. Being an AIDS carrier should not be a barrier to adopting a child.
16. In hospitals, AIDS carriers should not share a room with non-infected persons.
17. I would not like to work with an AIDS carrier.
18. Children who are AIDS carriers should attend special classes.
19. As a precautionary measure, we should avoid contact with AIDS patients and carriers.
20. We should always use gloves when touching someone infected with AIDS.
21. HIV testing should be voluntary and anonymous.

References

1. Tomás-Sábado, J.; Aradilla, H.A. Actitud ante el sida en estudiantes de enfermería. ¿Cuál es el papel de la formación académica? *Educ. Médic* **2003**, *6*, 87–92.
2. ONUSIDA. Monitoreo Global del Sida 2020. Available online: https://www.unaids.org/sites/default/files/media_asset/global-aids-monitoring_es.pdf (accessed on 1 June 2020).
3. ONUSIDA. Estadísticas Mundiales Sobre el VIH. Available online: https://www.unaids.org/es/resources/fact-sheet (accessed on 1 June 2020).
4. ONUSIDA. Estrategia ONUSIDA 2016–2021. Acción Acelerada Para Acabar Con el Sida. Available online: https://www.unaids.org/sites/default/files/media_asset/UNAIDS-strategy-2016-2021_es.pdf (accessed on 1 June 2020).
5. Gol-Montserrat, J.; del Llano, J.E.; del Amo, J.; Campbell, C.; Navarro, G.; Segura, F.; Suárez, I.; Teira, R.; Brañas, F.; Serrano-Villar, S.; et al. VIH en España 2017: Políticas para una nueva gestión de la cronicidad más allá del control virológico. *Rev. Esp. Salud Pública* **2018**, *92*, e201809062.
6. Serwaa, B.D.; Mavhandu-Mudzusi, A.H. Nurses knowledge, attitudes and practices towards patients with HIV and T AIDS in Kumasi, Ghana. *Int. J. Africa Nurs. Sci.* **2019**, *11*, 100147.
7. Diesel, H.; Ercole, P.; Taliaferro, D. Knowledge and Perceptions of HIV/AIDS among Cameroonian Nursing Students. *Int. J. Nurs. Educ. Scholarsh.* **2013**, *10*, 209–218. [CrossRef]
8. Frain, J.A. Preparing Every Nurse to become an HIV Nurse. *Nurse Educ. Today* **2017**, *48*, 129–133. [CrossRef]
9. Serrano-Gallardo, M.P.; Giménez-Maroto, A.M. Actitud ante el sida de los estudiantes de enfermería de la Escuela Puerta de Hierro (Universidad Autónoma de Madrid). *Enferm. Clin.* **2006**, *16*, 11–18.
10. Pickles, D.; King, L.; Belan, I. Attitudes of nursing students towards caring for people with VIH/AIDS: Thematic literature review. *J. Adv. Nurs.* **2009**, *65*, 2262–2273. [CrossRef]
11. Maina, G.; Sutankayo, L.; Chorney, R.; Caine, V. Living with and teaching about HIV: Engaging nursing students through body mapping. *Nurse Educ. Today* **2014**, *34*, 643–647. [CrossRef]
12. Som, P.; Bhattacherjee, S.; Guha, R.; Basu, M.; Datta, S. A study of knowledge and practice among nurses regarding care of human immunodeficiency virus positive patients in medical college and Hospitals of Kolkata, India. *Ann. Niger. Med.* **2015**, *9*, 15–19. [CrossRef]
13. González, P.C.; Durán, M.T.; Casals, B.I.; Lugones, B.M.; Castro, R.T. Reflexiones en torno a los problemas éticos y bioéticos en el tratamiento del paciente con VIH/SIDA. *Rev. Cubana Med. Gen. Integr.* **2009**, *25*, 105–112.
14. Relf, M.V.; Laverriere, K.; Devlin, C.; Salerno, T. Ethical beliefs related to HIV and AIDS among nursing students in South Africa and the United States: A cross-sectional analysis. *Int. J. Nurs. Stud.* **2009**, *46*, 1448–1456. [CrossRef] [PubMed]
15. Valencia-García, D.; Rao, D.; Strick, L.; Simoni, J.M. Women´s experiences with HIV-related stigma from health care providers in Lima, Peru: "I would rather die than go back for care". *Health Care Women Int.* **2017**, *38*, 144–158. [CrossRef] [PubMed]
16. Pickles, D.; King, L.; de Lacey, S. Culturally construed beliefs and perceptions of nursing students and the stigma impacting on people living with AIDS: A qualitative study. *Nurse Educ. Today* **2017**, *49*, 39–44. [CrossRef] [PubMed]
17. Shah, S.M.; Heylen, E.; Srinivasan, K.; Perumpil, S.; Ekstrand, M.L. Reductind HIV Stigma Among Nursing Students: A Brief Intervention. *West J. Nurs. Res.* **2014**, *36*, 1323–1337. [CrossRef] [PubMed]
18. West, A.M.; Davis-Lagrow, P.; Lesaure, R.; Allen, P. Low Versus High Prevalence of AIDS: Effect on Nursing Students' Attitudes and Knowledge. *AIDS Patient Care STDs* **1998**, *12*, 51–60. [CrossRef]

19. Lynette, S.B.A. An investigation into whether nursing students alter their attitudes and knowledge levels regarding HIV infection and AIDS following a 3-year programme leading to registration as a qualified nurse. *J. Adv. Nurs.* **1997**, *25*, 1167–1174.
20. Tomás, S.J. Actitud de enfermería ante el SIDA; construcción de una escala de Likert. *Enferm. Clínica* **1999**, *9*, 233–237.
21. Tomás, S.J.; Aradilla, H.A. Actitud ante el SIDA en enfermería. Un análisis de las diferencias entre estudiantes y profesionales. *Ago Enf.* **2006**, *10*, 956–959.
22. Fernández, D.L.; Fernández, N.P.; Tomás, S.J. Modificación de actitudes ante el Sida en estudiantes de enfermería. Resultados de una experiencia pedagógica. *Educ. Médic* **2006**, *9*, 84–90.
23. Kerby, D.S. The simple difference formula: An approach to teaching nonparametric correlation. *Innov. Teach.* **2014**, *3*, 1–9. [CrossRef]
24. Álvarez, V.M.; Guillem, P.C.; Navarro, P.R. Actitud ante el sida en estudiantes de enfermería. *Ago Enf.* **2008**, *12*, 13.
25. Leyva-Moral, J.M.; Terradas-Robledo, R.; Feijoo-Cid, M.; de Dios-Sánchez, R.; Mestres-Camps, L.; Lluva-Castaño, A.; Comas-Serrano, M. Attitudes to HIV and AIDS among students and faculty in a School of Nursing in Barcelona (Spain): A cross-sectional survey. *Collegian* **2017**, *24*, 593–601. [CrossRef]
26. Damrosch, S.; Abbey, S.; Warner, A.; Guy, S. Critical care nurses' attitudes toward, concerns about, and knowledge of the acquired immunodeficiency syndrome. *Heart Lung* **1990**, *19*, 395–400. [PubMed]
27. Young, D.M.; Garvin, B.J. Nurses' knowledge and attitudes and AIDS patients' perception of confirmation: A pilot study. *Appl. Nurs. Res.* **1990**, *3*, 105–111. [CrossRef]
28. Pérez, N.; García-Pérez, G.G. Conocimientos y actitudes de los estudiantes de enfermería para el abordaje de las personas VIH/SIDA seropositiva. *Rev. Cien. Cuidad.* **2014**, *11*, 7–18.
29. Fernández, S.; Rodríguez-Vázquez, B.; Pértega-Diaz, S. Attitudes of nursing and auxiliary hospital staff toward HIV infection and AIDS in Spain. *J. Assoc. Nurses AIDS Care* **2004**, *15*, 62–69. [CrossRef] [PubMed]
30. Rovira, V.M.D.; Uriz, S.E.; Rodríguez, S.C.; Vila Córcoles, A. Comportamiento y actitud de los profesionales de enfermería hospitalaria ante los pacientes VIH positivos. *Enferm. Clínica* **2004**, *3*, 135–141.
31. Conejeros, V.I.; Emig, S.H.; Ferrer, L.L.; Cabieses, V.B.; Acosta, R.C. Conocimientos, actitudes y percepciones de enfermeros y estudiantes de enfermería hacia VIH/Sida. *Invest Educ. Enferm.* **2010**, *28*, 345–354.
32. Waluyo, A.; Culbert, G.J.; Levy, J.; Norr, K.F. Understanding HIV-related stigma among indonesian nurses. *J. Assoc. Nurses AIDS Care* **2015**, *26*, 69–80. [CrossRef]
33. Tyer-Viola, L.A. Obstetric nurses' attitudes and nursing care intentions regarding care of HIV-positive pregnant women. *J. Obstet. Gynecol. Neonatal Nurs.* **2007**, *36*, 398–409. [CrossRef]
34. Korhonen, T.; Kylmä, J.; Houtsonen, J.; Välimäki, M.; Suominen, T. University students' knowledge of, and attitudes towards, HIV and AIDS, homosexuality and sexual risk behaviour: A questionnaire survey in two finnish universities. *J. Biosoc. Sci.* **2012**, *44*, 661–675. [CrossRef] [PubMed]
35. Zou, J.; Yamanaka, Y.; John, M.; Watt, M.; Ostermann, J.; Thielman, N. Religion and HIV in Tanzania: Influence of religious beliefs on HIV stigma, disclosure, and treatment attitudes. *BMC Public Health* **2009**, *9*, 75. [CrossRef] [PubMed]

Publisher's Note: MDPI stays neutral with regard to jurisdictional claims in published maps and institutional affiliations.

International Journal of
Environmental Research and Public Health

Article

Ethnic Disparities in Utilization of Maternal and Child Health Services in Rural Southwest China

Chaofang Yan [1], Charuwan Tadadej [1,*], Kanittha Chamroonsawasdi [2], Natkamol Chansatitporn [3] and John FC Sung [4]

[1] Department of Public Health Administration, Faculty of Public Health, Mahidol University, Bangkok 10400, Thailand; chaofangyan@hotmail.com
[2] Department of Family Health, Faculty of Public Health, Mahidol University, Bangkok 10400, Thailand; kanittha.cha@mahidol.ac.th
[3] Department of Biostatistics, Faculty of Public Health, Mahidol University, Bangkok 10400, Thailand; nutkamol.cha@mahidol.ac.th
[4] Institute of Health and Development Studies, School of Public Health, Kunming Medical University, Kunming 650500, China; fcsung1008@yahoo.com
* Correspondence: charuwan.tad@mahidol.ac.th; Tel.: +66-2-644-8833; Fax: +66-2-644-8554

Received: 4 November 2020; Accepted: 17 November 2020; Published: 19 November 2020

Abstract: Background: Studies in China on ethnic disparities in access to health care in remote and rural population remain insufficient. This study aimed to assess the disparities in utilization of maternal and child health (MCH) services, including antenatal care (ANC), hospital birth, child growth monitoring, and immunization compliance between Han and ethnic minority women in Yunnan Province. Methods: A multi-stage sampling scheme was used to randomly recruit women from 40 townships in 14 remote prefectures of extremely remote areas in Yunnan. From birth records, we identified and recruited 303 Han women and 222 ethnic minority women who had given birth to a child within 3 years for an interview. Results: Overall, 96% of women used the ANC checkups and more than 95% had infants born in hospitals. However, the proportion of women compliant with early ANC visits (having antenatal care in the first trimester) was 22.5% lower in minority women than in Han women (61.3% vs. 83.8%, $p < 0.001$) with an adjusted odds ratio (aOR) of 2.04 (95% confidence interval (CI) of 1.13–3.66) for the minority group. The proportion of children under one year old with immunizations completed in a timely manner was also lower in minority families than in Han families (80.2% vs. 86.8%, $p < 0.05$) with an aOR of 1.99 (95% CI = 1.16–3.40). Conclusions: Ethnic disparities remain in utilization of early ANC visits and timely immunization completion for newborns. Ethnic minority women tended to lag behind for both. Further intervention should focus on assisting minority women living in extremely rural areas to comply with the MCH policy. Culturally-sensitive policies and skills are needed, and priority should be given to improve utilization of early ANC and timely immunization completion.

Keywords: ethnic disparity; utilization; maternal and child health services; China

---

1. Introduction

Women and children are at a crucial position in social development. The United Nations (UN) has highlighted strategies to promote the health of women and children in the Millennium Development Goals (MDG) 4 and 5 from 2000 to 2015, and in Sustainable Development Goal (SDG) 3 from 2016 to 2030. Substantial progress in maternal and child survivals has been made since the 1990s [1]. Timely maternal and child health (MCH) services utilization is vital to reduce both the maternal and child deaths [2–5]. The mediation of the MDG has increased utilization of antenatal care (ANC) services to 44% and assisted delivery to 12%. The UN efforts have thus reduced maternal deaths by 44% from

283.2 per 100,000 live births in 1990 [4,6]. The coverage for priority interventions among disadvantaged populations in a country is an indication of the strength of the health system [7].

Reducing MCH disparity has been a crucial issue across the whole nation in China, a developing country with the largest population in the world [8–11]. The Chinese government has taken a series of actions in order to increase the coverage of the interventions of MCH among disadvantaged populations in less developed regions [12–14]. For example, from 1996 to 1999 the government started a World Bank loan program, the Poverty Alleviation Fund for Maternal and Child Health. The program aimed to reduce maternal mortality rate (MMR) and infant mortality rate (IMR) by assisting 5% of poor families with their payment of the MCH services in five western provinces [12,15]. From 2000 to 2008, the program of Reducing Maternal Mortality and Eliminating Neonatal Tetanus was launched to reduce MMR, mainly targeting 378 counties in rural areas [14]. The Basic Public Health Service (BPHS) project, launched in 2009, is one of the most effective policies for providing and promoting residents with equal access to basic public health services. This government-funded project includes nine service categories, including one reform to improve MCH services [13,16]. After years of effort, the urban–rural disparity of MMR in China has been greatly narrowed. The urban-to-rural ratio of MMR reduced from 1:2.2 in 1990 to 1:1.3 in 2018 (15.5 vs. 19.9 deaths per 100,000 livebirths) [10]. The western part of China is less developed than the eastern part of China; the eastern-to-western region ratio of MMR reduced from 1:4.7 in 1996 to 1:2.3 (10.9 vs. 25.2 per 100,000 livebirths) in 2018. The project also reduced the urban–rural ratio of the under-5 mortality from 1:3.4 in 1991 to 1:2.3 in 2018 (4.4 vs. 10.2 per 1000 livebirths). The gap in under-5 mortality rates between east and west regions was reduced to 8.5‰ in 2018 from 66.5‰ in 1991 [10].

The MCH disparities between east and west China, or between urban and rural, have been explored and well documented. However, studies evaluating the ethnic disparities with robust methods remain severely insufficient in China [17]. The Lancet–Lowitja Institute Global Collaboration study has shown that Yunnan province and Tibet Autonomous Region in China have not yet reached the UN SDG for underserved populations [18]. Some studies related to ethnic populations have reported health coverage and health outcomes based on the whole population from the Autonomous Region rather than classified by ethnic type [8,19]. No study has compared differences of MCH between Han and ethnic minorities at a provincial level, because the medical records and vital statistics were rarely classified by ethnicity in the official health statistics system. Most studies have emphasized the improvement of health data collection for ethnic minorities in China [17,18,20].

China is populated with multiple ethnic groups with high cultural and language diversities; these consist of the Han majority accounting for 91.5% of the total population and 55 minority groups accounting for the rest of the population (8.5%). The 2010 census estimated that 114 million people in China were ethnic minorities. Three-quarters (71.4%) of all ethnic minorities live in the remote southern and western regions of China, including Yunnan Province [21].

Yunnan Province is located in Southwestern China, bordering Myanmar in the west, and Laos and Vietnam in the south. It is one of the poorest remote provinces and has the highest ethnic diversity in China (Figure 1), including 8 ethnic autonomous prefectures and 29 ethnic autonomous counties. Among 56 recognized ethnic groups in China, Yunnan Province has recognized at least 25 ethnic minority groups, making up 33.6% of the 48 million population in the province in 2018. Yunnan is situated in mountainous land, with only 6% of plain areas suitable for cultivation [22]. The provincial authority has endeavored to carry out the ambitious provincial goal since 2010 for reducing MMR to 25/100,000 and IMR to 11 per 1000 live births by 2020, then reducing the MMR to 12/100,000 and the IMR to 5 per 1000 live births by 2030, based on the MMR 37.3/100,000 and the IMR 15.2 per 1000 live births in 2010 [23,24]. For better utilization of MCH services, pregnant women are officially encouraged to complete at least five ANC with the first ANC in the first trimester, hospital delivery, having their children under one year old checked for growth monitoring at least four times, and getting their children vaccinated in a timely manner according to the National Immunization Program [25].

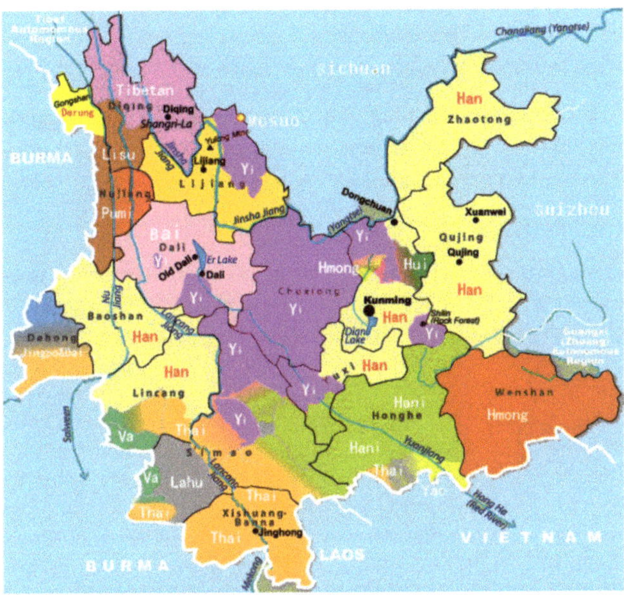

Figure 1. Ethnic diversity of Yunnan Province, China.

As a province with typical ethnic minorities, no study has yet compared the compliance of MCH care services between Hans and ethnic minorities in Yunnan. Therefore, we conducted a survey in remote rural areas in the province to assess the disparities in utilization of MCH services between Han and the ethnic minority mothers.

2. Materials and Methods

2.1. Study Design and Participants Recruitment

This study was a cross-sectional survey with a comparison between Han and ethnic minorities. Women who had given birth within the past three years at the time of the survey and had been residing in the selected township in Yunnan for at least three years were included in the study. This inclusion criteria ensured that these women could provide the health service information for both mothers and children, such as the information of growth monitoring and immunization. Women who worked and used health services out of the selected township were excluded to show the effect of ethnicity in the same context. A conceptual framework, adapted from the model of access to medical care by Andersen [26], was used to examine the access to MCH services and to analyze the factors affecting access to MCH services.

The sample size was calculated based on detecting a significant proportional difference of hospital births between ethnic minorities and the Han group at a 5% level with 80% power of a two-sided test. Previous studies in Yunnan Province have reported that the hospital birth rates were 81% (512/631) in minority women and 91% (476,132/523,223) in Han women [27,28]. These data were used to calculate the minimum sample size being at least 188 subjects in each group for this study.

A multi-stage sampling method was used to select the study sample. Firstly, we excluded the provincial capital Kunming to eliminate the effects of economic inequality and Nujiang prefectures that are difficult to reach by automobile transportation. For the rest of the 14 prefectures, 1 county was randomly selected from each prefecture. In each selected county, after excluding the township in which the county government was located, we randomly selected a township. Based on the birth records from the township hospital in the selected township, 40 women with a birth within the past three years at the time of the survey were randomly selected from each township, resulting in a total of 560 women selected as potential participants. For participants with more than one pregnancy and delivery, information from the last pregnancy was collected and used.

This study was approved by the ethics committee of the Faculty of Public Health, Mahidol University (COA No. MUPH 2014-214; 24 November 2014). All participants were clearly informed of their rights and any risks associated with participation. Verbal consent was obtained from all interview participants.

2.2. Study Procedure and Data Collection

We applied the Andersen health behavior model to establish a structured questionnaire for data collection of the survey to investigate MCH services associated with predisposing (demographic and social) factors, enabling (economic) factors, and need (health outcomes) factors [26]. The questionnaire was reviewed by experts in MCH, social medicine, and health management.

The questionnaire asked for four types of information. The first type regarded predisposing factors including the general characteristics of respondents, comprising age of the woman, age of the child, sex of the child, parity, ethnicity, language ability, education, and knowledge on MCH. The second part regarded enabling factors: the family income, family size and health insurance, the usual health facility, and travel time to the nearest health facility for childbirth. The third part regarded need factors: the woman's health status during the perinatal period, previous obstetric problems, child health status, preterm delivery, and birth weight of the child. The fourth part asked for information about MCH utilization, including ANC utilization, timing for ANC visits, hospital birth, delivery type, child growth monitoring services, timing for child growth monitoring visits, and immunization services.

Before the actual data collection, the questionnaire was pre-tested on 20 women from villages of the study area. The reliability of scaling variables was assessed by calculating Cronbach's coefficient alpha ($\alpha = 0.716$), which indicated that the measures had fairly high levels of reliability. The questionnaire was administered in person by trained interviewers at each participant's household. We trained 32 medical students who could speak the local dialog as interviewers. A two-day interview workshop was given to them before conducting the survey.

After excluding 22 women who refused interviews and 13 women who could not complete the questionnaires, information was obtained from 525 households in total. The ethnic minorities included 15 ethnicity groups of Yi, Hani, Thai, Zhuang, Hmong, Bai, Yao, Hui, Lisu, Va, Lahu, Nu, Jingpo, Achang, and Paijiao.

2.3. Statistical Analysis

Eligible questionnaires were edited and coded, and the data were entered and processed using SPSS version 19. Data analysis first compared distributions of participants' characteristics in predisposing factors, enabling factors, and need factors between Han and ethnic minority participants. Utilization of MCH care (ANC visit, infant delivery, growth monitoring, and immunization) was presented by numbers and proportions. Chi-square test was used to examine the difference between the two groups. Logistic regression analysis was further used to calculate odds ratio (OR) and 95% confidence interval (CI) of inadequate utilization of MCH care for the ethnic minority women, compared to Han women. The adjusted odds ratio (aOR) was estimated with a multivariable by three models: after controlling for predisposing factors, after controlling for predisposing and enabling factors, and after controlling for predisposing, enabling, and need factors.

3. Results

3.1. Socio-Demographic Profile

With response rates of 97.1% and 89.5% in Han and ethnic minorities, respectively, 303 Han women and 222 minority women completed the questionnaire interviews. The average ages of participants in both Han and ethnic minority groups were approximately similar at 26 years old. The Han group had higher education and better Mandarin language ability, with higher income, than the ethnic minority group. Ethnic minority women were more likely (69.4%) to be married to ethnic minority husbands (refer to Husband's Ethnicity type) than were Han women (8.3%), and they had a larger family size than Han women did. Ethnic minority women needed more travel time to reach their nearest health care facility and perceived that their children had a poor health status. (Table 1).

Table 1. Sociodemographic characteristics of study subjects in Yunnan.

Characteristics	Total N = 525	Han N = 303	Minorities N = 222	*p*-Value
	n (%)	n (%)	n (%)	
Predisposing factors				
Age, years	26 ± 4.6	26.1 ± 4.7	25.9 ± 4.6	0.48
Husband's Ethnicity				<0.001
Han	346 (65.9)	278 (91.7)	68 (30.6)	
Ethnic minority	179 (34.1)	25 (8.3)	154 (69.4)	
Family size	5.1 ± 1.4	4.9 ± 1.4	5.2 ± 1.4	0.04
Parity				0.49
1	297 (56.6)	174 (57.4)	123 (55.4)	
2	194 (37.0)	109 (36.0)	85 (38.3)	
≥3	34 (6.5)	20 (6.6)	14 (6.4)	
Enabling factors				
Women's Education				0.001
≤6 years	160 (30.5)	78 (25.8)	82 (36.9)	
7–9	241 (45.9)	138 (45.5)	103 (46.4)	
≥10	124 (23.6)	87 (28.7)	37 (16.7)	
Husband's Education				<0.001
≤6 years	112 (21.3)	51 (16.8)	61 (27.5)	
7–9	283 (53.9)	155 (51.2)	128 (57.7)	
≥10	130 (24.8)	97 (32.0)	33 (14.9)	
Fluent in Mandarin				<0.001
None/Partly	29 (5.5)	2 (0.7)	27 (12.2)	
Yes	496 (94.5)	301 (99.3)	195 (87.8)	
Average annual per capita income, CNY [1]				0.008
<2800	235 (44.8)	127 (41.9)	119 (53.6)	
≥2800	181 (34.5)	176 (58.1)	103 (46.4)	
Median	2800	3000	2500	
Health Insurance Status				0.88
Public (NCMS [2])	515 (98.1)	167 (98.0)	218 (98.2)	
Private/None	10 (1.9)	136 (2.0)	4 (1.8)	
Travel to nearest health facility				0.01
<30 min	176 (33.5)	94 (31.0)	82 (36.9)	
30–59 min	199 (37.9)	131 (43.2)	68 (30.6)	
≥60 min	150 (28.6)	78 (25.7)	72 (32.4)	
Need factors				
Perceived health status of mother				0.57
Poor/very poor	18 (3.4)	11 (3.7)	7 (3.2)	
Fair	125 (23.8)	72 (23.8)	53 (23.9)	
Good/Excellent	382 (72.8)	220 (72.6)	162 (73.0)	
Have any previous obstetric problem				0.60
Yes	48 (9.1)	26 (8.6)	22 (9.9)	
No	477 (90.9)	277 (91.4)	200 (90.1)	
Perceived health status of child				0.04
Poor/Very Poor	34 (6.5)	17 (5.7)	17 (7.7)	
Fair	98 (18.7)	54 (17.8)	44 (19.8)	
Good/Excellent	393 (74.8)	232 (76.6)	161 (72.5)	
Birth weight of Child	3210 ± 526	3230 ± 32	3183 ± 33	0.31

[1] CNY, Chinese yuan, 1 USD = 6.2 CNY; [2] NCMS, New rural cooperating medical scheme.

3.2. Comparison of MCH Utilization

More than 90% of women had at least one ANC visit and nearly 60% made five visits during their pregnancy (Table 2). The Han women had a higher adherence to the recommended 5-visit schedules than did ethnic minority women by making their first ANC visit in the first trimester (58.1% vs. 43.2%, *p* < 0.001). However, making an early ANC visit was 22.5% lower in minority women than in Han women (61.3 vs. 83.8%). Over 95% of participants had their infants delivered at health institutions, with a higher incidence of Caesarean delivery in Han women (26.1% vs. 20.3%, p 0.12). However, a lower proportion of ethnic minority women completed the required vaccinations for children within the first year of age than Han women (80.2% vs. 86.8%, $p < 0.05$).

Table 2. Comparison on utilization of maternal and child health care between Han and ethnic minority women.

Type of MCH [1] Services	Total N = 525	Han N = 303	Minorities N = 222	X2	*p*
	n (%)	n (%)	n (%)		
Antenatal care					
ANC [2] at least 1 visit	504 (96.0)	297 (98.0)	207 (93.2)	7.61	<0.05
ANC at least 5 visits	318 (60.6)	192 (63.4)	126 (56.8)	2.34	0.13
Early ANC visit [3]	390 (74.3)	254 (83.8)	136 (61.3)	27.38	<0.001
ANC at least 5 visits with an early ANC visit	272 (51.8)	176 (58.1)	96 (43.2)	11.3	<0.05
Hospital births					
Hospital birth	503 (95.8)	292 (96.4)	211 (95.0)	0.56	0.45
Caesarean Delivery	124 (23.6)	79 (26.1)	45 (20.3)	2.39	0.12
Growth monitoring services for children within the first year of age	231 (44.0)	126 (41.6)	105 (47.3)	1.70	0.19
Use at least 4 times	231 (44.0)	126 (41.6)	105 (47.3)	1.70	0.19
Immunization					
Vaccinated all doses	520 (99.0)	300 (99.0)	220 (99.1)	0.01	0.91
Vaccinated all doses timely	441 (84.0)	263 (86.8)	178 (80.2)	4.18	<0.05

[1] MCH, maternal and child care; [2] ANC, antenatal care; [3] Early ANC visit, having antenatal care in the first trimester.

3.3. Odds Ratio of MCH Services Utilization

Table 3 shows that, compared with the Han women, the minority women were at higher risks for inadequate antenatal care, with significant unadjusted ORs of 3.59 (95% CI = 1.37–9.40) for failing to make at least one ANC visit during the pregnancy, of 3.28 (95% CI = 2.00–4.75) for failing to make an early ANC visit in the first trimester, and of 1.82 (95% CI = 1.28–2.58) for failing to make at least 5 ANC visits with an early ANC visit in the first trimester. However, only the risk of failing to make an early ANC visit remained significant after controlling for predisposing, enabling, and need factors, with an aOR of 2.04 (95% CI = 1.13–3.66). The ethnic minorities were also at a higher risk of not receiving vaccinations for children with all doses in a timely manner, with an aOR of 1.99 (95% CI = 1.16–3.40) after controlling for all covariables.

Table 3. Ethnic minority women compared to Han women odds ratio (OR) of inadequate MCH utilization and adjusted odds ratio (aOR) controlling for predisposing, enabling, and need factors.

Type of MCH Services	Base Model, OR (95% CI)	Predisposing Factors Model, OR (95% CI)	Predisposing-Enabling Factors Model, OR (95% CI)	Predisposing-Enabling-Need Factors Model, OR (95% CI)
Antenatal care				
ANC at least 1 visit	3.59 (1.37–9.40) *	1.80 (0.51–6.31)	1.62 (0.36–7.38)	1.88 (0.42–8.51)
ANC at least 5 visits	1.32 (0.93–1.88)	0.96 (0.60–1.53)	0.97 (0.59–1.60)	1.01 (0.61–1.67)
Early ANC visit	3.28 (2.00–4.75) *	1.81 (1.05–3.15) *	1.91 (1.07–3.41) *	2.04 (1.13–3.66) *
ANC at least 5 visits with an early ANC visit	1.82 (1.28–2.58) *	1.27 (0.81–2.00)	1.34 (0.83–2.17)	1.42 (0.87–2.32)
Hospital birth				
Hospital birth	1.38 (0.59–3.25)	1.00 (0.32–3.14)	0.90 (0.23–3.90)	0.95 (0.25–3.66)
Caesarean Delivery	0.72 (0.48–1.09)	0.71 (0.42–1.23)	0.76 (0.44–1.31)	0.77 (0.44–1.33)
Growth monitoring services for children within the first year of age				
Use at least 4 times	0.79 (0.56–1.12)	0.70 (0.44–1.10)	0.65 (0.41–1.05)	0.65 (0.41–1.05)
Immunization services				
Vaccinated all doses	0.91 (0.15–5.49)	1.68 (0.21–13.6)	2.47 (0.19–31.4)	2.47 (0.20–31.0)
Vaccinated all doses timely	1.63 (1.02–2.60) *	1.73 (1.03–2.89) *	1.79 (1.06–3.04) *	1.99 (1.16–3.40) *

Note: The Predisposing Factors Model controlled for husband ethnicity and family size. Predisposing-Enabling Factors Model controlled for husband ethnicity, family size, education, husband education, Mandarin language ability, income, and travel time. The Predisposing-Enabling-Need Factors Model controlled for husband ethnicity, family size, education, husband education, Mandarin language ability, income, travel time, and perceived child health, which were significant predictors. * $p < 0.05$.

4. Discussion

This study aimed to assess the disparities in the utilization of MCH services between Han and ethnic minority mothers in remote rural areas. Theoretically, minority women and Han women could receive similar care even in the remote rural areas. We found that gaps of using MCH services still existed in ANC utilizations, significant for the early ANC visit in the mountainous rural region of Yunnan. The pregnant women from minority families were less likely than Han women to make the first ANC within the first 12 gestational weeks. Previous studies have rarely measured the disparities between the Han and ethnic minority populations in pregnancies, although some studies have reported the rates of the early ANC visits related to ethnic minority population in other provinces [29–33]. The timing of the first antenatal care visit is an important compliance measure to ensure optimal pregnancy outcomes for women and children [4,34]. The early antenatal care visit also has been used as an indicator to measure the quality of MCH services since launching the health care reform in 2009 in China [25]. This finding might imply the need of exploring the cultural factors for early ANC visits among ethnic minority women. Our study also showed the need to develop additional culture-sensitive policy to eliminate inequality and deliver cost effective MCH interventions.

Further, compared to Han mothers, the ethnic minority mothers were also less likely to have their children immunized in a timely manner, which is crucial for preventing childhood diseases. Studies have consistently reported a poor compliance with immunization services for children of ethnic minorities in China [17,35–37]. With an investment of USD $120 billion, the universal health insurance coverage increased to 95% of the population in 2019 from 30% in 2003 [38]. The economic barriers to accessing ANC, hospital births, and immunization services for rural populations have been removed, yet disparities were still observed.

This study found that hospital birth and caesarean delivery between Han and ethnic minority women were not significantly different. This was likely because of the compliance to the UN's MDG and SDG strategies to promote the health of women and children. In China, like many developing countries, the economic level plays an important role in health service accessibility, especially in remote rural areas where many ethnic minorities live. Over the past 20 years, the Chinese Government has introduced several strategies aimed specifically at reaching underserved populations in rural areas. In 2003, a voluntary health insurance program, New Cooperative Medical Scheme (NCMS), was launched for rural residents. This heavily subsidized program has increased outpatient and inpatient services utilization by reducing the health service cost with a sliding fee discount of 30% to 80%, varying by region annually [39,40]. In addition to the major support from the central authority, some of the local health administrations also received support from international organizations such as the United Nations International Children's Emergency Fund (UNICEF) [12,15,41]. The administrations were thus capable of conducting pilot studies using poverty alleviation funds to help low income pregnant women adhere to MCH services in a timely manner in some identified poverty-stricken counties.

Similarly, there were also no significant differences in using service of growth monitoring for children under 1 year old between Han and ethnic minority families in this study. The BPHS project of new medical reform policies has made a great contribution to promoting equal access to basic public health services since it was launched in 2009 [42]. This government-funded project included nine service categories, one of which was to improve infant cares. The project encouraged all families having their children under 1 year old to complete "four times-routine examination at the 3, 6, 8, and 12 months of age" to monitor growth and development free of charge. Further, all MCH physicians are required to have standardized training and to deliver the same maternal services for rural and urban populations [25].

As a whole, we demonstrated a substantial improvement in utilization of MCH services among all women in this study compared with the utilizations in the 1990s, providing valid information on the improvement of MCH services in remote rural areas. Overall, 96% of pregnant women used the antenatal checkups at least once in the present study, compared to 58.07% in 1992 in the whole Yunnan province. The compliance with at least five antenatal checkups improved to 60.6% from 23.78% in 1992. The greatest improvement is that the hospital birth increased to 95.8% from only 30.33% in 1990. More importantly, vital statistics data showed the MMR had a great decline from 115.29 per 100,000 livebirths in 1992 to 17.72 per 100,000 livebirths in 2018, and the infant mortality rates declined from 46.38‰ in 1992 to 5.85‰ in 2018 in Yunnan Province [43,44]. All the policies such as the NCMs and BPHS contribute to this great progress. The Targeted Poverty Alleviation Project was recently launched to further increase accessibility to local health care without charge for families officially registered as Poverty-Stricken Households [45]. We anticipate this effort will further reduce the ethnic disparities in MCH services utilization in rural China. However, most current policies tend to invest more money to reduce health inequality but have less emphasis on cultural barriers, transportation barriers, and development of the health workforce in extremely rural areas [46,47]. Minority women living in these areas deserve further intervention if the highest ethnic diversity province wants to reach the goals of MMR of 12/100,000 and the IMR of 5 per 100 live births by 2030 [24].

This study has a few limitations. This study excluded women who had experienced an abortion or had children who died under the age of 3 years. Their utilization of MCH services is not clear. Among the 560 women randomly invited to participate in this study, 35 women were not interviewed. These women were less likely to be compliant with required MCH checkups, and their information was not included in this study. Furthermore, information on MCH services was collected from women within 3 years of the childbirth at the time of survey, the self-reported data throughout the period might vary slightly among women. However, interviewers conducted the survey following strict protocols. The response bias was likely unintentional and would not adversely affect the survey outcome.

5. Conclusions

This study found that gaps of using MCH services between ethnic minorities and Han women exist in ANC utilization and timely immunization completion in China. Ethnic minority women tended to lag behind in utilization of early ANC, and in timely immunization completion for their children under one year old. However, there were no significant differences in hospital birth, caesarean delivery, and growth monitoring for children under 1 year old between Han and ethnic minority women in this study. China's political commitment to poverty reduction for compliance with the UN MDG and SDG might have contributed to the improvement. A heavily subsidized NCMS and the BPHS of new medical reforms might also play an important role in promoting equal access to MCH services. Minority women living in extremely rural areas deserve further intervention. Culturally-sensitive policies and skills are needed to target ethnic populations to improve their access to MCH services in order to continue decreasing the urban–rural and ethnic disparities. Priority should be given to improve utilization of early ANC and timely immunization completion.

Author Contributions: C.Y. designed the study, analyzed data, interpreted findings, and drafted the manuscript. C.T. provided supervision in the development of the study concept and design, advised on the scope of the paper, and provided critical revision of the manuscript. K.C. advised on the study theory and revised the manuscript critically. N.C. advised on sampling and data analysis methods and revised the manuscript critically. J.F.S. assisted in study design and data presentation and revised critically the manuscript for important intellectual content. All authors have read and agreed to the published version of the manuscript.

Funding: This research received no external funding.

Acknowledgments: We are thankful to health officers, rural physicians and medical students from Kunming Medical University in Yunnan, China, for their great assistance in this study. This article is a fulfillment of the requirements for a Doctoral study in Public Health, Mahidol University, Bangkok, Thailand, supported by the grant of International Fellowships Program of Ford Foundation (Grant ID:15090343).

Conflicts of Interest: The authors declare that they have no competing interests.

Abbreviations

aOR	adjusted odds ratio
ANC	antenatal care
BPHS	basic public health service
CI	confidence interval
IMR	infant mortality rate
MCH	maternal and child health
MDG	Millennium Development Goal
MMR	maternal mortality rate
NCMS	New Cooperative Medical Scheme
OR	odds ratio
SDG	Sustainable Development Goal
UNICEF	United Nations International Children's Emergency Fund

References

1. WHO; UNICEF; UNFPA; World Bank Group and the United Nations Population Division. *Trends in Maternal Mortality: 2000 to 2017*; World Health Organization: Geneva, Switzerland, 2019.
2. Spinelli, A.; Baglio, G.; Donati, S.; Grandolfo, M.E.; Osborn, J. Do antenatal classes benefit mother and her baby? *J. Matern. Fetal Neonatal Med.* **2003**, *13*, 94–101. [CrossRef] [PubMed]
3. WHO; UNICEF; UNFPA; The World Bank and the United Nations Population Division. *Trends in Maternal Mortality: 1990 to 2013*; World Health Organization: Geneva, Switzerland, 2014.
4. Moller, A.B.; Petzold, M.; Chou, D.; Say, L. Early antenatal care visit: A systematic analysis of regional and global levels and trends of coverage from 1990 to 2013. *Lancet Global Health* **2017**, *5*, e977–e983. [CrossRef]
5. Templeton, A. The Public Health Importance of Antenatal Care. *Facts Views Vis. Obgyn* **2015**, *7*, 5–6.
6. Alkema, L.; Chou, D.; Hogan, D.; Zhang, S.; Moller, A.-B.; Gemmill, A.; Fat, D.M.; Boerma, T.; Temmerman, M.; Mathers, C.; et al. Global, Regional, and National Levels and Trends in Maternal Mortality Between 1990 and 2015, With Scenario-based Projections to 2030. *Obstet. Anesthesia Dig.* **2016**, *36*, 191. [CrossRef]
7. Bryce, J.; Terreri, N.; Victora, C.G.; Mason, E.; Daelmans, B.; Bhutta, Z.A.; Bustreo, F.; Songane, F.; Salama, P.; Wardlaw, T. Countdown to 2015: Tracking intervention coverage for child survival. *Lancet* **2006**, *368*, 1067–1076. [CrossRef]
8. Ren, Y.; Zhao, Z.; Pan, J.; Qian, P.; Duan, Z.; Yang, M. Ethnic disparity in maternal and infant mortality and its health-system determinants in Sichuan province, China, 2002–2014: An observational study of cross-sectional data. *Lancet* **2017**, *390*, S7. [CrossRef]
9. Guo, Y.; Huang, Y. Realising equity in maternal health: China's successes and challenges. *Lancet* **2019**, *393*, 202–204. [CrossRef]
10. National Health Commission of the People's Republic of China. Report on the development of maternal and child health in China. *Chin. J. Women Child. Health* **2019**, *10*, 1–8.
11. Guo, Y.; Yin, H. Reducing child mortality in China: Successes and challenges. *Lancet* **2016**, *387*, 205–207. [CrossRef]
12. Liu, F.Y.; Zhu, X.X.; Yang, L.; Guo, G.P.; Zhou, H.; Li, Z. The Implementation and effect of World Bank Loan Health Project in Yunnan province. *Matern. Child Health Care China* **2002**, *17*, 403–406.
13. Zhao, P.; Diao, Y.; You, L.; Wu, S.; Liu, Y. The influence of basic public health service project on maternal health services: An interrupted time series study. *BMC Public Health* **2019**, *19*, 1–8. [CrossRef] [PubMed]
14. Liang, J.; Zhu, J.; Wang, Y.; Li, M. Factors of Reducting the Maternal Mortality in counties implementing project of "Reduce maternal mortality and eliminate neonatal tetanus". *Chin. J. Epidemiol.* **2007**, *28*, 746–748.
15. Chen, L.; Pei, H.; Huo, L. The Sustainable Development of Poverty Alleviation Fund for Maternal and Child Health in China. *Matern. Child Health Care China* **2002**, *23*, 87–89.
16. Wang, F.; Li, Y.; Ding, X.; Dai, T. The national essential public health services project in China: Progress and equity. *Chin. J. Health Policy* **2013**, *6*, 9–14.
17. Huang, Y.; Shallcross, D.; Pi, L.; Tian, F.; Pan, J.; Ronsmans, C. Ethnicity and maternal and child health outcomes and service coverage in western China: A systematic review and meta-analysis. *Lancet Glob. Health* **2018**, *6*, e39–e56. [CrossRef]

18. Anderson, I.; Robson, B.; Connolly, M.; Al-Yaman, F.; Bjertness, E.; King, A.; Tynan, M.; Madden, R.; Bang, A.; Coimbra, C.E., Jr.; et al. Indigenous and tribal peoples' health (The Lancet–Lowitja Institute Global Collaboration): A population study. *Lancet* **2016**, *388*, 131–157. [CrossRef]

19. Duan, M.F.; Zhou, Y.B.; Li, H.T.; Gao, Y.Q.; Zhang, Y.L.; Luo, S.S.; Kang, C.Y.; Liu, J.M. The rate of hospital births in regions inhabited by ethnic minorities China from 1996 to 2017. *Chin. Med. J.* **2019**, *99*, 2135–2140.

20. Stephens, C.; Porter, J.; Nettleton, C.; Willis, R. Disappearing, displaced, and undervalued: A call to action for Indigenous health worldwide. *Lancet* **2006**, *367*, 2019–2028. [CrossRef]

21. National Bureau of Statistics. *China Statistical Yearbook 2016*; China Statistics Press: Beijing, China, 2016.

22. Yunnan Provincial Government. Polulation and Ethnicity of Yunnan Province. Available online: http://www.yn.gov.cn/yngk/gk/201904/t20190403_96251.html (accessed on 4 November 2020).

23. National Working Committee on Children and Women under State Council of the People's Republic of China. A bulletin on "The Program of Yunan's Women Development (2011–2020)" and "The Program of Yunan's Children Development (2011–2020)". Available online: http://www.nwccw.gov.cn/node_2502.html (accessed on 4 November 2020).

24. People's Government of Yunnan Province. An Outline for the "Healthy Yunnan 2030" Initiative. Available online: http://www.yn.gov.cn/zwgk/zcwj/qtwj/201911/t20191101_183998.html (accessed on 4 November 2020).

25. Health Ministry of China. *The Outline of Basic Public Health Services Project in China*; People's Medical Publishing House: Beijing, China, 2012.

26. Andersen, R.M. Revisiting the Behavioral Model and Access to Medical Care: Does It Matter. *J. Health Soc. Behav.* **1995**, *36*, 1–10. [CrossRef]

27. He, G. Analysis of Hospital Delivery rate and it factors in DeQin County, Yunnan Province China. *Health Horiz. Med. Sci.* **2012**, *2*, 175.

28. Yu, Y. The rethinking about promoting safe childbirth in poor and remote rural Yunnan. *Soft Sci. Health* **2012**, *26*, 386–388.

29. An, L.; Gao, Y.; Zhang, J. Study on the influent factors of antenatal examination in West Poor Rural Areas of China. *Chin. J. Reprod. Health* **2004**, *15*, 146–149.

30. Lei, P.; Duan, C.G.; Zhuo-Chun, W.U.; Li, C.; Yu, L.J.; Li, X.H.; Xu, L.; Liu, Y.; Liu, M.W.; Liu, Y.G.; et al. Evaluation of Qinba Health Program: Its impact on antenatal care utilization. *Chin. Health Resour.* **2010**, *13*, 64–66.

31. Ren, Z. Utilisation of antenatal care in four counties in Ningxia, China. *Midwifery* **2011**, *27*, e260–e266. [CrossRef]

32. Liu, C.; Zhang, L.; Shi, Y.; Zhou, H.; Medina, A.; Rozelle, S. Maternal health services in China's western rural areas: Uptake and correlates. *China Agric. Econ. Rev.* **2016**, *8*, 250–276. [CrossRef]

33. Tan, L.F.; Huang, C.L.; Yang, R.R.; Zheng, X.Y. Study on the accessibility of maternal health services for Tibetan women in agricultural and pastoral areas in Yushu Autonomous Prefecture. *Matern. Child Health Care China* **2012**, *27*, 325–328.

34. Razum, O.; Breckenkamp, J.; Borde, T.; David, M.; Bozorgmehr, K. Early antenatal care visit as indicator for health equity monitoring. *Lancet Glob. Health* **2018**, *6*, e35. [CrossRef]

35. Qi, H.Y.; Fan, J.L.; Zhang, X.L.; Yao, K.L.; Zhang, H.S.; Shi, M.F.; Hou, C.X. Survey and analysis of vaccine inoculation rate of National Immunization Programs in Linxia Prefecture of Gansu Province. *Bull. Dis. Control Prev.* **2015**, *30*, 13–17.

36. Zhou, J.; Tang, G.P.; Li, Z.Y.; Deng, J.; Wu, G.G.; Tian, X.G. Investigation on children immunization coverage rate of hepatitis B vaccine and the influential factors. *Mod. Prev. Med.* **2007**, *34*, 1444–1447.

37. Lin, Z.; Wang, Q.; Zhang, S.; Wu, J.; Zhou, Y. Association between timely immunization of hepatitis B vaccine and the completion of the national immunization program vaccine. *China Trop. Med.* **2017**, *17*, 485–488.

38. National Healthcare Security Administration. Bulletin for HealthCare Statistics in 2019. Available online: http://www.nhsa.gov.cn/art/2020/3/30/art_7_2930.html (accessed on 4 November 2020).

39. Li, L. *The New Cooperation Medical System in China*; People's Publishing House: Beijing, China, 2009.

40. Xin, Y.; Xiang, L.; Jiang, J.; Lucas, H.; Tang, S.; Huang, F. The impact of increased reimbursement rates under the new cooperative medical scheme on the financial burden of tuberculosis patients. *Infect. Dis. Poverty* **2019**, *8*, 67. [CrossRef] [PubMed]

41. Ministry of Commerce of the People's Republic of China, The United Nations System in China. *Excellence in Poverty Reduction: Case Study on China-UN Poverty Reduction Collaboration in the Last Forty Years*; China Commerce and Trade Press: Beijing, China, 2019.
42. Health Ministry of China. A Bulletin on "National Standards for Basic Public Health Services". Available online: http://www.nhc.gov.cn/jws/s3581r/200910/fe1cdd87dcfa4622abca696c712d77e8.shtml (accessed on 19 November 2020).
43. People's Government of Yunnan Province. Report on Maternal and Child Health in Yunnan Province. Available online: http://m.yunnan.cn/system/2019/10/15/030399105.shtml (accessed on 4 November 2020).
44. Yu, Y.; Zhao, Z.; Liu, H.; Liu, J. The trend of maternal mortality ratio in yunnan province from 1987 to 2016 and the evaluation of the implementation of the millennium development goals. *Chin. Matern. Child Health Care* **2018**, *33*, 8–11.
45. National Health Commission of the People's Republic of China. Guidelines on the Implementation of the Health Poverty Alleviation Project. Available online: http://www.gov.cn/xinwen/2016-06/21/content_5084195.html (accessed on 4 November 2020).
46. Zhu, X.L.; Chen, Q.K.; Yang, S.X. The current situation and problems of primary health care personnel since the new round of China's health care reform. *Chin. J. Health Policy* **2015**, *8*, 57–62.
47. Gao, Y.; Zhou, H.; Singh, N.S.; Powell-Jackson, T.; Nash, M.S.; Yang, M.; Guo, S.; Fang, H.; Alvarez, M.M.; Liu, X.; et al. Progress and challenges in maternal health in western China: A Countdown to 2015 national case study. *Lancet Glob. Health* **2017**, *5*, e523–e536. [CrossRef]

Publisher's Note: MDPI stays neutral with regard to jurisdictional claims in published maps and institutional affiliations.

International Journal of
*Environmental Research
and Public Health*

Article

Maternal Dietary Patterns during Pregnancy and Their Association with Gestational Weight Gain and Nutrient Adequacy

Naomi Cano-Ibáñez [1,2,3], Juan Miguel Martínez-Galiano [2,4,*],
Miguel Angel Luque-Fernández [2,3,5], Sandra Martín-Peláez [1,2], Aurora Bueno-Cavanillas [1,2,3] and
Miguel Delgado-Rodríguez [2,6]

1 Department of Preventive Medicine and Public Health, University of Granada, 18071 Granada, Spain;
   ncaiba@ugr.es (N.C.-I.); sandramartin@ugr.es (S.M.-P.); abueno@ugr.es (A.B.-C.)
2 Consortium for Biomedical Research in Epidemiology and Public Health (CIBERESP), 28029 Madrid, Spain;
   miguel-angel.luque@lshtm.ac.uk (M.A.L.-F.); mdelgado@ujaen.es (M.D.-R.)
3 Instituto de Investigación Biosanitaria (ibs. GRANADA), Complejo Hospitales Universitarios de
   Granada/Universidad de Granada, 18071 Granada, Spain
4 Department of Nursing, University of Jaén, 23071 Jaén, Spain
5 Department of Noncommunicable Disease Epidemiology, London School of Hygiene and Tropical Medicine,
   London WC1H 9SH, UK
6 Division of Preventive Medicine and Public Health, University of Jaén, 23071 Jaén, Spain
* Correspondence: jgaliano@ujaen.es

Received: 2 October 2020; Accepted: 25 October 2020; Published: 28 October 2020

Abstract: Several epidemiologic studies have shown an association between Gestational Weight Gain (GWG) and offspring complications. The GWG is directly linked to maternal dietary intake and women's nutritional status during pregnancy. The aim of this study was (1) to assess, in a sample of Spanish pregnant women, the association between maternal dietary patterns and GWG and (2) to assess maternal dietary patterns and nutrient adequate intake according to GWG. A retrospective study was conducted in a sample of 503 adult pregnant women in five hospitals in Eastern Andalusia (Spain). Data on demographic characteristics, anthropometric values, and dietary intake were collected from clinical records by trained midwives. Usual food intake was gathered through a validated Food Frequency Questionnaire (FFQ), and dietary patterns were obtained by principal component analysis. Nutrient adequacy was defined according to European dietary intake recommendations for pregnant women. Regression models adjusted by confounding factors were constructed to study the association between maternal dietary pattern and GWG, and maternal dietary patterns and nutritional adequacy. A negative association was found between GWG and the Mediterranean dietary pattern (crude $\beta = -0.06$, 95% CI: -0.11, -0.04). Independent of maternal dietary pattern, nutrient adequacy of dietary fiber, vitamin B9, D, E, and iodine was related to a Mediterranean dietary pattern ($p < 0.05$). A Mediterranean dietary pattern is related to lower GWG and better nutrient adequacy. The promotion of healthy dietary behavior consistent with the general advice promoted by the Mediterranean Diet (based on legumes, vegetables, nuts, olive oil, and whole cereals) will offer healthful, sustainable, and practical strategies to control GWG and ensure adequate nutrient intake during pregnancy.

Keywords: maternal dietary patterns; pregnancy; gestational gain weight; offspring

---

1. Introduction

Excess body weight during pregnancy is a public health concern owing to its high prevalence; increased risk of maternal diseases, such as gestational diabetes; and delivery complications [1].

Even though the current maternal guidelines stress the need for Gestational Weight Gain (GWG) control [2], the 2011 US pregnancy nutrition surveillance system has shown that up to 47% of pregnant women had excessive GWG and 23% had inadequate GWG according to the recommendations provided by the Institute of Medicine (IOM) [3].

GWG, or the total amount of weight gained in pregnancy, is a complex physiologic response to accommodate the natural responses to pregnancy, such as gestational fat deposition and fetal growth [4]. Several epidemiological studies have shown an association between GWG and offspring complications. Women who gain excessive weight during pregnancy (i.e., more than the amount recommended in guidelines) are more likely to have infants with high birth weight, premature delivery, and infants with an increased risk of developing childhood obesity [5]. Furthermore, excessive GWG is associated with an increased risk of maternal diseases, such as gestational diabetes mellitus and preeclampsia [6]. On the other hand, women with insufficient GWG are more likely to have infants with low birth weight and intrauterine growth retardation [7].

Nevertheless, GWG is a potentially modifiable risk factor because it is directly linked to maternal nutritional habits during pregnancy. Current research evaluating GWG according to dietary intake has focused on isolated foods [8,9] or nutrients [10,11] instead of the dietary pattern of food consumption. The analysis of dietary patterns could provide a better understanding of maternal dietary food intake and thus of women's nutritional status during pregnancy. Although a correct diet is essential to maintain an adequate nutritional status at all stages of life, during pregnancy, nutritional needs are increased in order to meet fetal requirements, especially for some micronutrients. In particular, essential micronutrients are necessary to prevent maternal and perinatal adverse health conditions. However, a deficient intake of essential micronutrients is commonly reported [12] and is associated with an increased nutritional vulnerability in pregnant women [13,14], specifically among those women with worse dietary food intake. Usually, this deficient intake is linked to occidental dietary patterns characterized by a high intake of meat or meat products; snacks; baked desserts; and sugar-sweetened beverages, providing large amounts of saturated fatty acids and simple carbohydrates as added sugars [15].

There is a scarcity of findings regarding dietary factors, nutritional adequacy, and GWG. Furthermore, inadequate GWG and nutrient intake during pregnancy has a negative impact on maternal, perinatal, and fetal health [16]. Therefore, we aimed (1) to assess the association of maternal dietary patterns with GWG and (2) to analyze the association between maternal dietary patterns and nutrient adequate intake according to GWG among Spanish pregnant women.

2. Methods

2.1. Study Design, Settings, and Participants

We designed a retrospective, observational study. Data were obtained from a cohort of pregnant women attending five hospitals in Eastern Andalusia (Spain). Women were recruited from 15 May 2012 through 15 July 2015. Eligible participants were women who resided in the referral area of the five hospitals located in the provinces of Jaen (2 hospitals), Granada (2 hospitals), and Almeria (1 hospital), who understood the Spanish language, gave birth to a single live newborn, and agreed to complete and return the Food Frequency Questionnaire (FFQ) after delivery assessing their dietary intake during pregnancy. After applying the inclusion criteria, 533 women were included in the study. Fifteen women refused participation, and 15 women were excluded because they presented energy intakes outside of predefined limits [17]. A sample of 503 women was analyzed in the current analysis (Figure 1).

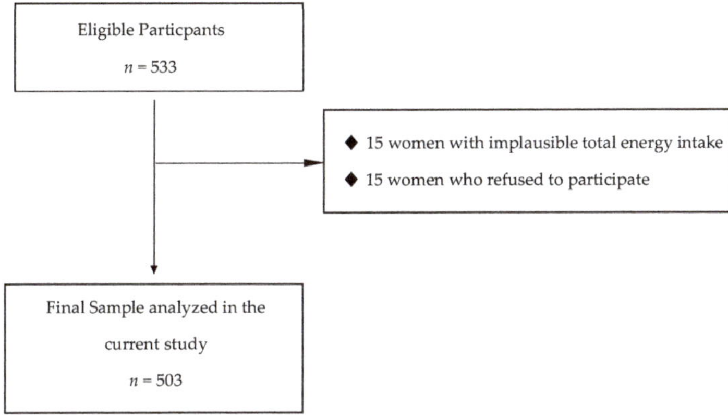

Figure 1. Study Flow-Chart.

The ethics committees from the hospitals approved the study protocol. All women included in this study filled out informed consent and data treatment forms to enroll in the study, following the ethical standards of institutions where they were identified.

2.2. Data Collection and Outcomes

Data were collected retrospectively on anthropometric measures and dietary food and energy intake. Deficient dietary patterns of food consumption and excessive GWG were identified based on IOM guidelines. Next, we evaluated the quality of diet, based on nutrient adequate intake according to recommendations for pregnant women. The primary outcome of the study was maternal GWG and adequacy of nutrient intake, with maternal dietary patterns as the main exposure.

2.2.1. Dietary Assessment

Trained midwives collected dietary intake information using a 137 item semiquantitative FFQ, which was given to women after delivery. This FFQ has been previously translated, adapted, and validated in Spanish women 18 to 74 years of age [18,19]. The FFQ provided a list of foods commonly used by the Spanish population and inquired about the consumption of these foods during the previous year. For each food item in the FFQ, women were asked to report the frequency of consumption and portion size. The FFQ included nine response options (never or almost never, 1–3 times a month, once a week, 2–4 times a week, 5–6 times a week, once a day, 2–3 times a weekday, 4–6 times a day, and more than six times a day). The dietary intake in grams per day was estimated using the indicated frequencies of consumption that were converted to intakes per day and multiplied by the weight of the standard serving size. Nutrient information, as well as total energy intake, was derived from Spanish food composition tables [20,21]. After computing total energy intake, 15 women were excluded because of implausible extreme energy intakes (<500 kcal/day and >3500 kcal/day) [17], leaving 503 women for analysis. Finally, food intake was adjusted for total energy intake using the residual method proposed by Willet et al. [22].

2.2.2. Dietary Pattern Construction

Factor analysis has been extensively used to detect common patterns among highly correlated variables through the use of FFQ [23]. This methodology is a tool commonly employed to extract posteriori dietary patterns [24]. We applied factor analysis with the Principal Components Method (PCA). First, the food items of the FFQ were combined into 16 groups by similar nutrient profile and culinary usage. A detailed description of each food group is reported in Table 1. The daily intake (in grams) for each food group, adjusted by total energy intake, was used in the factor analysis to identify maternal dietary patterns. On the basis of the values of the factor loadings, two main dietary patterns were defined, characterized by high factor loadings of specific food groups.

Table 1. Food groupings used in factor analysis.

Food Groups	Food Subgroups
Vegetables	(1) Green leafy vegetables: spinach, cruciferous, lettuce, green beans, eggplant, peppers, and asparagus; (2) Orange and yellow vegetables: tomatoes, carrots, and pumpkin; (3) Mushrooms.
Fruits	Dried fruit, canned fruit, and fresh fruit
Dairy Products	(1) Milk: low fat and high fat; (2) Yogurt: low fat and high fat; (3) Cheese: low fat and high fat.
Whole Cereals	Whole grain: bread, pasta, rice, and whole breakfast cereals
Refined Cereals	Refined grain: bread, pasta, and rice
Meat	(1) Red meats: beef, lamb, and organ meats; (2) White meats: poultry and rabbit.
Meat products	Hamburger, sausages, and other processed meats
Fish	White fish, oily fish, canned fish, and shellfish/seafood
Sweets and desserts	Biscuits, cakes, and cookies
Olive oil	Olive oil
Hydrogenated oil	Butter, margarine, and solid oil
Potatoes	Cooked and fried potato
Legumes	Peas, beans, lentils, and chickpeas
Nuts	Almonds, nuts, pistachios, and other nuts
Eggs	Eggs
Ready-mademeals	Pizza, soup, lasagna, meatballs, sauces, and other ready-made meals

2.2.3. Anthropometry

Pre-pregnancy weight was self-reported by pregnant women during the first appointment with the midwife. During each antenatal appointment, maternal weight (in kg) and height (in cm) were measured at the antenatal outpatient clinic, and maternal weight (in kg) before delivery was measured at delivery admission by a midwife. Total GWG (in kg) was obtained as the difference in maternal weight between pre-pregnancy weight and weight at delivery admission, as other authors have done previously [25]. Finally, according to the IOM guidelines, we defined the GWG as 12.5–18 kg for underweight, 11.5–16 kg for normal or adequate weight, 7–11.5 kg for overweight, and 5–9 kg for obese women. Weight gain below or above the recommended range was considered as inadequate or excessive GWG.

2.2.4. Diet Quality: Nutrient Adequate Intake

The dietary intake of a selection of nutrients, including dietary fiber, vitamins A, B_9, B_{12}, D, and E, and minerals such as calcium, phosphorus, magnesium, iron, iodine, potassium, selenium, and zinc, was compared with the recommended intakes for these nutrients in pregnant women according to the criteria established by the European Food Safety Agency (EFSA) [26]. Considering the Average Requirements (AR) and/or Average Intake (AI) and Upper-Level intake (UL), we built three categories: (i) deficit intake (intake below AR/AI), (ii) adequate intake (intake between AR/AI and UL), and (iii) excessive intake (intake higher UL). To decrease potential measurement errors derived from the use of the FFQ (overestimation bias), we calculated the proportion of women with intakes below two thirds (2/3) of the Dietary Reference Intakes (DRIs), as other authors have reported previously [14,27]. Results were based on dietary intake data only, excluding supplements.

2.2.5. Other Maternal Variables Related to Patient Characteristics

The assessment of data was obtained via three different sources: (i) personal interviews carried out by trained midwives in the hospital within the two days after giving birth, (ii) clinical records, and (iii) prenatal care records. Information was obtained on general sociodemographic characteristics, including maternal age at pregnancy, lifestyle habits, and personal characteristics. Social class ranged from I (the highest) to V (the lowest), coded according to the classification of the Spanish Society of Epidemiology [28], which is similar to that reported in the Black Report [29]. We categorized smoking status during pregnancy as a current smoker and non-smoker (including previous smokers). Finally, we also computed the number of prenatal appointment and the date of the first appointment. Prenatal care utilization was measured by using the Kessner index. This index considers the timing of entry in prenatal care, number of prenatal appointments, and gestational age at delivery [30].

2.3. Statistical Analysis

We described the study variables using proportions for qualitative variables and means and standard deviations (SD) for quantitative variables. We used the Pearson Chi-square and Kruskal–Wallis tests to assess differences in the distribution of means and percentages.

We conducted a PCA analysis to ascertain dietary group patterns [24]. According to the similarity of food items, we created 16 food groups (Table 1).

We used the Scree plot (Figure 2) and the eigenvalues >1 of the principal components to decide the number of factors to retain. We retained the factors with loadings showing an absolute value ≥0.3. We used these loadings to define the food groups that characterized each dietary pattern.

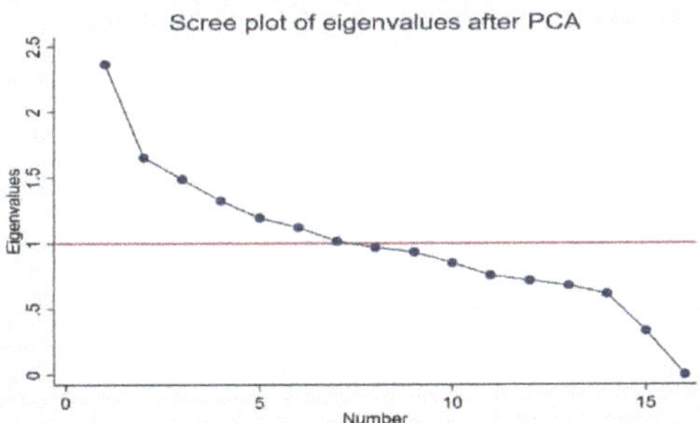

Figure 2. Scree plot of eigenvalues after Principal Components Method (PCA).

To explain sampling adequacy and inter-correlation of variables, we used the Kaiser–Meyer–Olkin value and Bartlett's test of sphericity. Finally, we identified three different dietary patterns explaining 30.6% of the total variance among the food groups included (Table 2).

We used multiple linear regression models to investigate the association between the different dietary patterns (independent variable) with GWG (dependent variable). Crude β-coefficients and adjusted β-coefficients, with their respective 95% confidence intervals (CI), were derived from the fitted univariate and multivariate models. Finally, we used logistic regression models to assess the association between dietary patterns (independent variable) and nutrient adequacy (dependent variable). To control for potential confounding factors in each of the models mentioned previously, multiple logistic and linear models were adjusted for maternal age at pregnancy, social class, Kessner index, and smoking status. Statistical analysis was performed using Stata (15.0, StataCorp L.P. College Station, TX, USA).

Table 2. Factor loadings for two main dietary patterns derived from a principal component analysis.

Foods/Food Groups	Occidental Dietary Pattern	Mediterranean Dietary Pattern
Meat	−0.147	0.237
Meat products	0.416	0.376
Fish	−0.352	0.263
Dairy Products	−0.075	0.063
Vegetables	−0.309	0.740
Whole Cereals	−0.183	0.504
Refined Cereals	0.054	−0.618
Fruits	−0.373	0.043
Nuts	−0.059	0.323
Legumes	−0.050	0.224
Potatoes	0.342	0.319
Olive oil	−0.005	0.315
Sweets and desserts	0.401	−0.128
Hydrogenated oil	0.314	−0.092
Eggs	0.096	0.032
Ready-made meals	0.373	0.207

The cumulative variance contribution rate is 30.6%. Values > 0.3 are factor loading of significant relevance.

3. Results

3.1. Characteristic of the Study Population

Based on pre-gestational BMI, the overall percentage of women with underweight, normal weight, overweight, and obesity were 12.1%, 57.3%, 22.7% and 8.0%, respectively. Regarding the GWG, 170 (33.8%) women had a reduced GWG. Meanwhile, 128 (25.5%) presented excessive GWG. Table 3 presents maternal sociodemographic, anthropometric, and lifestyle variables stratified by GWG. Women with an excessive GWG had a higher pre-pregnancy BMI ($p < 0.001$), higher mean birth weight, and length of gestation (<0.05) than women with a reduced and adequate GWG.

Table 3. Description of the study population characteristics in the study ($n = 503$).

	Reduced GWG		Adequate GWG		Excessive GWG		*p*-Value
	$n = 170$		$n = 205$		$n = 128$		
Age in years, mean (SD)	31.6	(5.5)	31.9	(5.3)	31.0	(4.8)	0.076
Pre-pregnancy BMI, mean (SD)	23.3	(3.9)	23.6	(4.0)	25.5	(4.2)	<0.001
Pre-pregnancy BMI, *n* (%)							<0.001
Underweight (<18.5 Kg/m^2)	19	(11.2)	31	(15.1)	11	(8.6)	
Normal weight (18.5–24.9 Kg/m^2)	121	(71.2)	120	(58.5)	47	(36.7)	
Overweight (25–29.9 Kg/m^2)	22	(12.9)	40	(19.5)	52	(40.6)	
Obesity (≥30 Kg/m^2)	8	(4.7)	14	(6.8)	18	(14.1)	
GWG (kg), mean (SD)	8.2	(2.9)	12.5	(2.5)	17.3	(3.6)	<0.001
Birth weight (g), mean (SD)	3310.5	(379.1)	3436.5	(384.8)	3465.2	(341.7)	<0.001
Length of gestation (weeks), mean (SD)	39.4	(1.2)	39.5	(1.2)	39.8	(1.2)	0.013
Marital status, *n* (%)							0.312
Singled, never married	15	(8.8)	11	(5.4)	14	(10.9)	
Married	115	(67.7)	147	(71.7)	80	(62.5)	
Couple	40	(23.5)	47	(22.9)	34	(26.6)	
Educational level, *n* (%)							0.173
Primary	31	(18.2)	33	(16.1)	26	(20.3)	
Secondary (unfinished)	10	(5.9)	12	(5.9)	4	(3.1)	
Secondary (completed)	50	(29.4)	81	(39.5)	54	(42.2)	
University	79	(46.5)	79	(38.5)	44	(34.4)	
Smoking during pregnancy, *n* (%)	14	(8.2)	35	(17.1)	29	(22.7)	0.002
Kessner index (prenatal care), *n* (%)							0.396
Adequate	80	(47.1)	95	(46.3)	71	(55.5)	
Intermediate	66	(38.9)	74	(36.1)	38	(29.7)	
Inadequate	24	(14.1)	36	(17.6)	19	(14.8)	

Abbreviations: (SD): standard deviation; BMI: body mass index. Pearson chi-square test and Kruskal–Wallis test were performed for evaluating differences in categorical and continuous variables, respectively.

3.2. Dietary Patterns and GWG

Table 4 presents the crude and adjusted beta (β) coefficients from the univariate and multivariate linear models evaluating the association between each of the two different dietary patterns with an increasing GWG. There was a negative association between GWG and Mediterranean dietary pattern (crude β = −0.06, 95% CI: −0.11, −0.04), whereas a positive association with GWG was found for the Occidental dietary pattern; nevertheless, this association is lost when adjusted for confounders.

Table 4. Multivariable regression models for the association between dietary patterns and gestational weight gain (GWG) (*n* = 503).

Dietary Pattern	GWG			
	Crude β-Coefficients	(95% CI)	Adjusted β-Coefficients [a]	(95% CI)
Occidental dietary pattern	0.02	(−0.05, 0.04)	0.08	(−0.04, 0.05)
Mediterranean dietary pattern	−0.06	(−0.11, −0.04)	−0.05	(−0.01, 0.01)

Crude β-coefficients: crude β-coefficients, adjusted β-coefficients [a]: adjusted β-coefficients. [a] Adjusted for age of parity, social class, Kessner index, and smoking habits.

3.3. Prevalence of Participants with Adequate, Deficient, or Excessive Nutrient Intake According to GWG

Table 5 shows the prevalence of participants with adequate, deficient, or excessive nutrient intake according to their GWG. The vitamins that exhibited the highest deficient intake for all the study participants were vitamins B9 and D. Women with a reduced GWG showed a lower prevalence of vitamin B9 deficient intake than women with an adequate or excessive GWG. In contrast, women with adequate GWG exhibited a higher intake deficit of Vitamin D than women with inadequate or excessive GWG (*p* < 0.05).

Table 5. Prevalence of participants with an adequate, deficient, or excessive intake of nutrients according to 2/3 EFSA DRIs stratified by GWG.

Nutrient	Reduced GWG *n* = 170			Adequate GWG *n* = 205			Excessive GWG *n* = 128			
	DI [a]	AI [b]	EI [c]	DI [a]	AI [b]	EI [c]	DI [a]	AI [b]	EI [c]	*p*-Value
Dietary fiber (g/day)	6.0	94.1	-	9.3	90.7	-	9.4	90.6	-	0.413
Vitamin A (µg/day)	0	58.2	41.8	0.5	59.5	40.0	0.8	62.5	36.7	0.756
Vitamin B9 (µg/day)	45.3	50.0	4.7	58.0	41.0	1.0	54.7	40.6	4.7	0.034
Vitamin B12 (µg/day)	0	100.0	-	0.5	99.5	-	1.6	98.4	-	0.215
Vitamin D (µg/day)	82.4	17.7	0	89.8	10.2	0	80.5	19.5	0	0.037
Vitamin E (mg/day)	5.9	94.1	0	2.9	97.1	0	3.1	96.9	0	0.293
Calcium (mg/day)	0.6	87.1	12.4	0.5	88.8	10.7	0.8	82.8	16.4	0.651
Magnesium (mg/day)	0	0	100.0	0	0	100.0	0	0	100.0	-
Iodine (µg/day)	9.4	60.6	30.0	7.8	53.7	38.5	10.9	50.0	39.1	0.299
Potassium (mg/day)	0.6	99.4	-	0	100.0	-	1.6	98.4	-	0.197
Selenium (µg/day)	1.2	98.8	0	1.5	98.5	0	3.1	96.1	0.8	0.314

Intake (I): [a] deficient intake (DI), [b] adequate intake (AI), and [c] excessive intake (EI). Values are % unless otherwise indicated. Pearson chi-square test was used in order to ascertain differences between groups. There is no Upper-Level intake (UL) in the micronutrient assessed.

3.4. Association between Maternal Dietary Patterns and Nutrient Adequate Intake According to GWG

Table 6 shows the association between nutritional adequacy and dietary patterns according to the GWG status. Independent of GWG, a Mediterranean dietary pattern showed moderate evidence of a higher probability of meeting an adequate intake for dietary fiber; vitamins B9, D, and E; and iodine (*p* < 0.05).

Table 6. Multivariate logistic regression of association between nutrient adequacy and dietary patterns according to GWG.

	Reduced GWG OR (95% CI)	Adequate GWG OR (95% CI)	Excessive GWG OR (95% CI)
	Dietary fiber		
Occidental dietary pattern	0.8 (0.42, 1.35)	0.6 (0.39, 1.03)	0.4 (0.15, 1.86)
Mediterranean dietary pattern	3.1 (1.37, 7.07)	1.6 (0.96, 2.59)	1.4 (0.72, 2.57)
	Vitamin A		
Occidental dietary pattern	0.9 (0.67, 1.10)	0.7 (0.54, 0.87)	0.6 (0.44, 0.82)
Mediterranean dietary pattern	1.1 (0.80, 1.36)	1.2 (0.94, 1.51)	1.1 (0.78, 1.43)
	Vitamin B9		
Occidental dietary pattern	0.5 (0.44, 0.81)	0.6 (0.44, 0.77)	0.8 (0.56, 1.08)
Mediterranean dietary pattern	1.6 (1.22, 2.21)	2.1 (1.56, 2.79)	1.7 (1.26, 2.36)
	Vitamin D		
Occidental dietary pattern	0.5 (0.32, 0.87)	0.74 (0.46, 1.19)	0.91 (0.61, 1.38)
Mediterranean dietary pattern	4.4 (2.50, 7.68)	4.89 (2.72, 8.77)	3.02 (1.84, 4.96)
	Vitamin E		
Occidental dietary pattern	1.06 (0.62, 1.82)	1.1 (0.54, 2.06)	1.00 (0.40, 2.51)
Mediterranean dietary pattern	2.7 (1.26, 5.72)	1.26 (0.64, 2.50)	2.75 (0.90, 8.34)
	Calcium		
Occidental dietary pattern	0.8 (0.56, 1.09)	0.8 (0.59, 1.15)	0.82 (0.65, 1.04)
Mediterranean dietary pattern	1.4 (0.92, 2.13)	1.20 (0.85, 1.70)	1.15 (0.80, 1.67)
	Iodine		
Occidental dietary pattern	0.8 (0.65, 1.12)	1.2 (0.93, 1.48)	1.0 (0.78, 1.35)
Mediterranean dietary pattern	1.3 (0.97, 1.68)	1.3 (1.02, 1.63)	1.1 (0.91, 1.43)

The multivariable model was adjusted for age of parity, social class, Kessner index, and smoking habits.

4. Discussion

This is the first study that evaluates the association of maternal dietary patterns and GWG and their association with adequacy nutrient intake in Spanish pregnant women. Independent of maternal GWG, we found that an adequate intake of dietary fiber; vitamins B9, D, and E; calcium; and iodine nutrients was directly related to a classical Mediterranean dietary pattern characterized by a high content of vegetables, olive oil, whole cereals, and nuts. Our findings showed moderate evidence for an association between this healthy dietary pattern and lower GWG trajectories.

Diet quality has been neglected as a risk factor and potential intervention target for inappropriate GWG. Traditionally, the classical nutritional epidemiological approach focused on isolated food groups and/or macronutrient intake instead of the assessment of dietary patterns [31]. However, the role of the overall diet versus individual foods or nutrients provides a more intuitive and objective holistic interpretation of the quality of a woman's dietary pattern [24].

Taking into account that women with healthy dietary intake also have a healthy lifestyle contributing to reducing excess GWG, the position of American Dietetic Association for pregnant women emphasized the adoption of healthy dietary patterns rich in fish and seafood, vegetables, legumes, and vegetable oils, along with engaging in physical activity [32], to prevent inadequate GWG. Accordingly, our study's main findings support the previous recommendation as we found an association between the classically traditional healthy Mediterranean dietary pattern richer in vegetables, olive oil, and nuts, and lower GWG. In line with our findings, Bassel et al. reported that a Mediterranean-style diet characterized by a high intake of extra virgin olive oil, vegetables, and legumes had a potential role in reducing GWG in British pregnant women [33]. Furthermore, the scientific literature has shown that an increased intake of whole-grain cereal has a positive effect on reducing weight gain not only in a general adult [34] but also in the pregnant women population [35]. Our findings highlight that the Mediterranean dietary pattern, which includes this food group, displays a similar effect. Surprisingly, in our study, dietary patterns dense in energy, such as processed meat and ready-made meals or sweets and dessert patterns, did not show any effect on weight gain.

This result contradicts those of other authors that report that a higher energy intake pattern is associated with higher GWG in European pregnant women [36]. Among the reasons that could explain these inconsistent results could be that women who presented a higher GWG (overweight/obese women) reported a healthier dietary intake, overreporting food groups considered as "healthy", resulting in an information bias, as other authors have extensively communicated [35,36]. Another explanation that could explain the discrepancy in research findings could be that the pregnant women included in our study reported a low intake of these non-healthy products, thus making the identification of this association more difficult.

Pre-pregnancy and pregnancy dietary patterns influence fetal health and the risk of fetal diseases, not only during intrauterine life but also into adulthood. Many nutrients, specifically vitamins and minerals such as vitamin B9 and iron, have been extensively investigated due to their relationship with the development of maternal morbidities and neurodevelopmental disease in babies [37,38]. Even though traditional pregnant counseling has emphasized the consumption of complex prenatal vitamins to provide the necessary amount of some micronutrients [39], Cano-Ibáñez et al. found a positive association between a healthy and diverse dietary pattern with nutrient adequacy in Spanish pregnant women [40]. Accordingly, we found a positive effect of the Mediterranean dietary pattern on nutrient adequacy intake in the present study, independent of GWG. These food groups are considered as "healthy food groups", typical of dietary patterns such as the Mediterranean diet [41] or Prudent diet [42]. In line with our results, several authors have demonstrated that adherence to both patterns might provide a balanced intake of micronutrients [43–45] instead of low energy quantity. The notion is that a food group provides the intake of several nutrients. For example, vitamin B9 is present in leafy vegetables, and dietary fiber can be found in legumes and vegetables.

Strengths and Limitations

We assessed food intake during pregnancy within 48 h after giving birth; thus, dietary intake during the last trimester may be remembered better than earlier dietary intake during the first trimester. Nevertheless, it has been reported that dietary patterns are stable across pregnancy despite an increased energy intake, with the exception of alcohol intake [46], although recent studies associate very moderate alcohol consumption with lower indices of small for gestational age (SGA) newborns [47]. For this reason, we decided not to include this group in our PCA analysis, as other authors have reported previously [48]. The semiquantitative FFQ has been validated for the Spanish population [18], but self-reporting questionnaires and memory problems could lead to information bias. However, this would more likely cause a non-differential misclassification bias and estimations would tend toward the null. To correct possible errors derived from the FFQ, we excluded participants with energy intakes outside of predefined limits [17], and we used the residual method to adjust for food intake for energy intake. However, although the FFQ specifies the usual portion size as part of the question on frequency, it might not be the ideal tool to measure micronutrient intake and is not validated for this specific population group, although it has been used previously in pregnant women [14]. For this reason, we considered that intake was adequate only when the intake reached at least 2/3 of the recommendations proposed by EFSA for pregnant women, correcting the possible bias introduced by the FFQ and assuming, in any case, that the inadequate micronutrient intake (deficit intake or excessive intake) would be higher than the estimated figures [27].

Furthermore, we cannot exclude a potential reverse causation bias in the association between the independent variable with the assessed outcomes due to the study design. Finally, residual confounding that might affect GWG and nutrient intake, such as physical activity or other unmeasured socioeconomic factors, cannot be discarded, even if we have gathered data on the relevant confounders in nutritional epidemiology and adjusted for them in the multivariate analyses.

Notwithstanding the above limitations, our study includes several strengths reinforcing the validity and consistency of the findings obtained. The inclusion of a large representative sample (533 healthy pregnant women), from a reference population of around 120,000 healthy pregnant women providing exhaustive and specific information on dietary intake, is a strength of this study. The use of dietary patterns instead of single food or nutrients is another strength. Dietary patterns are more exhaustive as specific potential factors related to GWG than just single isolated foods. Dietary patterns are more intuitive and objective to determine women's overall dietary intake during pregnancy [24].

Furthermore, our PCA results explained the 30.6% of the total variance among food groups, and the analysis to derive dietary patterns was based on well-established criteria. We also included a considerable amount of information collected using a standardized protocol that reduces information bias regarding reported food intakes, sociodemographic characteristics, and lifestyles. Finally, to the extent of our knowledge, this report is the first study in Spain assessing the association between GWG with maternal dietary patterns and adequate micronutrient intake.

5. Conclusions

In summary, we found that a healthy dietary food intake pattern, compatible with a Mediterranean diet, is associated with an adequate GWG during pregnancy and better nutrient adequacy. Pregnancy can be considered a window of opportunity for promoting healthy habits, as women are more willing to adopt healthier dietary habits during this time. Therefore, our findings support the promotion of healthy dietary habits based on a Mediterranean diet (characterized by a high intake of vegetables, nuts, whole cereals, and olive oil) during pregnancy. Furthermore, counseling and promoting the Mediterranean diet during antenatal visits could offer a sustainable and practical strategy in order to control GWG and ensure adequate nutrient intake during this critical fetal developmental period. Moreover, it could also be an important public health measure with implications that might span over a woman's life course.

Author Contributions: Conceptualization, N.C.-I., A.B.-C., J.M.M.-G., and M.D.-R.; methodology, N.C.-I. and M.A.L.-F.; data curation, M.D.-R., J.M.M.-G., and N.C.-I.; writing—original draft preparation, N.C.-I., S.M.-P., J.M.M.-G. and A.B.-C. writing—review and editing, all the authors; supervision, A.B.-C. and M.D.-R. All authors have read and agreed to the published version of the manuscript.

Funding: This work was supported by a grant from the National Institute of Health Carlos III (PI11/02199).

Conflicts of Interest: The authors declare no conflict of interest.

References

1. Guelinckx, I.; Devlieger, R.; Beckers, K.; Vansant, G. Maternal obesity: Pregnancy complications, gestational weight gain and nutrition. *Obes. Rev.* **2008**, *9*, 140–150. [CrossRef] [PubMed]
2. Institute of Medicine and National Research Council Committee to Reexamine IOMPWG. The National Academies Collection: Reports funded by National Institutes of Health. In *Weight Gain During Pregnancy: Reexamining the Guidelines*; Rasmussen, K.M., Yaktine, A.L., Eds.; National Academies Press (US) National Academy of Sciences: Washington, DC, USA, 2009.
3. Goldstein, R.F.; Abell, S.K.; Ranasinha, S.; Misso, M.; Boyle, J.A.; Black, M.H.; Li, N.; Hu, G.; Corrado, F.; Rode, L.; et al. Association of gestational weight gain with maternal and infant outcomes: A systematic review and meta-analysis. *JAMA* **2017**, *317*, 2207–2225. [CrossRef] [PubMed]
4. Herring, S.J.; Rose, M.Z.; Skouteris, H.; Oken, E. Optimizing weight gain in pregnancy to prevent obesity in women and children. *Diabetes Obes. Metab.* **2012**, *14*, 195–203. [CrossRef] [PubMed]
5. Su, W.J.; Chen, Y.L.; Huang, P.Y.; Shi, X.L.; Yan, F.F.; Chen, Z.; Yan, B.; Song, H.-Q.; Lin, M.-Z.; Li, X.-J. Effects of Prepregnancy Body Mass Index, Weight Gain, and Gestational Diabetes Mellitus on Pregnancy Outcomes: A Population-Based Study in Xiamen, China, 2011–2018. *Ann. Nutr. Metab.* **2019**, *75*, 31–38. [CrossRef] [PubMed]
6. Gilmore, L.A.; Klempel-Donchenko, M.; Redman, L.M. Pregnancy as a window to future health: Excessive gestational weight gain and obesity. *Semin. Perinatol.* **2015**, *39*, 296–303. [CrossRef] [PubMed]
7. Guan, P.; Tang, F.; Sun, G.; Ren, W. Effect of maternal weight gain according to the Institute of Medicine recommendations on pregnancy outcomes in a Chinese population. *J. Int. Med Res.* **2019**, *47*, 4397–4412. [CrossRef] [PubMed]

8. Tovar, A.; Kaar, J.L.; McCurdy, K.; Field, A.E.; Dabelea, D.; Vadiveloo, M. Maternal vegetable intake during and after pregnancy. *BMC Pregnancy Childbirth* **2019**, *19*, 267. [CrossRef] [PubMed]

9. Lai, J.S.; Soh, S.E.; Loy, S.L.; Colega, M.; Kramer, M.S.; Chan, J.K.Y.; Tan, T.C.; Shek, L.P.C.; Yap, F.K.P.; Tan, K.H. Macronutrient composition and food groups associated with gestational weight gain: The GUSTO study. *Eur. J. Nutr.* **2019**, *58*, 1081–1094. [CrossRef] [PubMed]

10. Tebbani, F.; Oulamara, H.; Agli, A. Factors associated with low maternal weight gain during pregnancy. *Revue d'Epidemiologie et de Sante Publique* **2019**, *67*, 253–260. [CrossRef]

11. Campos, C.A.S.; Malta, M.B.; Neves, P.A.R.; Lourenço, B.H.; Castro, M.C.; Cardoso, M.A. Gestational weight gain, nutritional status and blood pressure in pregnant women. *Revista de Saude Publica* **2019**, *53*, 57. [CrossRef]

12. Gernand, A.D.; Schulze, K.J.; Stewart, C.P.; West, K.P.; Christian, P. Micronutrient deficiencies in pregnancy worldwide: Health effects and prevention. *Nat. Rev. Endocrinol.* **2016**, *12*, 274–289. [CrossRef] [PubMed]

13. Black, R.E. Micronutrients in pregnancy. *Br. J. Nutr.* **2001**, *85* (Suppl. S2), S193–S197. [CrossRef]

14. Cano-Ibáñez, N.; Martínez-Galiano, J.M.; Amezcua-Prieto, C.; Olmedo-Requena, R.; Bueno-Cavanillas, A.; Delgado-Rodríguez, M. Maternal dietary diversity and risk of small for gestational age newborn: Findings from a case–control study. *Clin. Nutr.* **2019**, *39*, 1943–1950. [CrossRef]

15. Martínez-González, M.A.; Martín-Calvo, N. The major European dietary patterns and metabolic syndrome. *Rev. Endocr. Metab. Disord.* **2013**, *14*, 265–271. [CrossRef]

16. Plante, A.S.; Lemieux, S.; Labrecque, M.; Morisset, A.S. Relationship Between Psychosocial Factors, Dietary Intake and Gestational Weight Gain: A Narrative Review. *J. Obstet. Gynaecol. Can.* **2019**, *41*, 495–504. [CrossRef] [PubMed]

17. Willet, W. *Nutritional Epidemiology*, 3rd ed.; Oxford University Press: New York, NY, USA, 2013.

18. Martin-Moreno, J.M.; Boyle, P.; Gorgojo, L.; Maisonneuve, P.; Fernandez-Rodriguez, J.C.; Salvini, S.; Willett, W.C. Development and validation of a food frequency questionnaire in Spain. *Int. J. Epidemiol.* **1993**, *22*, 512–519. [CrossRef] [PubMed]

19. Fernandez-Ballarth, J.; Pinol, J.; Zazpe, I.; Corella, D.; Carrasco, P.; Toledo, E.; Perez-Bauer, M.; Martinez-Gonzalez, M.; Salas-Salvado, J.; Martin-Moreno, J. Relative validity of a semi-quantitative food-frequency questionnaire in an elderly Mediterranean population of Spain. *Br. J. Nutr.* **2010**, *103*, 1808–1816. [CrossRef]

20. Moreiras, O.C.A.; Cabrera, L.; Cuadrado, C. *Tablas de Composición de Alimentos. (Spanish Food Composition Tables)*; Pirámide: Madrid, Spain, 2003.

21. Mataix, J.; Manas, M.; Llopis, J.; Martínez de Victoria, E.; Juan, J.; Borregón, A. *Tabla de Composición de Alimentos Españoles. (Spanish Food Composition Tables)*; Universidad de Granada: Granada, Spain, 2003.

22. Willett, W.; Stampfer, M. Implications of Total Energy Intake for Epidemiologic Analyses. In *Nutritional Epidemiology*; Oxford University Press: New York, NY, USA, 2009.

23. Hu, F.B.; Rimm, E.; A Smith-Warner, S.; Feskanich, D.; Stampfer, M.J.; Ascherio, A.; Sampson, L.; Willett, W.C. Reproducibility and validity of dietary patterns assessed with a food- frequency questionnaire. *Am. J. Clin. Nutr.* **1999**, *69*, 243–249. [CrossRef]

24. Hu, F.B. Dietary pattern analysis: A new direction in nutritional epidemiology. *Curr. Opin. Lipidol.* **2002**, *13*, 3–9. [CrossRef]

25. Wei, X.; He, J.-R.; Lin, Y.; Lu, M.; Zhou, Q.; Li, S.; Lu, J.; Yuan, M.; Chen, N.; Zhang, L.; et al. The influence of maternal dietary patterns on gestational weight gain: A large prospective cohort study in China. *Nutrition* **2019**, *59*, 90–95. [CrossRef]

26. EFSA. Dietary Reference Values and Dietary Guidelines. 2018. Available online: https://www.efsa.europa. eu/en/topics/topic/dietary-reference-values (accessed on 10 June 2019).

27. Aranceta-Bartrina, J.; Serra-Majem, L.; Pérez-Rodrigo, C.; Llopis, J.; Mataix, J.; Ribas, L.; Tojo, R.; Tur, J.A. Vitamins in Spanish food patterns: The eVe study. *Public Health Nutr.* **2001**, *4*, 1317–1323. [CrossRef] [PubMed]

28. Álvarez-Dardet, C.A.J.; Domingo, A.; Regidor, E. *La Medicina de la Clase Social en Ciencias de la Salud, Informe de un Grupo de Trabajo de la Sociedad Española de Epidemiología*; SG Editores: Barcelona, Spain, 1995.

29. Black, D.T.P. *Inequalities in Health*; Penguin: Harmondsworth, UK, 1983.

30. Kessner, D.M.S.J.; Kalk, C.E.; Schlesinger, E.R. *Infant Death: An Analysis by Maternal Risk and Health Care, Contrasts in Health Status*; Institute of Medicine and National Academy of Sciences: Washington, DC, USA, 1973.

31. Tielemans, M.J.; Garcia, A.H.; Santos, A.P.; Bramer, W.M.; Luksa, N.; Luvizotto, M.J.; Moreira, E.; Topi, G.; Jonge, E.A.L.D.; Visser, T.L.; et al. Macronutrient composition and gestational weight gain: A systematic review. *Am. J. Clin. Nutr.* **2016**, *103*, 83–99. [CrossRef] [PubMed]

32. Kaiser, L.; Allen, L.H. Position of the American Dietetic Association: Nutrition and lifestyle for a healthy pregnancy outcome. *J. Am. Diet. Assoc.* **2008**, *108*, 553–561.

33. Al Wattar, B.H.; Dodds, J.; Placzek, A.; Beresford, L.; Spyreli, E.; Moore, A.; Carreras, F.J.G.; Austin, F.; Murugesu, N.; Roseboom, T.J.; et al. Mediterranean-style diet in pregnant women with metabolic risk factors (ESTEEM): A pragmatic multicentre randomised trial. *PLoS Med.* **2019**, *16*, e1002857. [CrossRef]

34. Mozaffarian, D.; Hao, T.; Rimm, E.B.; Willett, W.C.; Hu, F.B. Changes in diet and lifestyle and long-term weight gain in women and men. *New Engl. J. Med.* **2011**, *364*, 2392–2404. [CrossRef] [PubMed]

35. Wrottesley, S.V.; Pisa, P.T.; Norris, S.A. The influence of maternal dietary patterns on body mass index and gestational weight gain in urban black South African women. *Nutrients* **2017**, *9*, 732. [CrossRef] [PubMed]

36. Maugeri, A.; Barchitta, M.; Favara, G.; La Rosa, M.C.; La Mastra, C.; Lio, R.M.S.; Agodi, A. Maternal dietary patterns are associated with pre-pregnancy body mass index and gestational weight gain: Results from the "mamma & bambino" cohort. *Nutrients* **2019**, *11*, 1308.

37. Yang, J.; Cheng, Y.; Pei, L.; Jiang, Y.; Lei, F.; Zeng, L.; Wang, Q.; Li, Q.; Kang, Y.; Shen, Y.; et al. Maternal iron intake during pregnancy and birth outcomes: A cross-sectional study in Northwest China. *Br. J. Nutr.* **2017**, *117*, 862–871. [CrossRef]

38. Simpson, J.L.; Bailey, L.B.; Pietrzik, K.; Shane, B.; Holzgreve, W. Micronutrients and women of reproductive potential: Required dietary intake and consequences of dietary deficiency or excess. Part i Folate, Vitamin B12, Vitamin B6. *J. Matern. Fetal Neonatal Med.* **2010**, *23*, 1323–1343. [CrossRef]

39. Haider, B.A.; Yakoob, M.Y.; Bhutta, Z.A. Effect of multiple micronutrient supplementation during pregnancy on maternal and birth outcomes. *BMC Public Health* **2011**, *11* (Suppl. S3), S19. [CrossRef]

40. Cano-Ibáñez, N.; Gea, A.; Martínez-González, M.A.; Salas-Salvadó, J.; Corella, D.; Zomeño, M.D.; Romaguera, D.; Vioque, J.; Aros, F.; Wärnberg, J.; et al. Dietary diversity and nutritional adequacy among an older Spanish population with metabolic syndrome in the PREDIMED-plus study: A cross-sectional analysis. *Nutrients* **2019**, *11*, 958. [CrossRef] [PubMed]

41. Trichopoulou, A.; Martínez-González, M.A.; Tong, T.Y.; Forouhi, N.G.; Khandelwal, S.; Prabhakaran, D.; Mozaffarian, D.; De Lorgeril, M. Definitions and potential health benefits of the Mediterranean diet: Views from experts around the world. *BMC Med.* **2014**, *12*, 112. [CrossRef] [PubMed]

42. Fung, T.T.; Willett, W.C.; Stampfer, M.J.; Manson, J.E.; Hu, F.B. Dietary patterns and the risk of coronary heart disease in women. *Arch. Intern. Med.* **2001**, *161*, 1857–1862. [CrossRef]

43. Serra-Majem, L.; Bes-Rastrollo, M.; Román-Viñas, B.; Pfrimer, K.; Sánchez-Villegas, A.; Martínez-González, M.A. Dietary patterns and nutritional adequacy in a Mediterranean country. *Br. J. Nutr.* **2009**, *101*, S21–S28. [CrossRef] [PubMed]

44. Phillips, S.M.; Fulgoni, V.L.; Heaney, R.P.; Nicklas, T.A.; Slavin, J.L.; Weaver, C.M. Commonly consumed protein foods contribute to nutrient intake, diet quality, and nutrient adequacy. *Am. J. Clin. Nutr.* **2015**, *101*, 1346S–1352S. [CrossRef]

45. Cano-Ibáñez, N.; Bueno-Cavanillas, A.; Martínez-González, M.Á.; Salas-Salvadó, J.; Corella, D.; Freixer, G.-L.; Romaguera, D.; Vioque, J.; Alonso-Gómez, Á.M.; Wärnberg, J.; et al. Effect of changes in adherence to Mediterranean diet on nutrient density after 1-year of follow-up: Results from the PREDIMED-Plus Study. *Eur. J. Nutr.* **2019**, *59*, 2395–2409.

46. Crozier, S.R.; Robinson, S.M.; Godfrey, K.M.; Cooper, C.; Inskip, H.M. Women's dietary patterns change little from before to during pregnancy. *J. Nutr.* **2009**, *139*, 1956–1963. [CrossRef]

47. Martínez-Galiano, J.M.; Amezcua-Prieto, C.; Salcedo-Bellido, I.; Olmedo-Requena, R.; Bueno-Cavanillas, A.; Delgado-Rodriguez, M. Alcohol consumption during pregnancy and risk of small-for-gestational-age newborn. *Women Birth* **2019**, *32*, 284–288. [CrossRef]
48. Rifas-Shiman, S.L.; Rich-Edwards, J.W.; Kleinman, K.P.; Oken, E.; Gillman, M.W. Dietary Quality during Pregnancy Varies by Maternal Characteristics in Project Viva: A US Cohort. *J. Am. Diet. Assoc.* **2009**, *109*, 1004–1011. [CrossRef]

Publisher's Note: MDPI stays neutral with regard to jurisdictional claims in published maps and institutional affiliations.

International Journal of
*Environmental Research
and Public Health*

Article

Sexually Transmitted Infections and Associated Factors in Southeast Spain: A Retrospective Study from 2000 to 2014

María Ángeles Pérez-Morente [1], María Gázquez-López [2], María Adelaida Álvarez-Serrano [2,*], Encarnación Martínez-García [3,4,*], Pedro Femia-Marzo [5], María Dolores Pozo-Cano [3] and Adelina Martín-Salvador [6]

[1] Department of Nursing, Faculty of Health Sciences, University of Jaén, 23071 Jaén, Spain; mmorente@ujaen.es
[2] Department of Nursing, Faculty of Health Sciences, University of Granada, 51001 Ceuta, Spain; mgazquez@ugr.es
[3] Department of Nursing, Faculty of Health Sciences, University of Granada, 18016 Granada, Spain; pozocano@ugr.es
[4] Guadix High Resolution Hospital, 18500 Guadix, Granada, Spain
[5] Department of Statistics, Faculty of Health Sciences, University of Granada, 18016 Granada, Spain; pfemia@ugr.es
[6] Department of Nursing, Faculty of Health Sciences, University of Granada, 52005 Melilla, Spain; ademartin@ugr.es
* Correspondence: adealvarez@ugr.es (M.A.Á.-S.); emartinez@ugr.es (E.M.-G.)

Received: 19 September 2020; Accepted: 12 October 2020; Published: 13 October 2020

Abstract: The World Health Organization estimates that more than one million people acquire a Sexually Transmitted Infection (STI) every day, compromising quality of life, sexual and reproductive health, and the health of newborns and children. It is an objective of this study to identify the factors related to a Sexually Transmitted Infection diagnosis in the province of Granada (Spain), as well as those better predicting the risk of acquiring such infections. In this study, 678 cases were analyzed on a retrospective basis, which were treated at the Centre for Sexually Transmitted Diseases and Sexual Orientation in Granada, between 2000–2014. Descriptive statistics were applied, and by means of binary logistic regression, employing the forward stepwise-likelihood ratio, a predictive model was estimated for the risk of acquiring an STI. Sex, age, occupation, economic crisis period, drug use, number of days in which no condoms were used, number of sexual partners in the last month and in the last year, and number of subsequent visits and new subsequent episodes were associated with an STI diagnosis ($p < 0.05$). The risk of being diagnosed with an STI increased during the economic crisis period (OR: 1.88; 95%-CI: 1.28–2.76); during the economic crisis and if they were women (OR:2.35, 95%- CI: 1.24–4.44); and if they were women and immigrants (OR: 2.09; 95%- CI:1.22–3.57), while it decreased with age (OR: 0.97, 95%-CI: 0.95–0.98). Identification of the group comprised of immigrant women as an especially vulnerable group regarding the acquisition of an STI in our province reflects the need to incorporate the gender perspective into preventive strategies and STI primary health care.

Keywords: sexually transmitted diseases; public health; risk groups

1. Introduction

Sexually Transmitted Infections (STIs) constitute a significant public health issue on a worldwide basis. The World Health Organization (WHO) estimates that more than one million people acquire an STI every day, compromising quality of life, sexual and reproductive health, and newborns' and children's health [1].

Such infections are caused by more than 30 different bacteria, viruses, and parasites, the most frequent being syphilis, gonorrhea, chlamydia, human papillomavirus (HPV), hepatitis B and C, and human immunodeficiency virus (HIV), the first three of which being curable and the last four being incurable viral infections, although treatments exist to attenuate or modify symptoms and the disease [1].

STIs are more frequent in people with risky sexual behaviors, such as not using condoms, having multiple sexual partners, having a highly risky partner (individual with many sexual partners or other risk factors), having anal sex or a partner who practices it, or having sexual intercourse with a partner who injects or has injected drugs before [2–4]. Teenagers, immigrants, sex workers, men who have sex with men (MSM), and bisexual individuals are particularly vulnerable groups regarding the acquisition of STIs [5,6].

Furthermore, the current financial situation impacts on the incidence of STIs, mainly in the most vulnerable population [1,7–10]. In the recent economic crisis period (2008–2014) [11–14], syphilis and gonorrhoea reappeared in Spain, and the incidence of HIV, hepatitis, and HPV increased. During this period, the probability of the appearance of an STI was higher than in the non-crisis period [15], due to the financial difficulties to access contraceptive methods, thus increasing the absence of protection against STIs [15–17].

At the world level, the prevalence of STIs in men and women is similar, with some regional differences. Nevertheless, complications in these processes do affect disproportionately women and generate a strong impact, mainly on their sexual and reproductive health. The WHO points out that the lack of data on STIs at the local level and, in particular, of data classified by sex, compromises the solution to this problem [18]. Moreover, today, despite the outstanding advance in medical knowledge and development of primary health care, sexual prevention programs seem to continue being inefficient, since the number of STIs continues to increase [1]. In this sense, an increase in STIs may be observed in Spain in general [19] and in the province of Granada in particular [10,15], which becomes an especially alarming issue.

The objective of this study was to identify any factors associated with STI diagnoses in the province of Granada (Spain), as well as those that better predict the risk of acquiring such infections.

2. Materials and Methods

The group of cases involving users attended at the Center for Sexually Transmitted Diseases and Sexual Orientation in Granada (Spain) with a new STI diagnosis from 2000–2014 was analyzed on a retrospective basis. This center is attached to the Andalusian Health Service (AHS), which offers universal and free healthcare to people residing in the province regardless of their nationality and income level, providing healthcare to 730,000 people older than 15 years old [20].

The main source of information was the patient's medical history in which health professionals completed sociodemographic data, symptoms, and medicals signs, results of diagnostic tests, therapeutic evolution, and final diagnosis. The study population consists of users with a positive or negative STI diagnosis clearly recorded on the medical histories contained in a database of 1437 histories created in a previous project from which this study is derived, and whose details on the sampling process are described in another paper [15]. Exclusion criteria included being younger than 18 years old and having a cognitive deficit. Investigators collected the information and attended a training course to reduce possible variations among them.

STI diagnosis was included as the dependent variable and coded as binary (yes/no), following the pattern established by other studies in this line of research [21]. Independent variables were classified in three groups: 1) Sociodemographic characteristics such as sex (female/male), age, nationality (Spanish, foreigner), occupation (sex worker/former worker/others), working status (active/non-active), education level (with no studies or primary/secondary/higher-level education); crisis period, with the year 2008 being considered the commencement of the financial crisis in Spain [11–14] (yes/no); 2) characteristics regarding the healthcare received, such as reason for consultation (HIV/another); previous treatment (yes/no), number of subsequent consultations, number of new subsequent episodes, and 3) STI risk indicators such as sexual behaviour (heterosexual, homosexual, bisexual); stable relationship (yes/no); stable relationship with symptoms (yes/no); days elapsed since the last contact without using a condom; number of partners in the last month; number of partners in the last year; sex life understood as the total number of partners in their whole life; drug use (yes/no); previous STI (yes/no) and age of first sexual intercourse. Information was recorded on an ad hoc designed computerized data collection sheet, which was then exported to the statistical program for analysis.

Variables were analyzed on a descriptive basis calculating measures of central tendency (mean and standard deviation) for quantitative variables and absolute and relative frequencies for categorical variables. Comparison among groups was carried out using the chi-square test and the *t*-test according to the variable nature. The level of statistical significance was established at 0.05.

Finally, through binary logistic regression, the optimal model that best predicted a positive STI diagnosis was selected, employing the forward stepwise-likelihood ratio. At first, a model with no explanatory variable (only the constant) was used, and, in every step, a variable including a minor *p* was introduced. The potential predictors group comprised first-order interactions among explanatory variables. For each variable included in the model, the odds ratio (OR) and its confidence interval (CI) were calculated at 95%. The validity of the model was verified by means of the Hosmer-Lemeshow goodness-of-fit test. Analyses were performed using the Statistical Package for the Social Sciences (SPSS) program, version 24, (IBM, New York, NY, USA).

Before this study was carried out, approval was obtained from the Biomedical Research Ethics Committee of the province of Granada and from the Management Directorate of the Granada-Metropolitan Health District, which is responsible for the STI clinic where the research was approved. Patient data were handled with the utmost confidentiality and in compliance with the Spanish Organic Law 15/1999, of December 13, on Personal Data Protection, and the Spanish Organic Law 3/2018, of December 5, on Personal Data Protection and a guarantee of digital rights.

3. Results

Inclusion criteria were fulfilled by 678 cases, out of which 440 (65.6%) had a positive STI diagnosis and 230 (34.2%), a negative one. Results show that STIs spread homogeneously across the population in terms of nationality, working status, and education level, establishing, with statistically significant differences, that persons diagnosed with an STI are younger ($p < 0.001$), mostly men ($p < 0.001$), who have not been or are not sex workers ($p < 0.001$), and who were diagnosed during the economic crisis period (2008–2014) ($p < 0.001$) (Table 1).

With respect to the healthcare received, differences were found in the number of subsequent visits ($p = 0.001$) and of new subsequent episodes ($p < 0.001$), being more frequent in cases diagnosed with STIs (Table 2).

As regards risk indicators, the group diagnosed with STIs used drugs more frequently, mentioned a higher number of days elapsed since the last unprotected sexual intercourse, and had more sexual partners in the last month and in the last year (Table 3).

Table 4 shows the logistic regression model adjusted that better predicts a positive STI diagnosis. It is noted that the probability of acquiring an STI is higher in a crisis period, if it is to be a woman in the said period, and if it is to be a woman and an immigrant. In addition, this probability diminishes drastically with age (Table 4).

Table 1. STI diagnosis vs. sociodemographic characteristics.

Variables	Negative STI Diagnosis		Positive STI Diagnosis		*p*
	n	Mean (SD)	*n*	Mean (SD)	
Age (*n* = 678)	234	31.21 (10.62)	444	28.16 (8.27)	<0.001
	n	%	*n*	%	
Sex (*n* = 678)					
Female	76	32.5%	213	48%	<0.001
Male	158	67.5%	231	52%	
Nationality (*n* = 663)					
Spanish	178	78.8%	326	74.6%	0.234
Non-Spanish	48	21.2%	111	25.4%	
Occupation (*n* = 622)					
Sex worker/Former sex worker	13	6.4%	72	17.2%	<0.001
Other occupations	191	93.6%	346	82.8%	
Employment status (*n* = 607)					
Employed	97	46.4%	204	51.3%	0.257
Unemployed	112	53.6%	194	48.7%	
Level of education (*n* = 632)					
No education/Primary education	43	19.8%	77	18.6%	0.926
Secondary education	72	33.2%	141	34%	
Higher education	102	47%	197	47.5%	
Economic crisis (*n* = 678)					
Yes (2008–2014)	80	34.2%	243	54.7%	<0.001
No (2000–2007)	154	65.8%	201	45.3%	

Note: STI—Sexually Transmitted Infection; *n*—sample size; SD—standard deviation; *p*—*p* value.

Table 2. STI diagnosis vs. healthcare received.

Variables	Negative STI Diagnosis		Positive STI Diagnosis		*p*
	n	%	*n*	%	
Reason for visit (*n* = 678)					
HIV	67	28.7%	151	34%	0.156
Other reasons	167	71.3%	293	66%	
Previous treatment (*n* = 456)					
Yes	53	40.8%	107	32.8%	0.108
No	77	59.2%	219	67.2%	
	n	Mean (SD)	*n*	Mean (SD)	
No. of subsequent visits (*n* = 667)	229	0.95 (1.01)	438	1.30 (1.41)	0.001
No. of new subsequent episodes (*n* = 668)	229	0.34 (0.67)	439	0.79 (1.24)	<0.001

Note: *n*—sample size; SD—standard deviation; *p*—*p* value.

Table 3. STI diagnosis vs. risk indicators.

Variables	Negative STI Diagnosis		Positive STI Diagnosis		*p*
	n	%	*n*	%	
Sexual orientation identity (*n* = 649)					0.250
Heterosexual	190	85.6	351	82.2	
Bisexual	11	5	17	4	
Homosexual	21	9.5	59	13.8	
Regular partner (*n* = 627)					
Yes	141	66.8	282	67.8	0.808
No	70	33.2	134	32.2	
Regular partner having symptoms (*n* = 242)					
Yes	32	42.1	68	41	0.867
No	44	57.9	98	59	
Drug use (*n* = 284)					
Yes	24	26.4	78	40.4	0.021
No	67	73.6	115	59.6	
Previous STIs (*n* = 534)					0.415
Yes	40	22.2	90	25.4	
No	140	77.8	264	74.6	
	n	Mean (SD)	*n*	Mean (SD)	*p*
No. of days since last sexual contact without a condom (*n* = 382)	123	2.71 (0.87)	259	3.29 (0.83)	0.001
No. of partners in the last month (*n* = 629)	209	1.40 (1.01)	420	1.87 (1.49)	<0.001
No. of partners in the last year (*n* = 626)	208	2.33 (1.6)	418	3.05 (2.06)	<0.001
Sex life (*n* = 114)	45	1.84 (0.90)	99	1.97 (0.88)	0.436
Age of first sexual intercourse (*n* = 318)	86	18.23 (2.9)	232	17.49 (3.1)	0.054

Note: *n*—sample size; SD—standard deviation; *p*—*p* value.

Table 4. Logistic regression for STI diagnosis [†].

Variables (Reference Category or Units)	OR	(95% CI)
Crisis (yes)	1.88	(1.28–2.76)
Crisis (yes) × Sex (women)	2.35	(1.24–4.44)
Sex (women) × Nationality (Non-Spanish)	2.09	(1.22–3.57)
Age (years)	0.97	(0.95–0.98)

Note: [†]—Hosmer-Lemeshow goodness-of-fit test: χ^2 (8) = 3.89; *p* = 0.867. OR—odds ratio; CI: confidence interval.

4. Discussion

In relation to the sociodemographic characteristics of the sample analyzed, it is noted that the mean age of individuals diagnosed with an STI was lower than in the case of those who did not have them. This result coincides with the publications in the scientific literature, considering the young population as one of the groups most exposed to STIs [22–24] along with other vulnerable groups such as gay men [5,25–28], men who have sex with men (MSM), transgender people, injecting drug users, women, sex workers, and immigrants, especially irregular immigrants [18,29–33]. Likewise, in a recent study on HIV and other STI epidemiology, it is observed that in 2006 the higher number of cases of some STIs such as syphilis, gonorrhea, chlamydia trachomatis, and lymphogranuloma venereum occurred in young adults between 25–44 years old [34], the age range in which the mean age of persons positively diagnosed with STIs is found in our study.

We have found no relationship between sex work and a positive STI diagnosis, despite the already-known vulnerability of such a group [35–37]. Although sexual orientation plays a relevant role, gay men sex workers assume more risky practices compared to heterosexuals [38], they were not well represented in our sample. In addition, safe sexual practices are often more present in commercial sexual relations, while they are relaxed in non-commercial sexual relations [38,39], which may have an impact on this study. However, our results coincide with the ones published regarding Spain by other authors, who found a low seroprevalence of some STIs [40], such as HIV, among sex workers [41–43].

The association found between the period of economic crisis and the increased risk of being diagnosed with STIs is in line with the fact that a crisis debilitates educational and health systems as well as prevention and promotion measures related to sexual health [44]. Therefore, such a situation may lead to an increase in the prevalence of infectious diseases like STIs, in particular to an increase in new HIV diagnoses [45,46].

Upon analyzing the healthcare received, the significant association found between the number of subsequent visits and of new episodes, among people who were positively diagnosed with an STI, could be justified by the need for a more intense follow-up of such new diagnoses.

In relation to risk indicators, we have found that drug use becomes a risky practice for transmission of STIs, due to the limitation imposed on the individual's decision-making capability, making them more vulnerable, as pointed out by other authors [33,47,48]. Evidence associates the low use of condoms with a higher risk of STI contagion [49–52], in line with our results, despite some nuances. We have analyzed the time period elapsed since the last time a condom was used and not its frequency of use, but literature identifies both factors as strongly associated with a significant risk of acquiring an STI [22,24,53]. Moreover, upon evaluating jointly two of the abovementioned variables, namely, the time elapsed since the last sexual intercourse without using a condom and the economic crisis period, it is observed that, in Spain, as from 2007, there has been a decrease in the use of contraceptive methods and STI prevention measures in one-fifth of casual or sporadic relationships [45]. Therefore, it should be noted that, in the crisis period, financial problems pose obstacles to access to contraception [16], with the consequent risk not only of unintended pregnancy but also of contagion of venereal diseases. Regarding the number of sexual partners in the last month and in the last year, people having a positive diagnosis had more partners, which finding is in line with the literature consulted [22,31].

In the predictive model designed for the diagnosis of STIs in our province, the variables of the economic crisis period and interactions between crisis and women, and between women and non-Spanish nationality remained, regardless of the other variables, showing an increase in risk, while age acted as a protective factor (Table 4). That is, structural determinants of health inequalities are the ones that have had a direct impact on positive STI diagnoses, reinforcing the knowledge that health is not just an individual issue, but it depends, to a great extent, on the surrounding environment [54,55].

As previously stated, this study is within the group of studies finding an association between an adverse financial context and bad health results, which becomes worse in women. The fact that women suffer in a different way than men do, the impacts of every financial crisis, has been widely reported [56–61]. And this is so because of the different and unequal opportunities men and women have regarding access to powerful, prestigious positions and to available resources, where women are usually at a disadvantage, even in developed countries like ours [62]. Furthermore, it should be pointed out that gender inequality occurring in the context of relationships affects the sexual and reproductive health of women, exposing them to a higher number of risks [45].

If to the fact of being a woman, the fact of being a foreigner is added, the likelihood of being diagnosed with a STI in our study increased even more. This result confirms the findings in previous Spanish studies where immigrant women were more exposed to HIV infection [32,33]. In this sense, the last report on Epidemiologic Surveillance of HIV and AIDS in Spain states that among the foreign population with new HIV diagnoses, 56% were women, had a worse immunological response to antiretroviral therapy, showed less follow-up, and less time for therapeutic failure [63]. Although the fact of being a foreigner, especially if coming from low-income countries, usually entails some disadvantages in the destination country, in financial, working, administrative and legal terms, that affect their health in the medium and long term, the austerity and exclusion policies carried out during the economic crisis seem to have sped up the process [64,65], particularly in relation to women's sexual health.

Finally, the predictive model results confirm the protective effect of age in relation to the appearance of STIs. The fact that only a small group of European young people state they have access to information on STI prevention, while most of them have erroneous concepts on it and are in favor of causal sexual intercourse and with multiple partners, shows the existing lack of knowledge on these issues and the underestimation of the risk of acquiring STIs to which they are exposed [22].

Limitations

This study presents some limitations and strengths. On the one hand, due to its cross-sectional nature, associations detected shall not be interpreted as causal relations, but, in any case, they allow for hypotheses that shall require confirmation in subsequent research with more complex designs. It should also be taken into account that the sample of the population analyzed comes from a single center and may not be fully representative of subjects vulnerable to an STI diagnosis in the province, which affects the external validity of the study. Specifically, it is fairly likely that the immigrant population is overrepresented, since, for instance, the native population may have more access to private healthcare for the treatment of these infections, as it occurs with other health/disease processes [66]. Such a fact would minimize the magnitude of the associations found. Nevertheless, the WHO recommendations are in line with the promotion of knowledge at the local level in order to address STIs, which justifies the need for this type of analysis [18]. On the other hand, although we may think people coming to the center may have had risky sexual behavior, information taken from clinical interviews in-person may be influenced by a social desirability bias, which, even in such a case, would not be discriminatory among the groups analyzed.

Finally, we believe the power of the study to evaluate factors related to STI diagnosis is high for several reasons. A long time period was evaluated; cases with an accurate (yes/no) STI diagnosis were selected, and individual and clinical data taken from medical histories were analyzed which, to the best of our knowledge, contribute more information and of better quality than any population database available in our context.

5. Conclusions

Our findings have identified immigrant women as a risk group regarding the acquisition of STIs in the province of Granada, and especially during an economic crisis. However, there is an evident need for further research, with a gender-oriented perspective, of sexual behavior within such a group, as well as any potential limitations to access the health system, in order to understand differences in behavior and, therefore, in health results between men and women.

Public health policies should be aimed at a re-evaluation of prevention strategies present in our province, as well as of services available, reduced during the crisis. Designing and implementing specific, more effective preventive measures that shall prepare immigrant women to face these infections, as well as detecting non-reported cases and controlling and treating existing cases at a sustainable human and economic cost, entail the first-level challenge in the context of STIs.

Author Contributions: Conceptualization, A.M.-S; Methodology, E.M.-G. and P.F.-M.; Investigation, M.Á.P.-M., M.G.-L, M.A.Á.-S., E.M.-G., P.F.-M., M.D.P.-C, and A.M.-S.; Data curation, E.M.-G. and P.F.-M.; Writing—original draft preparation, M.Á.P.-M., M.G.-L, M.A.Á.-S., E.M.-G., P.F.-M., M.D.P.-C, and A.M.-S.; Writing—review and editing, M.Á.P.-M., M.G.-L, M.A.Á.-S., E.M.-G., P.F.-M., M.D.P.-C, and A.M.-S.; Supervision, M.Á.P.-M., E.M.-G., and A.M.-S. All authors have read and agreed to the published version of the manuscript.

Funding: This research received no external funding.

Conflicts of Interest: The authors declare no conflict of interest.

References

1. Organización Mundial de la Salud. Infecciones de Transmisión Sexual. 2019. Available online: https://www.who.int/es/news-room/fact-sheets/detail/sexually-transmitted-infections-(stis) (accessed on 19 August 2020).
2. An, Q.; Wejnert, C.; Bernstein, K.; Paz-Bailey, G. Syphilis screening and diagnosis among men who have sex with men, 2008–2014, 20 U.S. cities. *J. Acquir. Immune Defic. Syndr.* **2017**, *75*, S363–S369. [CrossRef]
3. De Munain, J.L. Epidemiología y control actual de las infecciones de transmisión sexual. Papel de las unidades de ITS. *Enferm. Infecc. Microbiol. Clin.* **2019**, *37*, 45–49. [CrossRef] [PubMed]
4. William, H.; Blahd, J. MF-M de Emergencia & AHM-M Familiar & HMOMMmeF-M de Emergencia. Comportamiento Sexual de Alto Riesgo. Cigna. 2019. Available online: https://www.cigna.com/individuals-families/health-wellness/hw-en-espanol/temas-desalud/comportamiento-sexual-de-alto-riesgo-tw9064 (accessed on 19 August 2020).
5. Johnston, L.G.; Alami, K.; El Rhilani, M.H.; Karkouri, M.; Mellouk, O.; Abadie, A.; Rafif, N.; Ouarsas, L.; Bennani, A.; Omari, B.E. HIV, syphilis and sexual risk behaviours among men who have sex with men in Agadir and Marrakesh, Morocco. *Sex. Transm. Infect.* **2013**, *89*, 45–48. [CrossRef]
6. Godoy, P. La vigilancia y el control de las infecciones de transmisión sexual: Todavía un problema pendiente. *Gac. Sanit.* **2011**, *25*, 263–266. [CrossRef]
7. Llácer, A.; Fernández-Cuenca, R.; Navarro, J.F.M. Crisis económica y patología infecciosa. Informe SESPAS 2014. *Gac. Sanit.* **2014**, *28*, 97–103. [CrossRef]
8. Suhrcke, M.; Stuckler, D.; Suk, J.E.; Desai, M.; Senek, M.; McKee, M.; Tsolova, S.; Basu, S.; Abubakar, I.; Hunter, P.R.; et al. The impact of economic crises on communicable disease transmission and control: A systematic review of the evidence. *PLoS ONE* **2011**, *6*, e20724. [CrossRef]
9. Martínez García, M.F.; Sánchez, A.; Martínez, J. Crisis económica, salud e intervención psicosocial en España. *Apunt. Psicol.* **2017**, *35*, 5–24.
10. Pérez-Morente, M.A.; Martín-Salvador, A.; Gázquez-López, M.; Femia-Marzo, P.; Pozo-Cano, M.D.; Hueso-Montoro, C.; Martínez-García, E. Economic Crisis and Sexually Transmitted Infections: A Comparison Between Native and Immigrant Populations in a Specialised Centre in Granada, Spain. *Int. J. Environ. Res. Public Health* **2020**, *17*, 2480. [CrossRef] [PubMed]
11. Dávila Quintana, C.D.; González López-Valcárcel, B. Crisis económica y salud. *Gac. Sanit.* **2009**, *23*, 261–265. [CrossRef] [PubMed]
12. Meléndez, E.S. El mercado de trabajo español en la crisis económica (2008–2012): Desempleo y reforma laboral. *Rev. Estud. Empresariales Segunda Época* **2012**, *2*, 29–57.
13. Valls, A.D.I.; Coll, A.S.; Rivera, E.O. ¿Migración neohispánica? El impacto de la crisis económica en la emigración española. *Empiria Rev. Metodol. Las Cienc. Soc.* **2014**, *29*, 39–66.
14. Ministerio de Sanidad, Consumo y Bienestar Social. Crisis Económica y Salud en España. *Madrid.* 2018. Available online: https://www.mscbs.gob.es/estadEstudios/estadisticas/docs/CRISIS_ECONOMICA_Y_SALUD.pdf (accessed on 3 September 2020).
15. Pérez-Morente, M.A.; Sánchez-Ocón, M.T.; Martínez-García, E.; Martín-Salvador, A.; Hueso-Montoro, C.; García-García, I. Differences in Sexually Transmitted Infections between the Precrisis Period (2000–2007) and the Crisis Period (2008–2014) in Granada, Spain. *J. Clin. Med.* **2019**, *8*, 277. [CrossRef] [PubMed]
16. McBride, O.; Morgan, K.; McGee, H. Irish Contraception and Crisis Pregnancy Study 2010 (ICCP-2010) A Survey of the General Population. Crisis Pregnancy Programme Report 2012, 24. Available online: https://www.sexualwellbeing.ie/for-professionals/research/research-reports/iccp-2010_report.pdf (accessed on 3 September 2020).
17. Vallejo, S. El Virus de la Crisis Hace Estragos en la Salud. 2013. Available online: http://www.granadahoy.com/article/granada/1268438/sos/virus/la/crisis/hace/estragos/la/salud.htm (accessed on 25 August 2020).
18. World Health Organization. Global Health Sector Strategy on Sexually Transmitted Infections, 2016–2021. Available online: https://www.who.int/reproductivehealth/publications/rtis/ghss-stis/en/ (accessed on 25 August 2020).

19. Unidad de Vigilancia del VIH y Conductas de Riesgo. Vigilancia Epidemiológica de las Infecciones de Transmisión Sexual, 2017. Madrid: Centro Nacional de Epidemiología, Instituto de Salud Carlos III/Plan Nacional sobre el Sida, Dirección General de Salud Pública, Calidad e Innovación. 2019. Available online: https://www.mscbs.gob.es/ciudadanos/enfLesiones/enfTransmisibles/sida/vigilancia/Vigilancia_ITS_1995_2017_def.pdf (accessed on 25 August 2020).
20. INE: Instituto Nacional de Estadística. Principales Series de Población Desde 1998. *Población por Provincias, Edad (3 Grupos de Edad), Españoles/Extranjeros, Sexo y año. INEbase.* 2019. Available online: http://www.ine.es/jaxi/Tabla.htm?path=/t20/e245/p08/l0/&file=03001.px&L=0 (accessed on 25 August 2020).
21. Llavero-Molino, I.; Sánchez-Ocón, M.T.; Pérez-Morente, M.A.; Espadafor-López, B.; Martín-Salvador, A.; Martínez-García, E.; Hueso-Montoro, C. Sexually Transmitted Infections and Associated Factors in Homosexuals and Bisexuals in Granada (Spain) during the Period 2000–2015. *Int. J. Environ. Res. Public Health* **2019**, *16*, 2958. [CrossRef] [PubMed]
22. Calatrava, M.; López-del Burgo, C.; Irala, J. Factores de riesgo relacionados con la salud sexual en los jóvenes europeos. *Med. Clin.* **2012**, *138*, 534–540. [CrossRef]
23. Faílde Garrido, J.M.; Lameiras Fernández, M.; Bimbela Pedrola, J.L. Prácticas sexuales de chicos y chicas españoles de 14–24 años de edad. *Gac. Sanit.* **2008**, *22*, 511–519. [CrossRef]
24. Leoni, A.F.; Martelloto, G.I.; Jakob, E.; Cohen, J.E.; Aranega, C.I. Sexual behavior and risk of sexually transmitted infections in medicine students of the National University of Cordoba. *Dst. J. Bras. Doenças Sex. Transm.* **2005**, *17*, 93–98.
25. Catallozzi, M.; Bell, D.L.; Short, M.B.; Marcell, A.V.; Ebel, S.C.; Rosenthal, S.L. Does perception of relationship type impact sexual health risk? *Sex. Transm. Dis.* **2013**, *40*, 473–475. [CrossRef]
26. Senn, T.E.; Carey, M.P. Age of partner at first adolescent intercourse and adult sexual risk behavior among women. *J. Womens Health (Larchmt.)* **2011**, *20*, 61–66. [CrossRef]
27. Eaton, D.K.; Kann, L.; Kinchen, S.; Shanklin, S.; Flint, K.H.; Hawkins, J.; Harris, W.A.; Lowry, R.; McManus, T.; Chyen, D.; et al. Youth risk behavior surveillance—United States, 2011. *Surveill. Summ.* **2012**, *61*, 1–162.
28. Conde Gutiérrez Del Álamo, F.; Santoro Domingo, P.; Grupo de Asesores en Adherencia al Tratamiento Antirretrovírico de Sida. Tipología, valores y preferencias de las personas con VIH e imaginarios de la infección: Resultados de un estudio cualitativo. *Rev. Esp. Salud Publica* **2012**, *86*, 139–152. [CrossRef]
29. Organista, K.C.; Worby, P.A.; Quesada, J.; Arreola, S.G.; Kral, A.H.; Khoury, S. Sexual health of Latino migrant day labourers under conditions of structural vulnerability. *Cult. Health Sex.* **2013**, *15*, 58–72. [CrossRef]
30. Platt, L.; Grenfell, P.; Fletcher, A.; Sorhaindo, A.; Jolley, E.; Rhodes, T.; Bonell, C. Systematic review examining differences in HIV, sexually transmitted infections and health-related harms between migrant and nonmigrant female sex workers. *Sex. Transm. Infect.* **2013**, *89*, 311–319. [CrossRef] [PubMed]
31. Gesink, D.C.; Sullivan, A.B.; Miller, W.C.; Bernstein, K.T. Sexually transmitted disease core theory: Roles of person, place, and time. *Am. J. Epidemiol.* **2011**, *174*, 81–89. [CrossRef]
32. Hernando Rovirola, C.; Ortiz-Barreda, G.; Galán Montemayor, J.C.; Sabidó Espin, M.; Casabona Barbarà, J. Infección VIH/Sida y otras infecciones de transmisión sexual en la población inmigrante en España: Revisión bibliográfica. *Rev. Esp. Salud Publica* **2014**, *88*, 763–781. [CrossRef]
33. Morán Arribas, M.; Rivero, A.; Fernández, E.; Poveda, T.; Caylá, J.A. Burden of HIV infection, vulnerable populations and access barriers to health care. *Enferm. Infecc. Microbiol. Clin.* **2018**, *36*, 3–9. [CrossRef]
34. Área de Vigilancia del VIH y Conductas de Riesgo. Epidemiología del VIH y de Otras Infecciones de Transmisión Sexual en Mujeres. España, Diciembre 2018. Madrid: Centro Nacional de Epidemiología-Instituto de Salud Carlos III/Plan Nacional sobre el Sida—Dirección General de Salud Pública, Calidad e Innovación. 2018. Available online: https://www.isciii.es/QueHacemos/Servicios/VigilanciaSaludPublicaRENAVE/EnfermedadesTransmisibles/Documents/archivos%20A-Z/INFECCION%20GONOCOCICA/InformeMujeres2018.pdf#search=infecciones%20de%20transmision%20sexual (accessed on 3 September 2020).
35. Yáñez Álvarez, I.; Sánchez Alemán, M.A.; Conde González, C.J. Migration and sexual behavior effects on HIV infection among Mexican migrants to the USA. *Enf. Inf. Microbiol.* **2011**, *31*, 98–104.
36. Folch, C.; Sanclemente, C.; Esteve, A.; Martró, E.; Molinos, S.; Casabona, J. Social characteristics, risk behaviours and differences in the prevalence of HIV/sexually transmitted infections between Spanish and immigrant female sex workers in Catalonia, Spain. *Med. Clin.* **2009**, *132*, 385–388. [CrossRef] [PubMed]

37. del Romero Guerrero, J.; Rojas Castro, D.; Ballesteros Martín, J.; Clavo Escribano, P.; Menéndez Prieto, B. Prostitución: Un colectivo de riesgo. *JANO* **2004**, *LXVII*, 106–108.

38. Jacques-Aviñó, C.; de Andrés, A.; Roldán, L.; Fernández-Quevedo, M.; García de Olalla, P.; Díez, E.; Romaní, O.; Caylà, J.A. Trabajadores sexuales masculinos: Entre el sexo seguro y el riesgo. Etnografía en una sauna gay de Barcelona, España. *Ciênc. Saúde Coletiva* **2019**, *24*, 4707–4716. [CrossRef] [PubMed]

39. Folch, C.; Casabonaa, J.; Sanclemente, C.; Esteve, A.; González, V.; Grupo HIVITS-TS1. Trends in HIV prevalence and associated risk behaviors in female sex workers in Catalonia (Spain). *Gac. Sanit.* **2014**, *28*, 196–202. [CrossRef] [PubMed]

40. Cabrerizo Egea, M.-J.; Barroso García, M.P.; Rodríguez-Contreras Pelayo, R. Infecciones de transmisión sexual en mujeres que ejercen la prostitución en Almería. *Actual. Med.* **2013**, *789*, 74–77.

41. Belza, M.J.; Clavo, P.; Ballesteros, J.; Menéndez, B.; Castilla, J.; Sanz, S.; Jerez, N.; Rodríguez, C.; Sánchez, F.; Romero, J. Condiciones sociolaborales, conductas de riesgo y prevalencia de infecciones de transmisión sexual en mujeres inmigrantes que ejercen la prostitución en Madrid. *Gac. Sanit.* **2004**, *18*, 177–183. [CrossRef] [PubMed]

42. Vall-Mayans, M.; Villa, M.; Saravanya, M.; Loureiro, E.; Meroño, M.; Arellano, E.; Sanz, B.; Saladié, P.; Andreu, A.; Codina, M. Sexually transmitted Chlamydia trachomatis, Neisseria gonorrhoeae, and HIV-1 infections in two at-risk populations in Barcelona: Female street prostitutes and STI clinic attendees. *Int. J. Infect. Dis.* **2007**, *11*, 115–122. [CrossRef] [PubMed]

43. del Amo, J.; Gonzalez, C.; Losana, J.; Clavo, P.; Munoz, L.; Ballesteros, J.; Garcia-Saiz, A.; Belza, M.; Ortiz, M.; Menendez, B.; et al. Influence of age and geographical origin in the prevalence of high risk human papilloma virus inmigrant female sex workers in Spain. *Sex. Transm. Infect.* **2005**, *81*, 79–84. [CrossRef] [PubMed]

44. Vázquez, M.L.; Vargas, I.; Aller, M.B. The impact of the economic crisis on the health and healthcare of the immigrant population. SESPAS report 2014. *Gac. Sanit.* **2014**, *28*, 142–146. [CrossRef] [PubMed]

45. Larrañaga, I.; Martín, U.; Bacigalupe, A. Sexual and reproductive health and the economic crisis in Spain. SESPAS report 2014. *Gac. Sanit.* **2014**, *28*, 109–115. [CrossRef] [PubMed]

46. European Centre for Disease Prevention and Control. *Joint Technical Mission: HIV in Greece 28–29 May 2012*; ECDC: Stockholm, Sweden, 2013. Available online: https://ecdc.europa.eu/sites/portal/files/media/en/publications/Publications/hiv-joint-technical%20mission-HIV-in-Greece.pdf (accessed on 3 September 2020).

47. Gilbert, P.A.; Rhodes, S.D. HIV Testing Among Immigrant Sexual and Gender Minority Latinos in a US Region with Little Historical Latino Presence. *AIDS Patient Care STDs* **2013**, *27*, 628–636. [CrossRef]

48. Pinedo, M.; Burgos, J.L.; Ojeda, V.D. A critical review of social and structural conditions that influence HIV risk among Mexican deportees. *Microbes Infect.* **2014**, *16*, 379–390. [CrossRef]

49. Blanc Molina, A.; Rojas Tejada, A.J. Uso del preservativo, número de parejas y debut sexual en jóvenes en coito vaginal, sexo oral y sexo anal. *Rev. Int.* **2018**, *16*, 8–14. [CrossRef]

50. Ward, J.; Wand, H.; Bryant, J.; Delaney-Thiele, D.; Worth, H.; Pitts, M.; Byron, K.; Moore, E.; Donovan, B.; Kaldo, J.M. Prevalence and Correlates of a Diagnosis of Sexually Transmitted Infection Among Young Aboriginal and Torres Strait Islander People: A National Survey. *Sex. Transm. Dis.* **2016**, *43*, 177–184. [CrossRef]

51. Velo-Higueras, C.; Cuéllar-Flores, I.; Sainz-Costa, T.; Navarro-Gómez, M.L.; García-Navarro, C.; Fernández-McPhee, C.; Ramírez, A.; Bisbal, O.; Blazquez-Gameroa, D.; Ramos-Amadorb, J.T.; et al. Jóvenes y VIH. Conocimiento y conductas de riesgo de un grupo residente en España. *Enferm. Infecc. Microbiol. Clin.* **2019**, *37*, 176–182. [CrossRef] [PubMed]

52. Pérez-Morente, M.A.; Campos-Escudero, A.; Sánchez-Ocon, M.T.; Hueso-Montoro, C. Características sociodemográficas, indicadores de riesgo y atención sanitaria en relación a infecciones de transmisión sexual en población inmigrante de Granada. *Rev. Esp. Salud Pública* **2019**, *93*, 1–13.

53. Acién Alvarez, P. *Planificación Familiar: Fisiología Reproductiva. Sexualidad, Anticoncepción y ETS*; Molloy: Alicante, Spain, 2002; p. 522.

54. CSDH. *Closing the Gap in a Generation: Health Equity through Action on the social Determinants of Health. Final Report of the Commission on Social Determinants of Health*; World Health Organization: Geneva, Switzerland, 2008; Available online: https://www.who.int/social_determinants/final_report/csdh_finalreport_2008.pdf (accessed on 1 September 2020).

55. Ministerio de Sanidad, Servicios Sociales e Igualdad. Comisión Para Reducir las Desigualdades Sociales en Salud en España. Avanzando Hacia la Equidad. Propuestas de Políticas E Intervenciones Para Reducir las Desigualdades Sociales en Salud en España. 2015. Available online: https://www.mscbs.gob.es/profesionales/saludPublica/prevPromocion/promocion/desigualdadSalud/docs/Propuesta_Politicas_Reducir_Desigualdades.pdf (accessed on 1 September 2020).

56. Smith, J. Overcoming the 'tyranny of the urgent': Integrating gender into disease outbreak preparedness and response. *Gend. Dev.* **2019**, *27*, 355–369. [CrossRef]

57. Advisory Committee on Equal Opportunities for Women and Men. European Commision Employment, Social Affairs and Equal Opportunities. Opinion on the Gender Perspective on the Response to the Economic and Financial Crisis. 2009. Available online: file:///C:/Users/USUARIO/Downloads/Opinion%20economic%20crisis.pdf (accessed on 1 September 2020).

58. Walby, S. Gender and the Financial Crisis. Paper for UNESCO Proyect on Gender and the Financial Crisis. 2009. Available online: https://www.lancaster.ac.uk/fass/doc_library/sociology/Gender_and_financial_crisis_Sylvia_Walby.pdf (accessed on 1 September 2020).

59. Seguino, S. The global economic crisis, its gender implications and policy responses. *Gend. Dev.* **2010**, *18*, 179–199. [CrossRef]

60. Gálvez-Muñoz, L.; Rodríguez-Madroño, P. La desigualdad de género en las crisis económicas. *Investig. Fem.* **2011**, *2*, 113–132. [CrossRef]

61. Gálvez-Muñoz, L. La brecha de género en la crisis económica. *USTEA* **2012**, 4–5.

62. Borrel, C.; Rodríguez-Sanz, M.; Bartoll, X.; Malmusi, D.; Novoa, A.M. El sufrimiento de la población en la crisis económica del Estado español. *Salud Colect.* **2014**, *10*, 95–98. [CrossRef]

63. Unidad de Vigilancia de VIH y Comportamientos de Riesgo. Vigilancia Epidemiológica del VIH y Sida en España 2019: Sistema de Información Sobre Nuevos Diagnósticos de VIH y Registro Nacional de Casos de Sida. Plan Nacional sobre el Sida. D.G de Salud Pública, Calidad e Innovación/Centro Nacional de Epidemiología. ISCIII. Madrid. 2019. Available online: https://www.mscbs.gob.es/ciudadanos (accessed on 1 September 2020).

64. Robert, G.; Martínez, J.M.; García, A.M.; Benavides, F.G.; Ronda, E. From the boom to the crisis: Changes in employment conditions of immigrants in Spain and their effects on mental health. *Eur. J. Public Health* **2014**, *24*, 404–409. [CrossRef]

65. Malmusi, D.; Borrel, C.; Benach, J. Migration-related health inequalities: Showing the complex interactions between gender, social class and place of origin. *Soc. Sci. Med.* **2010**, *71*, 1610–1690. [CrossRef]

66. Dahlen, H.G.; Schmied, V.; Dennis, C.L.; Thornton, C. Rates of obstetric intervention during birth for low risk women born in Australia compared to those born overseas. *BMC Pregnancy Child* **2013**, *13*, 100.

International Journal of
*Environmental Research
and Public Health*

Article

The Quality of Counselling for Oral Emergency Contraceptive Pills—A Simulated Patient Study in German Community Pharmacies

Bernhard Langer *, Sophia Grimm, Gwenda Lungfiel, Franca Mandlmeier and Vanessa Wenig

Department of Health, Nursing, Management, University of Applied Sciences Neubrandenburg, 17033 Neubrandenburg, Germany; sg2807@t-online.de (S.G.); Gwenda.lungfiel@gmx.de (G.L.); franca_m@web.de (F.M.); vanessawenig@gmx.de (V.W.)
* Correspondence: langer@hs-nb.de

Received: 8 August 2020; Accepted: 11 September 2020; Published: 15 September 2020

Abstract: Background: In Germany, there are two different active substances, levonorgestrel (LNG) and ulipristal acetate (UPA), available as emergency contraception (the "morning after pill") with UPA still effective even 72 to 120 h after unprotected sexual intercourse, unlike LNG. Emergency contraceptive pills have been available without a medical prescription since March 2015 but are still only dispensed by community pharmacies. The aim of this study was to determine the counselling and dispensing behaviour of pharmacy staff and the factors that may influence this behaviour in a scenario that intends that only the emergency contraceptive pill containing the active substance UPA is dispensed (appropriate outcome). Methods: A cross-sectional study was carried out in the form of a covert simulated patient study in a random sample of community pharmacies stratified by location in the German state of Mecklenburg-Vorpommern and reported in accordance with the STROBE statement. Each pharmacy was visited once at random by one of four trained test buyers. They simulated a product-based request for an emergency contraceptive pill, stating contraceptive failure 3.5 days prior as the reason. The test scenario and the evaluation forms are based on the recommended actions, including the checklist from the Federal Chamber of Pharmacies. Results: All 199 planned pharmacy visits were carried out. The appropriate outcome (dispensing of UPA) was achieved in 78.9% of the test purchases (157/199). A significant correlation was identified between the use of the counselling room and the use of a checklist ($p < 0.001$). The use of a checklist led to a significantly higher questioning score ($p < 0.001$). In a multivariate binary logistic regression analysis, a higher questioning score (adjusted odds ratio [AOR] = 1.41; 95% CI = 1.22–1.63; $p < 0.001$) and a time between 12:01 and 4:00 p.m. (AOR = 2.54; 95% CI = 1.13–5.73; $p = 0.024$) compared to 8:00 to 12:00 a.m. were significantly associated with achieving the appropriate outcome. Conclusions: In a little over one-fifth of all test purchases, the required dispensing of UPA did not occur. The use of a counselling room and a checklist, the use of a checklist and the questioning score as well as the questioning score and achieving the appropriate outcome are all significantly correlated. A target regulation for the use of a counselling room, an explicit guideline recommendation about the use of a checklist, an obligation for keeping UPA in stock and appropriate mandatory continuing education programmes should be considered.

Keywords: non-prescription drugs; community pharmacies; consultation; patient simulation; emergency contraception; ulipristal acetate; Germany

---

1. Introduction

Sexual intercourse without contraception, an improperly used or broken condom, improperly applied regular contraception or sexual assaults can lead to unintended pregnancies [1].

The consequences of unintended pregnancies or the need to terminate a pregnancy affect not only the physical but also the social and emotional health of the women concerned and often do so over the long term [2]. It is estimated that in Germany about 34% [3] of all pregnancies are unintended (globally, about 44% [4]), of which about 43% [3] (globally, about 56% [4]) end in termination of a pregnancy. Emergency contraception (EC) plays an important role in preventing this situation. EC can be divided into copper intrauterine devices (Cu-IUD) and oral hormonal methods [5].

The Cu-IUD is the most reliable EC method with a failure rate of <0.1%, and it is also effective after ovulation has occurred [6]. However, access to this form of EC is made difficult in that a gynaecologist must insert the Cu-IUD, meaning that this is not the preferred method for many users [5]. In Germany, it is only used in isolated cases [7]. Of the oral hormonal methods, the Yuzpe method (oestrogen–progesterone combination comprising two doses of 50 μg ethinylestradiol and 0.25 mg levonorgestrel) is also no longer recommended in Germany for post-coital contraception [7] because it is less effective than other EC methods and is also associated with more adverse events [1,6,7]. Because mifepristone is not available in Germany for this indication [7,8]—unlike some other countries [5,9]—the oral hormonal methods are concentrated on emergency contraceptive pills (or "morning after pills") containing the active substances levonorgestrel (LNG) and ulipristal acetate (UPA).

LNG is a synthetic gestagen that reduces the surge in luteinising hormone (LH), thus delaying ovulation. However, it must be taken before the surge in LH. UPA is a selective progesterone receptor modulator that also inhibits the LH surge but, unlike LNG, is still effective during the LH surge up to the LH peak [5,6,10]. Therefore, UPA is effective for up to 120 h after unprotected sexual intercourse or contraceptive failure. In contrast to UPA, LNG is effective up to 72 h after unprotected sexual intercourse or contraceptive failure, even though a moderate efficacy of up to 120 h is discussed in the international literature [5,6]. The superior effectiveness of UPA compared to LNG is also apparent in the first 24 to 72 h [11–13]. There are differences in terms of the statistical significance, however [11–13], which in Germany has led to different recommendations from gynaecology associations (generally UPA is considered superior) and the Federal Chamber of Pharmacies (BAK) (less than 72 h LNG or UPA) [14,15]. Both substances are more effective the sooner they are taken after unprotected sexual intercourse [7], which is why prompt access to the medication is of great importance. Analogous to many other countries [16], it has therefore been possible to obtain such preparations in Germany without a prescription since March 2015 [17], but the dispensing of these medications is still restricted to community pharmacies. Gynaecology associations and the German Medical Association (BÄK) argued against no-prescription dispensing of emergency contraceptive pills, citing the need for a medical consultation [18,19]. Not least, given this background [20], the BAK published corresponding recommendations including a checklist for quality assurance of the counselling provided when dispensing emergency contraceptive pills, whereby the use of this checklist is not explicitly advised in the recommendations [15]. The recommendations, which were developed with input from many experts across a range of organisations in various disciplines (e.g., BÄK, professional organisations and associations of gynaecologists, pharmaceutical OTC industry, government-controlled, private and church-based organisations and centres providing advice on sex education and family planning) [20], include the prerequisites for dispensing emergency contraception containing the active substances LNG or UPA for self-medication. Information about counselling and dispensing are also provided as well as the criteria regarding the limits of self-medication and referral to a doctor. Not only pharmacists but pharmacy technicians and pharmaceutical technical assistants are authorised to provide counselling for dispensing medicinal products such as the emergency contraceptive pill in Germany. Considering that the issue is still a sensitive one for many consumers, maintaining the privacy of the customer for this indication is a particularly important counselling criterion [21–24]. Nevertheless, both the Ordinance on the Operation of Pharmacies in Germany and the BAK recommendations expect that the counselling is conducted in private to maintain confidentiality and prevent other customers overhearing the counselling as far as possible [15,25]. There is as yet no legal obligation or a recommendation based on BAK guidelines to keep a separate counselling room, however [15,25].

The BAK recommendations stipulate the dispensing of only UPA 72 to 120 h after unprotected sexual intercourse or contraceptive failure, but it is not currently known to what extent pharmacies comply with these recommendations. The three studies available to date in Germany on the counselling and dispensing behaviour for emergency contraception have also not investigated this issue [26–28]. The study results are also based on self-assessment by pharmacies and may therefore be distorted, for example, due to social desirability [29,30]. Applying the simulated patient method (SPM), which is considered to be the "gold standard" [31] even if the relatively high administrative and financial costs and the comparably smaller sample sizes [31] as well as any intra- and inter-observer variabilities [32] are taken into account, can avoid such weaknesses because a realistic (counselling) situation can be simulated [33]. Unlike the many studies conducted in other countries [21,23,30,34–45], in Germany, there have not been any investigations to date on the counselling and dispensing of an emergency contraceptive pill using an SPM approach because the few scientific SP studies available for Germany on the counselling quality in pharmacies are always based on indications other than emergency contraception. Clear deficits in the quality of counselling have been identified in these studies [46–51].

The aim of this study was to identify the counselling and dispensing behaviour of pharmacy staff and the factors that may influence this behaviour in a scenario that intends that the emergency contraceptive pill containing only the active substance UPA is dispensed (appropriate outcome).

2. Methods

2.1. Design

A cross-sectional study design was chosen in accordance with the "STROBE Statement—Checklist of items that should be included in reports of cross-sectional studies" [52]. The quality of the counselling provided in the pharmacies visited is determined using the SPM, which has often been applied in international studies in the community pharmacy (CP) setting [33,53,54]. The SPM is a covert participatory observation [55] by a person, who in an ideal case cannot be differentiated from a real customer (simulated patient, SP), who visits a CP to simulate a real-life counselling situation based on a previously defined scenario. The data are then collected according to previously defined criteria using an assessment form and the CP is provided with performance feedback, if applicable [33].

2.2. Setting and Participation

The test purchases took place between 1 July and 30 September, 2019 in the German state of Mecklenburg-Vorpommern (31 December, 2019: approx. 1.61 million residents, 23,216 km^2 area, low population density of 69.3 residents/km^2 [56]). Upon request by telephone, the Pharmacy Association of Mecklenburg-Vorpommern did not want to provide a list of all registered community pharmacies in the state. Consequently, the list of these pharmacies was identified using the pharmacy finder of the website Apotheken-Umschau.de [57]. All community pharmacies that had a postcode in the state of Mecklenburg-Vorpommern on the reference date of 1 June, 2019 using the postcode search of the pharmacy finder were included in the study. These hits were validated with a corresponding Google search. As a result, a basic population of N = 395 pharmacies was formed. A comparison with the most recently available information from the German Federal Chamber of Pharmacies (ABDA; Bundesvereinigung Deutscher Apothekerverbände) for the end of 2018 regarding the total number of pharmacies in Mecklenburg-Vorpommern [58] showed a 98% agreement.

In Germany there have been no studies conducted to date on the dispensing of emergency contraceptive pills for a scenario in which dispensing UPA is mandatory. The degree of variability is therefore unknown. The minimum necessary sample size (n) was determined for a population size (N) of 395 and an error margin (e) of 0.05 using the following formula, which is based on a degree of variability of $p = 0.5$ and a 95% confidence interval [59]:

$$n = \frac{N}{1 + N(e)^2} = \frac{395}{1 + 395(0.05)^2} = \frac{395}{1.9875} = 198.74 \, Population \ size = N \mid Margin \ of \ error = e \quad (1)$$

The assumed degree of variability of $p = 0.5$ maximises the required sample size. The 395 CPs were stratified by location to indicate whether they are urban or rural. Using the MS Excel random number generator, they were each then assigned a random number and the 199 participating CPs were then selected by simple random sampling from each stratum.

2.3. Scenario and Assessment

The recommendations, including the checklist published by the BAK, formed the basis of the test scenario that was developed (see Table 1) and the evaluation form used (see Table 2). The request for the "morning after pill" because of a broken condom that was simulated in the scenario is realistic because this reason was cited most frequently in requests for emergency contraception in pharmacies in a recent German study [26]. Dispensing the emergency contraception ellaOne®, the only preparation with the active substance UPA available on the German market [60], was defined as the appropriate outcome. According to the recommendation from the BAK, unprotected sexual intercourse 3.5 days ago must be considered the criterion for dispensing UPA instead of LNG [15].

The evaluation form includes 14 items, whereby the first nine evaluate whether appropriate questions were asked analogous to the BAK checklist. Based on the form, the decision was subsequently made by pharmacy staff whether the emergency contraceptive pill ellaOne® with the active substance UPA is dispensed (item 10). It was also recorded whether LNG was erroneously dispensed instead of UPA (item 11). For the situation in which a medicinal product was dispensed, it was also evaluated according to the BAK recommendations whether the test buyers were informed about common adverse events such as headaches, nausea, dizziness, abdominal pain and cramps, dysmenorrhoea, vomiting, tiredness or breast tenderness [15] (item 12). Finally, the overall conditions of the counselling were also recorded, that is, whether the counselling was conducted in a separate room (item 13) and whether a checklist was visibly used to ensure the quality of the counselling (item 14).

For the evaluation form, only objective items were used to avoid a subjective assessment and thus latitude in the evaluation by the SPs (for example, on the friendliness of the pharmacy staff). Therefore, to complete the individual items, only dichotomous scales were used (closed yes/no questions).

Table 1. Test scenario.

Scenario
The test buyers enter the pharmacy and ask for oral emergency contraception without having a specific product in mind (product-based query).
When questioned by the pharmacy staff, the following information is provided: - Real age of the test buyer - "Morning after pill" is needed because of a broken condom - Unprotected sexual intercourse was 3.5 days ago - Last period was 11 days ago - An existing pregnancy is not suspected - No nausea with urge to vomit, no vomiting, no medical conditions - Not breastfeeding - Not taking other medications - Not repeated application

Table 2. Evaluation form.

	Items	Yes	No
1.	Age of the patient?	1	0
2.	Why was the "morning after pill" requested?	1	0
3.	When did the unprotected sexual intercourse take place?	1	0
4.	When was the customer's last period?	1	0
5.	Are there indications of an existing pregnancy?	1	0
6.	Are there existing health problems or are nausea with an urge to vomit or vomiting present?	1	0
7.	Is the customer breastfeeding?	1	0
8.	Current or regular use of other medications?	1	0
9.	Is this a repeated application?	1	0
10.	Dispensing the emergency contraceptive pill (UPA) ('appropriate outcome')	1	0
11.	Dispensing the emergency contraceptive pill (LNG)	1	0
12.	Explanation of side effects	1	0
13.	Counselling room used	1	0
14.	Visible use of a checklist	1	0

2.4. Data Collection

Four female Master's students from the Department of Health, Nursing, Management of the Neubrandenburg University of Applied Sciences acted as test buyers. They were selected on the basis of their participation in a 3-semester student research project. The students are all in the age group of 18 to 29 years, which corresponds to the age group of the primary users of emergency contraception [61]. Each pharmacy was visited once (a total of 199 test purchases), whereby the pharmacies were distributed randomly across the four SPs.

Before the data collection was started, the test buyers underwent comprehensive training. Each test buyer first familiarised herself with the theoretical principles of the SP method, the test scenario itself and the contents of the evaluation form. The test buyers then each carried out four pre-tests for the scenario to confirm the functionality of the evaluation form and the test scenario and to train in the practical application of the SP method. The total of 16 pre-tests were carried out in pharmacies in Mecklenburg-Vorpommern that were not included in the stratified random sample. The pre-tests indicated that no changes to the test scenario or the evaluation form were required (see Figure 1).

The test purchases were carried out on different days of the week and at different times of the day. The SPs made their request to the pharmacy staff who first approached them. The SPs only provided additional information if they were then asked by the pharmacy staff to ensure that the information provided is consistent.

Along with the items on the evaluation form, the SPs planned, analogous to the international literature (see Table 3), to also collect a number of variables during and after the test purchases that may possibly affect the appropriate outcome. Unlike other international SP studies, the variable "pharmacy type" (chain vs. independent) was not recorded [35,37,44] because in Germany a pharmacy may only be operated by one pharmacist as the owner and in addition ownership is restricted to 1 primary pharmacy with up to 3 subsidiary pharmacies in the immediate neighbourhood [62].

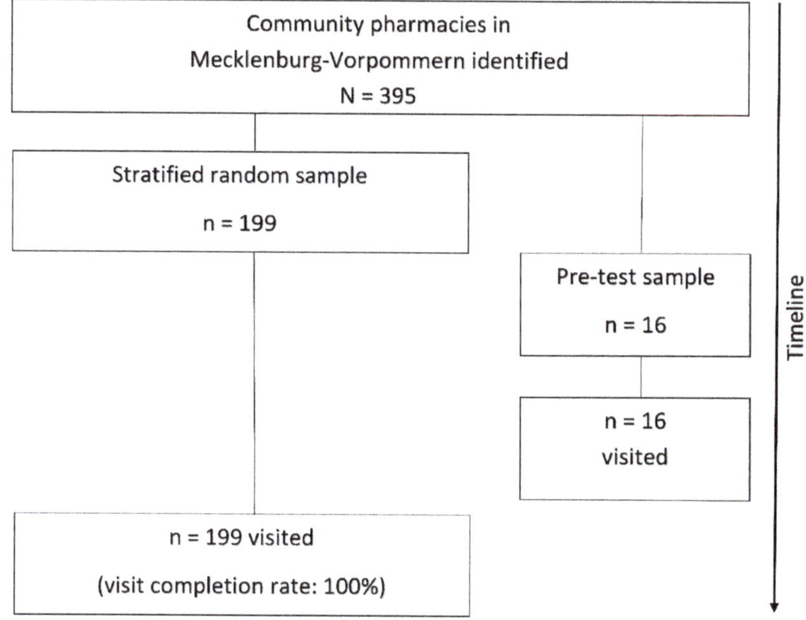

Figure 1. The design and flow of the study.

Table 3. Possible influencing factors as well as time and type of data collection.

Possible Influencing Factors [Literature Source *]	Time of Data Collection	Type of Data Collection
Location of the pharmacy [47] as an indicator for urban/rural	Before the test purchase because stratification variable	Precise measurement by allocating the number of pharmacies identified in the particular area
SP number [63]	During the test purchase	Exact measurement by assigning a number to each SP
Age of the pharmacy staff [64]	During the test purchase	Estimate based on visual impression by SP
Gender of the pharmacy staff [64]	During the test purchase	Exact measurement using visual impression of the SP
Queue—patients waiting after the SP [65]	During the test purchase	Exact measurement using visual impression of the SP
Time of the test purchase [66]	During the test purchase	Exact measurement using the SP's watch
Professional group of the pharmacy staff [67]	During and after the test purchase	Exact measurement based on the name tag and, if necessary, using a telephone query by the SP after completing all the test purchases
Pharmacy quality certificate [68]	After the test purchase	Precise measurement using a telephone query by the SP after completing all the test purchases
Questioning score [69]	After the test purchase	Precise measurement by summing the dichotomous evaluation of the nine individual questions (minimum possible score of 0 points and maximum possible score of 9 points)

Note: * The possible influencing factors were taken from the specific literature sources.

So that, on the one hand, the most realistic counselling situation possible could be simulated and to identify any planned dispensing of medicinal products while at the same time avoiding unnecessary purchasing of medicinal products and thus waste (because the test buyers do not actually need the emergency contraceptive pill), in the cases in which a medicinal product was going to be dispensed, the test buyers informed the pharmacy staff just before making payment that they had left their wallet at home.

In the literature, audio recordings are recommended for quality assurance of the test purchases [70]. For data privacy reasons, however, this was omitted because otherwise the pharmacies must be informed beforehand about the audio recordings and thus the study design is no longer covert. However, the corresponding evaluation form was completed by the test buyers immediately after visiting the pharmacy so that any distortions in the study results due to faulty recall by the test buyers could be minimised.

After evaluating the data collection, each pharmacy received a written performance feedback specific for each pharmacy including graphically edited benchmarking, whereby for each pharmacy any improvement or deterioration regarding the individual criteria was shown compared to the other anonymised pharmacies in the stratified random sample. In this way, the pharmacies were informed about the position of their competitors, so that ideally appropriate optimisation processes by the pharmacies investigated could be initiated based on the feedback with the aim of sustainably improving the quality of counselling provided.

2.5. Ethical Approval

According to the "Guideline for the use of mystery research in market and social research" [71], the data collected were anonymised and recorded in such a way that the pharmacies or the personnel involved could not be identified. To avoid a possible Hawthorne effect [72] and also a possible selection bias [73], the test purchases were carried out covertly—that is, without informing the pharmacies in advance—analogous to some other national [48–51] and international studies [21,33,38] and therefore pharmacies were not asked for consent to participate in advance. The lack of informed consent in advance was—analogous to the international literature [40,74,75]—resolved in that all pharmacies were informed about the procedure and the background of the study upon completion of the study. Recruited students provided their written informed consent to act as SPs. The study protocol was approved by the institutional ethics committee of the Neubrandenburg University of Applied Sciences (registration number: HSNB/KHM/152/19).

2.6. Data Analysis

The data were entered using the four-eyes principle and analysed with SPSS Version 25 for Windows (IBM, Armonk, NY, USA). As part of the descriptive statistics, frequencies and percentages were determined. 95% confidence intervals for categorical data using bootstrapping were also reported. A Pearson's chi-square test was performed to determine if interactions involving "counselling room used" were more likely to result in "visible use of a checklist". Cramer's V was reported as a measure of effect size. Applying the Kolmogorov–Smirnov test as well as the Shapiro–Wilk test indicated that the data did not have a normal distribution. Therefore, the median, interquartile range (IQR), minimum and maximum were calculated for continuous variables. With the help of the non-parametric Mann–Whitney U test, it was analysed whether "visible use of a checklist" led to a significantly higher median questioning score. Pearson's r was reported as a measure of the effect size. A binomial logistic regression model was also developed to identify the influence of various independent variables (see Table 3) on the appropriate outcome (dispensing of the emergency contraceptive pill with the active substance UPA). All independent variables were checked for outliers and multicollinearity. Variables with a p-value less than 0.05 in the univariate analysis were included in the multivariate analysis. Odds ratios (OR), 95% confidence intervals, p values and as a measure of the effect size Cohen's f^2 were reported. A p value of less than 0.05 was considered to be significant in all analyses.

3. Results

All 199 planned test purchases could be carried out (visit completion rate: 100%). Due to the exit strategy developed that resulted in the ending of the purchase, there were no direct costs. The total of

5080 km between the individual test purchases were driven in private vehicles and led to indirect costs totalling €1150 that were financed from the primary author's own resources. Socio-demographic data for the pharmacies or the advising pharmacy staff are shown in Table 4. Most of the pharmacies tested were in local competition with one another and often did not have a quality certificate (62.8%, 125/199). The advising pharmacy staff were in most cases female (87.4%, 174/199) and aged between 30 and 49 years (57.3%, 114/199). In the cases in which it could be determined, the proportion of pharmacists (37.7%, 75/199) compared to non-pharmacists (pharmacy technicians and pharmaceutical technical assistants) (40.7%, 81/199) was almost equal.

Table 4. Socio-demographic data for the pharmacies or the advising pharmacy staff.

	Frequency (*n*)	Percentage (%)
All pharmacies	199	100
Location of the pharmacy		
• 1 pharmacy in the area	37	18.6
• 2–4 pharmacies in the area	58	29.1
• 5–19 pharmacies in the area	46	23.1
• ≥20 pharmacies in the area	58	29.2
Pharmacy quality certificate		
• No	125	62.8
• Yes	51	25.6
• Not able to be determined	23	11.6
Age of the pharmacy staff		
• <30	27	13.6
• 30–49	114	57.3
• ≥50	58	29.1
Gender of the pharmacy staff		
• Male	25	12.6
• Female	174	87.4
Professional group of the pharmacy staff		
• Pharmacist	75	37.7
• Non-pharmacist	81	40.7
• Not able to be determined	43	21.6

The appropriate outcome—dispensing of UPA—was achieved in 78.9% of the test purchases (157/199) (see Table 5). In 3.0% of the test purchases LNG (6/199) was dispensed while in 18.1% (36/199) no preparation was dispensed. There was a median questioning score of 5.0 (IQR 2.0–9.0) with a minimum score of 0 in 2.5% (5/199) of test purchases and a maximum score of 9 in 27.6% (55/199) of test purchases. The question about the time since the unprotected sexual intercourse took place was asked most frequently (93.5%, 186/199). Regarding the test purchases in which a medication was dispensed, information was provided about possible side effects by the pharmacy staff in 59.5% (97/163) of test purchases.

Table 5. Assessment items ($n = 199$).

		Yes		
		Frequency (n)	Percentage (%)	95% CI
1.	Age of the patient?	118	59.3	52.3–66.3
2.	Why was the "morning after pill" requested?	103	51.8	45.2–58.8
3.	When did the unprotected sexual intercourse take place?	186	93.5	89.9–96.5
4.	When was the customer's last period?	133	66.8	59.8–73.4
5.	Are there indications of an existing pregnancy?	92	46.2	39.2–53.3
6.	Are there existing health problems or are nausea with an urge to vomit or vomiting present?	94	47.2	39.7–54.3
7.	Is the customer breastfeeding?	82	41.2	34.2–48.2
8.	Current or regular use of other medications?	125	62.8	55.3–70.4
9.	Is this a repeated application?	110	55.3	48.2–62.8
10.	Dispensing the emergency contraceptive pill (UPA)('appropriate outcome')	157	78.9	72.9–84.4
11.	Dispensing the emergency contraceptive pill (LNG)	6	3.0	1.0–5.5
12.	Explanation of side effects	97	59.5	52.0–67.1
13.	Counselling room used	88	44.2	37.2–51.8
14.	Visible use of a checklist	107	53.8	46.7–61.3

Counselling was provided in a counselling room in slightly less than half of all the test purchases (44.2%, 88/199). In just over half of all test purchases, a checklist was visibly used for the counselling (53.8%, 107/199). A significant correlation could be found between the use of a counselling room and the use of a checklist (Pearson's chi-square test; $\chi 2(1) = 104.355$, $p < 0.001$, V = 0.724) whereby the effect size V according to Cohen [76] corresponded to a 'large' effect. In 43.7% (87/199) of the test purchases, the counselling was not carried out in a counselling room nor was a checklist used. In contrast to this, in 41.7% (83/199) of the test purchases, a counselling room was sought out, and a checklist was used for the counselling.

In addition, the use of a checklist led to a significantly higher questioning score (Mann–Whitney U test; U = 385.500, $p < 0.001$, r = 0.806), whereby the effect size r according to Cohen [76] corresponded to a 'large' effect. If a checklist was visibly used, this led to a median questioning score of 9.0 (IQR 7.0–9.0) with a minimum score of 2 in 0.9% (1/107) and a maximum score of 9 in 51.4% (55/107) of test purchases. If a checklist was not visibly used, the median questioning score was 2.0 (IQR 1.0–3.0) with a minimum score of 0 in 5.4% (5/92) and a maximum score of 7 in 2.2% (2/92) of test purchases.

Table 6 shows the binary logistic regression model. As part of the bivariate analysis, three (age of the pharmacy staff, time of the test purchase, questioning score) of nine predictor variables had a p-value of <0.05 and were included in the multivariate logistic regression model. A time between 12:01 and 4:00 p.m. (AOR = 2.54; 95% CI = 1.13–5.73; p = 0.024) compared to 8:00 to 12:00 a.m. and a higher questioning score (AOR = 1.41; 95% CI = 1.22–1.63; p < 0.001) were significantly associated with dispensing of UPA (appropriate outcome). The location of the pharmacy, the presence of a quality certificate, a queue and the age, gender and professional group of the pharmacy staff as well as the SP number did not have any significant effect on the appropriate outcome. The model yielded a Nagelkerke R^2 value of 0.255, which corresponds to a Cohen's f^2 of 0.342 and thus to a 'medium' effect size [76].

Table 6. Possible influencing factors on the recommendation of UPA.

Possible Influencing Factors and Categories	n (%) Total 199 (100)	n (%) Recommendation 157 (78.9)	n (%) No Recommendation 42 (21.1)	COR (95% CI)	p-Value	AOR (95% CI)	p-Value
Location of the pharmacy							
• 1 pharmacy in the area	37 (100)	27 (73.0)	10 (27.0)	1			
• 2–4 pharmacies in the area	58 (100)	49 (84.5)	9 (15.5)	2.02 (0.73–5.57)	0.176		
• 5–19 pharmacies in the area	46 (100)	37 (80.4)	9 (19.6)	1.52 (0.55–4.26)	0.423		
• ≥ 20 pharmacies in the area	58 (100)	44 (75.9)	14 (24.1)	1.16 (0.45–2.99)	0.752		
Pharmacy quality certificate							
• No	125 (100)	102 (81.6)	23 (18.4)	1			
• Yes	51 (100)	40 (78.4)	11 (21.6)	0.82 (0.37–1.84)	0.629		
• Not able to be determined	23 (100)	15 (65.2)	8 (34.8)	0.42 (0.16–1.12)	0.082		
Age of the pharmacy staff							
• <30	27 (100)	17 (63.0)	10 (37.0)	1		1	
• 30–49	114 (100)	91 (79.8)	23 (20.2)	2.33 (0.94–5.75)	0.067	2.05 (0.74–5.64)	0.166
• ≥50	58 (100)	49 (84.5)	9 (15.5)	3.20 (1.11–9.21)	0.031	2.68 (0.84–8.54)	0.095
Gender of the pharmacy staff							
• Male	25 (100)	18 (72.0)	7 (28.0)	1			
• Female	174 (100)	139 (79.9)	35 (20.1)	1.54 (0.60–3.99)	0.369		
Professional group of the pharmacy staff							
• Pharmacist	75 (100)	63 (84.0)	12 (16.0)	1			
• Non-pharmacist	81 (100)	63 (77.8)	18 (22.2)	0.67 (0.30–1.50)	0.326		
• Not able to be determined	43 (100)	31 (72.1)	12 (27.9)	0.49 (0.20–1.22)	0.126		
Time of the test purchase							
• 8:00 a.m.–12:00 p.m.	85 (100)	61 (71.8)	24 (28.2)	1		1	
• 12:01 p.m.–4:00 p.m.	92 (100)	78 (84.8)	14 (15.2)	2.19 (1.05–4.59)	0.037	2.54 (1.13–5.73)	0.024 *
• 4:01 p.m.–8:00 p.m.	22 (100)	18 (81.8)	4 (18.2)	1.77 (0.54–5.77)	0.343	1.51 (0.41–5.56)	0.533
Queue							
• No	144 (100)	115 (79.9)	29 (20.1)	1			
• Yes	55 (100)	42 (76.4)	13 (23.6)	0.82 (0.39–1.71)	0.589		
SP number							
• 1	46 (100)	37 (80.4)	9 (19.6)	1			
• 2	55 (100)	40 (72.7)	15 (27.3)	0.65 (0.25–1.66)	0.367		
• 3	49 (100)	42 (85.7)	7 (14.3)	1.46 (0.50–4.31)	0.493		
• 4	49 (100)	38 (77.6)	11 (22.4)	0.84 (0.31–2.26)	0.731		
Questioning score [a]	5.0 (2.0–9.0)	7.0 (3.0–9.0)	2.5 (1.0–4.0)	1.39 (1.21–1.60)	<0.001	1.41 (1.22–1.63)	<0.001 *

[a] Median (Interquartile range; IQR); Abbreviations: COR = Crude Odds Ratio; AOR = Adjusted Odds Ratio. * significant at $p < 0.05$.

4. Discussion

The appropriate outcome, that is, the dispensing of UPA, was achieved in almost 80% of all test purchases. International SP studies on the quality of counselling provided for "morning after pills" in the Democratic Republic of Congo, Kenya and India show a similarly high case resolution of 74% [38], 82% [34] and 86% [30], while in Australia and Switzerland values of 95% [21] and even 100% [23] were achieved. In contrast, in one Australian SP study, this value turned out to be considerably lower with 24% [35]. The highly divergent values in some cases must be viewed and interpreted in light of the very different scenarios used.

The fact that in slightly more than 20% of all test purchases, a medicinal product was not dispensed, or the wrong medicinal product was dispensed is problematic [77]. This situation is all the more significant for such an indication in which the consequence of providing the wrong advice or not dispensing the correct medicinal product—possibly an unwanted pregnancy—must be considered to be very high.

Possible reasons for not dispensing UPA that were also seen in the international SP studies of oral emergency contraceptive pills is the unavailability of the preparation UPA [78,79] and a lack of knowledge on behalf of the pharmacy staff [80]. Appropriate mandatory continuing education programs could reduce the issue of a lack of knowledge [81]. The unavailability of the UPA preparation could be minimised by mandating keeping the preparation in stock, especially as UPA is the only recommended active substance on the market for a specific time window and also because it should be taken as soon as possible.

In contrast to a Saudi Arabian study [64], the time proved to be a significant influencing factor on the appropriate outcome even though the type of outcome between the two studies is not comparable (UPA dispensed vs. antibiotics dispensed without a prescription). This creates a need for further research about the possible reasons, especially as the best counselling was provided not in the morning as expected on the basis of the model of the performance curve [82] but instead from midday to the late afternoon.

Analogous to the international literature [67,83] the questioning score had a significant effect on the appropriate outcome. Regarding the individual questions, the question about when the unprotected sexual intercourse took place proved to be the "master question" because only then could the correct decision—dispensing UPA—be made at all. The SP studies conducted in Switzerland, Scotland and Australia showed similarly high values to those in this investigation with values of 100% [23,45], 93% [35] and 88% [36] respectively. In contrast, in one SP study in the Democratic Republic of Congo, this question was only asked in 7% of the test purchases [38], and in a Turkish SP study it was asked in only one of 155 test purchases [41]. In an Indian and a Brazilian SP study this question was not asked once in 70 [30] and 122 [44] test purchases respectively.

A significant correlation between the questioning score and the visible use of a checklist was apparent, analogous to the international literature [23,35]. However, a checklist was only used in a little over half of all test purchases. In contrast, in a qualitative German study, it was reported that almost all pharmacies stated that they used a checklist [27]. Such discrepancies—worse results in SP studies than in self-reported studies of the same issue—are also seen in the international mixed-methods literature for requests for medicinal products other than oral emergency contraceptive pills [84,85]. Internationally, a checklist was only used in 10% of all test purchases in an Australian SP study [21] whereas this value increased to 83% in another Australian SP study that was conducted because a checklist had since been developed and recommended by the Pharmaceutical Society of Australia [35]. However, in a more recent Australian SP study, a checklist was not used in a single test purchase [36]. The use of a checklist is also recommended in the guidelines for Switzerland, which led to an application rate of 99% in a recent Swiss SP study [23]. In contrast to an Australian SP study [35], our study showed not only a significant correlation between "checklist" and "questioning score" but the questioning score also has a significant effect on the dispensing of UPA. It, therefore, appears to be advisable—despite criticism from pharmacy staff in a qualitative German study that the BAK checklist, in particular, is overly detailed and thus requires considerable time [27]—to introduce an explicit guideline recommendation about the use of a checklist in Germany as well. However, optimising or shortening the checklist should be considered to increase its acceptance.

Although there is a significant correlation between the use of a checklist and the use of a counselling room, in this study, a counselling room was only used in a little less than half of the test purchases. In contrast, two-thirds of the interviewed pharmacy staff in the qualitative German study indicated using a counselling room [27]. Internationally, there is an inconsistent picture regarding the use of a counselling room. For example, in an Australian SP study, only 9% of the test purchases were conducted in a counselling room [36] while in an Indian SP study this value was as high as 47% [30]. In contrast to this, an extra counselling room or a quiet area in the pharmacy was used in 90% of the consultations in another Australian SP study [21]. In Switzerland, pharmacists are supposed to carry out the consultation for the "morning after pill" in a separate counselling room, which was again implemented in 94% of test purchases in a corresponding SP study—analogous to the recommended use of a checklist [23]. To protect the privacy of patients for this sensitive issue, such a target regulation should therefore be considered for Germany as well. Furthermore, the flow of information between the consumer and the pharmacy staff could also be improved as a result, which was established in a qualitative Australian study [22]. Due to the correlation between the use of a counselling room and the use of a checklist, it is also likely that the use of a checklist could be increased in this way [21].

There is a need for further research in Germany regarding access to and the quality of counselling provided for emergency contraceptive pills for minors [30], male consumers [39] and victims of sexual assault [43]. It should also be analysed in future whether pharmacy staff provide additional important information—about sexually transmissible infections, for example—during the counselling [35]. Future studies should also analyse the size of any differences between SP results and self-reported results by pharmacy staff by using a mixed-methods approach regarding the counselling provided for oral emergency contraception.

Strengths and Limitations

This study is the first in Germany that investigated the counselling and dispensing behaviour of pharmacies for the "morning after pill" using an SP method. The SP study design used is particularly advantageous because unlike other data collection methods, both the effect of social desirability and the Hawthorne effect can be avoided. The results—at least for one German state—can also be generalised because of the stratified random sampling used. There is also no selection bias because there is no option for opting out. The study could also be fully implemented as planned due to the 100% visit completion rate.

Because the study results only refer to one German state, future studies should be expanded to additional states or all of Germany. However, this would be an ambitious undertaking in light of the relatively high data collection costs. Furthermore, only one specific counselling and dispensing scenario (mandatory UPA dispensing) for oral emergency contraceptive pills were used for the data collection. Other scenarios could therefore lead to different results [48,67]. Because only students in a certain age range were used as SPs, it also cannot be ruled out that the counselling and dispensing behaviour would differ for other age groups (such as adolescents) or women with different levels of education (such as customers with secondary school education only). The study also did not differentiate between non-dispensing of UPA due to a lack of availability and non-dispensing due to incorrect counselling. Not dispensing UPA due to incorrect counselling would be classified as particularly serious because the test buyers would not receive UPA as a result, even though it would be the suitable active substance. However, not dispensing UPA due to a lack of availability would be problematic because the medication is supposed to be taken as soon as possible. Future studies should, therefore, record this differentiation because it would provide important information about possible optimisation measures. Furthermore, this is a cross-sectional study, and therefore, no causal relationship can be created between the variables studied. Because the study design (no audiotaping and no second observer) meant that the test buyers had to remember the recorded items until they left the pharmacy and many items had to be recorded (14 items for counselling as well as several influencing factor items), collecting additional important items (such as information about taking the medication as soon as possible or about taking the medication again in case of vomiting) was also omitted to prevent any recall bias. Immediate feedback after the test purchase would also have been desirable because then the pharmacy staff's memory of a specific counselling situation would have still been fresh. However, there is a risk that the pharmacy staff inform their colleagues in the neighbourhood about the test purchases. As a result, it was also not possible to ask the counselling pharmacy staff about their age, meaning that there may be certain distortions in the estimations of the age variable.

5. Conclusions

In almost four-fifths of all test purchases, the appropriate outcome was achieved. This meant, however, that in a little over one-fifth of all test purchases, the required dispensing of UPA did not occur, which could have led to serious consequences such as an unwanted pregnancy. The use of a counselling room and a checklist, the use of a checklist and the questioning score as well as the questioning score and achieving the appropriate outcome are all significantly correlated. A target regulation for the use of a counselling room, an explicit guideline recommendation about the use of a checklist, an obligation for keeping UPA in stock and appropriate mandatory continuing education programs should be considered. Surprisingly, the time of day proved to be a significant influencing factor on the appropriate outcome, which requires further investigation.

Author Contributions: Conceptualisation: B.L., S.G., G.L., F.M. and V.W.; data curation, B.L., S.G., G.L., F.M. and V.W.; formal analysis, B.L., S.G., G.L., F.M. and V.W.; methodology, B.L., S.G., G.L., F.M. and V.W.; project administration, B.L.; validation, B.L., S.G., G.L., F.M. and V.W.; visualisation, B.L., S.G., G.L., F.M. and V.W.; writing—original draft preparation, B.L., S.G., G.L., F.M. and V.W.; writing—review and editing, B.L., S.G., G.L., F.M. and V.W.; All authors have read and agreed to the published version of the manuscript.

Funding: We acknowledge support for the article processing charge from the Open Access Publication Fund of the Hochschule Neubrandenburg (Neubrandenburg University of Applied Sciences).

Conflicts of Interest: The authors declare no conflict of interest.

References

1. Matyanga, C.M.J.; Dzingirai, B. Clinical pharmacology of hormonal emergency contraceptive pills. *Int. J. Reprod. Med.* **2018**, *2018*, 2785839. [CrossRef]
2. Association of Women's Health, Obstetric and Neonatal Nurses (AWHONN). Emergency contraception. *J. Obs. Gynecol. Neonatal. Nurs.* **2017**, *46*, 886–888. [CrossRef] [PubMed]
3. Helfferich, C.; Klindworth, H.; Heine, Y.; Wlosnewski, I. Frauen leben 3—Familienplanung im Lebenslauf von Frauen. Schwerpunkt: Ungewollte Schwangerschaften. Studie im Auftrag des Federal Centre for Health Education (BZgA). 2016. Available online: https://service.bzga.de/pdf.php?id=34c03dbb54311efa0d4aa9ad5bb9bd9c (accessed on 29 March 2020).
4. Bearak, J.; Popinchalk, A.; Alkema, L.; Sedgh, G. Global, regional, and subregional trends in unintended pregnancy and its outcomes from 1990 to 2014: Estimates from a Bayesian hierarchical model. *Lancet Glob. Health* **2018**, *6*, e380–e389. [CrossRef]
5. Cameron, S.T.; Li, H.; Gemzell-Danielsson, K. Current controversies with oral emergency contraception. *BJOG* **2017**, *124*, 1948–1956. [CrossRef] [PubMed]
6. Haeger, K.O.; Lamme, J.; Cleland, K. State of emergency contraception in the U.S., 2018. *Contracept. Reprod. Med.* **2018**, *3*, 20. [CrossRef]
7. Rabe, T.; Goeckenjan, M.; Ahrendt, H.J.; Ludwig, M.; Merkle, E.; König, K.; Merki Feld, G.; Albring, C. Postkoitale Kontrazeption—Gemeinsame Stellungnahme der Deutschen Gesellschaft für *Gynäkologische Endokrinologie* und Fortpflanzungsmedizin (DGGEF) e.V. und des Berufsverbands der Frauenärzte (BVF) e.V. *J. Reprod. Endokrinol.* **2011**, *8*, 390–414. Available online: https://www.kup.at/kup/pdf/10330.pdf (accessed on 29 March 2020).

8. International Consortium on Emergency Contraception. EC Status and Availability: Germany. Available online: https://www.cecinfo.org/country-by-country-information/status-availability-database/countries/germany/ (accessed on 29 March 2020).

9. Rosato, E.; Farris, M.; Bastianelli, C. Mechanism of action of ulipristal acetate for emergency contraception: A systematic review. *Front. Pharm.* **2016**, *6*, 315. [CrossRef]

10. Black, K.I.; Hussainy, S.Y. Emergency contraception: Oral and intrauterine options. *Aust. Fam. Phys.* **2017**, *46*, 722–726. Available online: https://www.racgp.org.au/download/Documents/AFP/2017/October/V2/AFP-2017-10-Focus-Emergency-Contraception.pdf (accessed on 29 March 2020).

11. Creinin, M.D.; Schlaff, W.; Archer, D.F.; Wan, L.; Frezieres, R.; Thomas, M.; Rosenberg, M.; Higgins, J. Progesterone receptor modulator for emergency contraception: A randomized controlled trial. *Obs. Gynecol.* **2006**, *108*, 1089–1097. [CrossRef]

12. Glasier, A.F.; Cameron, S.T.; Fine, P.M.; Logan, S.J.; Casale, W.; Van Horn, J.; Sogor, L.; Blithe, D.L.; Scherrer, B.; Mathe, H.; et al. Ulipristal acetate versus levonorgestrel for emergency contraception: A randomised non-inferiority trial and meta-analysis. *Lancet* **2010**, *375*, 555–562. [CrossRef]

13. Shen, J.; Che, Y.; Showell, E.; Chen, K.; Cheng, L. Interventions for emergency contraception. *Cochrane Database Syst. Rev.* **2019**, *1*, CD001324. [CrossRef]

14. Rabe, T.; Albring, C. Arbeitskreis "Postkoitale Kontrazeption": Ahrendt HJ, Mueck A, Merkle E, König K, Merki, G. Notfallkontrazeption—Ein Update: Gemeinsame Steellungnahme der Deutschen Gesellschaft für *Gynäkologische Endokrinologie* und Fortpflanzungsmedizin (DGGEF) e.V. und des Berufsverbands der Frauenärzte (BVF) e.V. *Gynäkologische Endokrinol.* **2013**, *11*, 197–202. [CrossRef]

15. Bundesapothekerkammer (BAK). Rezeptfreie Abgabe von oralen Notfallkontrazeptiva ("Pille danach"). Handlungsempfehlungen der BAK. 2018. Available online: https://www.abda.de/fileadmin/assets/Praktische_Hilfen/Leitlinien/Selbstmedikation/BAK_Handlungsempfehlungen-Checkliste-NFK_20180228.pdf (accessed on 29 March 2019).

16. International Consortium on Emergency Contraception. EC Status and Availability: Countries with Non-prescription Access to EC. Available online: https://www.cecinfo.org/country-by-country-information/status-availability-database/countries-with-non-prescription-access-to-ec/ (accessed on 29 March 2020).

17. Bundesrat. Beschluss des Bundesrates. Vierzehnte Verordnung zur Änderung der Arzneimittelverschreibung sverordnung. 2015. Available online: https://www.bundesrat.de/SharedDocs/drucksachen/2015/0001-0100/28-15(B).pdf?__blob=publicationFile&v=4 (accessed on 29 March 2020).

18. Berufsverband der Frauenärzte (BVF) und Deutsche Gesellschaft für Gynäkologie und Geburtshilfe (DGGG). Stellungnahme des Berufsverbandes der Frauenärzte (BVF) und der Deutschen Gesellschaft für Gynäkologie und Geburtshilfe (DGGG) zu den Anträgen: Abgeordnete Mechthild Rawert et al., und die Fraktion der SPD, Rezeptfreiheit von Notfallkontrazeptiva Pille danach gewährleisten' Drucksache 17/11039 und Abgeordnete Yvonne Ploetz et al., und die Fraktion DIE LINKE ,Die Pille danach rezeptfrei machen' Drucksache 17/12102 vom 17.04.2013. Available online: https://www.dggg.de/fileadmin/documents/stellungnahmen/aktuell/2013/180_Rezeptfreiheit_von_Notfallkontrazeptiva_Pille_danach_gewaehrleisten.pdf (accessed on 5 February 2020).

19. Bundesärztekammer. Stellungnahme der Bundesärztekammer zum Entwurf einer Vierzehnten Verordnung zur Änderung der Arzneimittelverschreibungsverordnung vom 14.01.2015. 2015. Available online: https://www.bundesaerztekammer.de/fileadmin/user_upload/downloads/BAeK-Stn_und_AkdAe_AMVV-AendVO_15.01.2015.pdf (accessed on 5 February 2020).

20. Schulz, M.; Goebel, R.; Schumann, C.; Zagermann-Muncke, P. Non-prescription dispensing of emergency oral contraceptives: Recommendations from the German Federal Chamber of Pharmacists [Bundesapothekerkammer]. *Pharm. Pr.* **2016**, *14*, 828. [CrossRef]

21. Queddeng, K.; Chaar, B.; Williams, K. Emergency contraception in Australian community pharmacies: A simulated patient study. *Contraception* **2011**, *83*, 176–182. [CrossRef]

22. Seubert, L.J.; Whitelaw, K.; Boeni, F.; Hattingh, L.; Watson, M.C.; Clifford, R.M. Barriers and facilitators for information exchange during over-the-counter consultations in community pharmacy: A focus group study. *Pharmacy* **2017**, *5*, 65. [CrossRef]

23. Haag, M.; Gudka, S.; Hersberger, K.E.; Arnet, I. Do Swiss community pharmacists address the risk of sexually transmitted infections during a consultation on emergency contraception? A simulated patient study. *Eur. J. Contracept. Reprod. Health Care* **2019**, *24*, 407–412. [CrossRef]

24. Khojah, H.M.J. Privacy level in private community pharmacies in Saudi Arabia: A simulated client survey. *Pharm. Pharm.* **2019**, *10*, 445–455. [CrossRef]

25. Ordinance on the Operation of Pharmacies. Available online: https://www.abda.de/fileadmin/assets/Gesetze/ApBetrO_engl_Stand-2016-12.pdf (accessed on 11 May 2020).

26. Said, A.; Ganso, M.; Freudewald, L.; Schulz, M. Trends in dispensing oral emergency contraceptives and safety issues: A survey of German community pharmacists. *Int. J. Clin. Pharm.* **2019**, *41*, 1499–1506. [CrossRef]

27. Bruhns, C. Ergebnisse einer bundesweiten Befragung zur aktuellen Abgabepraxis der Pille danach. Rezeptfreie Pille danach—Abgabepraxis und Information. *Pro Fam. Dok.* **2015**, 14–20. Available online: https://www.profamilia.de/fileadmin/publikationen/Fachpublikationen/doku_pille__danach-2016_web.pdf (accessed on 12 April 2020).

28. Dierolf, V.; Freytag, S. Zugang zur Pille danach in den Apotheken nach der Rezeptfreigabe. Verhütungsberatung. *Pro Fam. Mag.* **2017**, *4*, 9–12.

29. Callegaro, M. Social desirability. *Encycl. Surv. Res. Methods* **2008**, 825–826. [CrossRef]

30. Saxena, P.; Mishra, A.; Nigam, A. Evaluation of pharmacists' services for dispensing emergency contraceptive pills in Delhi, India: A mystery shopper study. *Indian J. Community Med.* **2016**, *41*, 198–202. [CrossRef] [PubMed]

31. Converse, L.; Barrett, K.; Rich, E.; Reschovsky, J. Methods of observing variations in physicians' decisions: The opportunities of clinical vignettes. *J. Gen. Intern. Med.* **2015**, *30*, S586–S594. [CrossRef]

32. Bardage, C.; Westerlund, T.; Barzi, S.; Bernsten, C. Non-prescription medicines for pain and fever—A comparison of recommendations and counseling from staff in pharmacy and general sales stores. *Health Policy* **2013**, *110*, 76–83. [CrossRef] [PubMed]

33. Xu, T.; de Almeida Neto, A.C.; Moles, R.J. A systematic review of simulated-patient methods used in community pharmacy to assess the provision of non-prescription medicines. *Int. J. Pharm. Pract.* **2012**, *20*, 307–319. [CrossRef]

34. Obare, F.; Liambila, W. The provision of emergency contraceptives in private sector pharmacies in urban Kenya: Experiences of mystery clients. *Afr. Popul. Stud.* **2013**, *24*, 42–52. [CrossRef]

35. Schneider, C.R.; Gudka, S.; Fleischer, L.; Clifford, R.M. The use of a written assessment checklist for the provision of emergency contraception via community pharmacies: A simulated patient study. *Pharm. Pract.* **2013**, *11*, 127–131. [CrossRef]

36. Collins, J.C.; Schneider, C.R.; Moles, R.J. Emergency contraception supply in Australian pharmacies after the introduction of ulipristal acetate: A mystery shopping mixed-methods study. *Contraception* **2018**, *98*, 243–246. [CrossRef]

37. Cleland, K.; Bass, J.; Doci, F.; Foster, A.M. Access to emergency contraception in the over-the-counter era. *Womens Health Issues* **2016**, *26*, 622–627. [CrossRef]

38. Hernandez, J.H.; Mbadu, M.F.; Garcia, M.; Glover, A. The provision of emergency contraception in Kinshasa's private sector pharmacies: Experiences of mystery clients. *Contraception* **2018**, *97*, 57–61. [CrossRef]

39. Bell, D.L.; Camacho, E.J.; Velasquez, A.B. Male access to emergency contraception in pharmacies: A mystery shopper survey. *Contraception* **2014**, *90*, 413–415. [CrossRef]

40. Smith, H.; Whyte, S.; Chan, H.F.; Kyle, G.; Lau, E.T.L.; Nissen, L.M.; Torgler, B.; Dulleck, U. Pharmacist compliance with therapeutic guidelines on diagnosis and treatment provision. *JAMA Netw. Open* **2019**, *2*, e197168. [CrossRef]

41. Uzun, G.D.; Sancar, M.; Okuyan, B. Evaluation of knowledge and attitude of pharmacist and pharmacy technicians on emergency contraception method in Istanbul, Turkey: A simulated patient study. *J. Res. Pharm.* **2018**, *23*, 395–402. [CrossRef]

42. Kagashe, G.A.B.; Maregesi, S.M.; Mashaka, A. Availability, awareness, attitude and knowledge of emergency contraceptives in Dar Es Salaam. *J. Pharm. Sci. Res.* **2013**, *5*, 216–219. Available online: https://www.jpsr.pharmainfo.in/Documents/Volumes/vol5issue11/jpsr05111302.pdf (accessed on 12 April 2020).

43. Higgins, S.J.; Hattingh, H.L. Requests for emergency contraception in community pharmacy: An evaluation of services provided to mystery patients. *Res. Soc. Adm. Pharm.* **2013**, *9*, 114–119. [CrossRef]

44. Tavares, M.P.; Foster, A.M. Emergency contraception in a public health emergency: Exploring pharmacy availability in Brazil. *Contraception* **2016**, *94*, 109–114. [CrossRef]

45. Glasier, A.; Manners, R.; Loudon, J.C.; Muir, A. Community pharmacists providing emergency contraception give little advice about future contraceptive use: A mystery shopper study. *Contraception* **2010**, *82*, 538–542. [CrossRef]

46. Berger, K.; Eickhoff, C.; Schulz, M. Counselling quality in community pharmacies: Implementation of the pseudo customer methodology in Germany. *J. Clin. Pharm.* **2005**, *30*, 45–57. [CrossRef] [PubMed]

47. Alte, D.; Weitschies, W.; Ritter, C.A. Evaluation of consultation in community pharmacies with mystery shoppers. *Ann. Pharm.* **2007**, *41*, 1023–1030. [CrossRef]

48. Langer, B.; Bull, E.; Burgsthaler, T.; Glawe, J.; Schwobeda, M.; Simon, K. Assessment of counselling for acute diarrhoea in German pharmacies: A simulated patient study. *Int. J. Pharm. Pr.* **2018**, *26*, 310–317. [CrossRef]

49. Langer, B.; Kieper, M.; Laube, S.; Schramm, J.; Weber, S.; Werwath, A. Assessment of counselling for acute diarrhoea in North-Eastern German pharmacies—A follow-up study using the simulated patient methodology. *Pharm. Pharm.* **2018**, *9*, 257–269. [CrossRef]

50. Langer, B.; Kunow, C. Medication dispensing, additional therapeutic recommendations, and pricing practices for acute diarrhoea by community pharmacies in Germany: A simulated patient study. *Pharm. Pract.* **2019**, *17*, 1579. [CrossRef]

51. Langer, B.; Kunow, C. Do north-eastern German pharmacies recommend a necessary medical consultation for acute diarrhoea? Magnitude and determinants using a simulated patient approach. *F1000Research* **2020**, *8*, 1841. [CrossRef]

52. STROBE Statement—Checklist of Items That Should Be Included in Reports of Cross-Sectional Studies. Available online: https://www.strobe-statement.org/fileadmin/Strobe/uploads/checklists/STROBE_checklist_v4_combined.pdf (accessed on 8 February 2020).

53. Watson, M.C.; Norris, P.; Granas, A.G. A systematic review of the use of simulated patients and pharmacy practice research. *Int. J. Pharm. Pr.* **2006**, *14*, 83–93. [CrossRef]

54. Björnsdottir, I.; Granas, A.G.; Bradley, A.; Norris, P. A systematic review of the use of simulated patient methodology in pharmacy practice research from 2006 to 2016. *Int. J. Pharm. Pr.* **2020**, *28*, 13–25. [CrossRef]

55. da Costa, F.A. Covert and overt observations in pharmacy practice. In *Pharmacy Practice Research Methods*; Babar, Z.U.D., Ed.; Springer: Singapore, 2020. [CrossRef]

56. Statistisches Amt Mecklenburg-Vorpommern. Statistisches Jahrbuch Mecklenburg-Vorpommern 2019. 2019. Available online: https://www.laiv-mv.de/static/LAIV/Statistik/Dateien/Publikationen/Statistisches%20Jahrbuch/Z011%202019%2000.pdf (accessed on 24 April 2020).

57. Apotheken Umschau. Apothekenfinder. 2019. Available online: https://www.apotheken-umschau.de/Apothekenfinder (accessed on 11 April 2019).

58. ABDA—Bundesvereinigung Deutscher Apothekerverbände. Die Apotheke—Zahlen, Daten, Fakten 2019. Berlin: ABDA. 2019. Available online: https://www.abda.de/fileadmin/user_upload/assets/ZDF/ZDF_2019/ABDA_ZDF_2019_Brosch.pdf (accessed on 30 June 2019).

59. Israel, G.D. *Determining Sample Size*; University of Florida: Gainesville, FL, USA, 2012; Available online: http://www.psychosphere.com/Determining%20sample%20size%20by%20Glen%20Israel.pdf (accessed on 4 April 2019).

60. Bundesapothekerkammer (BAK). Rezeptfreie Abgabe von oralen Notfallkontrazeptiva ("Pille danach"). Handlungsempfehlungen der BAK. 2018. Anhang 1. Available online: https://www.abda.de/fileadmin/assets/Praktische_Hilfen/Leitlinien/Selbstmedikation/Anhang_1_Vergleich-NFK_20180228.pdf (accessed on 28 April 2019).

61. Federal Centre for Health Education (BZgA). Contraceptive Behaviour of Adults 2011. 2011. Available online: https://www.bzga.de/infomaterialien/sexualaufklaerung/sexualaufklaerung/contraceptive-behaviour-of-adults-2011/ (accessed on 28 April 2020).

62. German Pharmacies Act. Available online: https://www.abda.de/fileadmin/assets/Gesetze/Apothekengesetz_engl-Stand_2012-10-26.pdf (accessed on 11 May 2020).

63. Zapata-Cachafeiro, M.; Piñeiro-Lamas, M.; Guinovart, M.C.; López-Vázquez, P.; Vázquez-Lago, J.M.; Figueiras, A. Magnitude and determinants of antibiotic dispensing without prescription in Spain: A simulated patient study. *J. Antimicrob. Chemother.* **2019**, *74*, 511–514. [CrossRef]

64. Saba, M.; Diep, J.; Bittoun, R.; Saini, B. Provision of smoking cessation services in Australian community pharmacies: A simulated patient study. *Int. J. Clin. Pharm.* **2014**, *36*, 604–614. [CrossRef]

65. Langer, B.; Bull, E.; Burgsthaler, T.; Glawe, J.; Schwobeda, M.; Simon, K. Die Beratungsqualität von Apotheken bei akutem Durchfall—Eine Analyse unter Verwendung der "Simulated Patient"-Methode [Using the simulated patient methodology to assess counselling for acute diarrhoea—Evidence from Germany]. *Z. Evid. Qual. Gesundhwes.* **2016**, *112*, 19–26. [CrossRef]

66. Kashour, T.S.; Joury, A.; Alotaibi, A.M.; Althagafi, M.; Almufleh, A.S.; Hersi, A.; Thalib, L. Quality of assessment and counselling offered by community pharmacists and medication sale without prescription to patients presenting with acute cardiac symptoms: A simulated client study. *Eur. J. Clin. Pharm.* **2016**, *72*, 321–328. [CrossRef]

67. Collins, J.C.; Schneider, C.R.; Faraj, R.; Wilson, F.; de Almeida Neto, A.C.; Moles, R.J. Management of common ailments requiring referral in the pharmacy: A mystery shopping intervention study. *Int. J. Clin. Pharm.* **2017**, *39*, 697–703. [CrossRef]

68. Kippist, C.; Wong, K.; Bartlett, D.; Saini, B. How do pharmacists respond to complaints of acute insomnia? A simulated patient study. *Int. J. Clin. Pharm.* **2011**, *33*, 237–245. [CrossRef]

69. Collins, J.C.; Schneider, C.R.; Naughtin, C.L.; Wilson, F.; de Almeida Neto, A.C.; Moles, R.J. Mystery shopping and coaching as a form of audit and feedback to improve community pharmacy management of non-prescription medicine requests: An intervention study. *BMJ Open* **2017**, *7*, e019462. [CrossRef]

70. Werner, J.B.; Benrimoj, S.I. Audio taping simulated patient encounters in community pharmacy to enhance the reliability of assessments. *Am. J. Pharm. Educ.* **2008**, *72*, 136. [CrossRef]

71. BVM—Berufsverband Deutscher Markt- und Sozialforscher e.V. Richtlinie für den Einsatz von Mystery Research in der Markt- und Sozialforschung [Guideline for the Use of Mystery Research in Market and Social Research]. 2006. Available online: http://bvm.org/fileadmin/pdf/Recht_Berufskodizes/Richtlinien/RL_2006_Mystery.pdf (accessed on 4 April 2020).

72. McCambridge, J.; Witton, J.; Elbourne, D.R. Systematic review of the Hawthorne effect: New concepts are needed to study research participation effects. *J. Clin. Epidemiol.* **2014**, *67*, 267–277. [CrossRef]

73. Rhodes, K.V.; Miller, F.G. Simulated patient studies: An ethical analysis. *Milbank Q.* **2012**, *90*, 706–724. [CrossRef]

74. Fitzpatrick, A.; Tumlinson, K. Strategies for optimal implementation of simulated clients for measuring quality of care in low- and middle-income countries. *Glob. Health Sci. Pr.* **2017**, *5*, 108–114. [CrossRef]

75. Ibrahim, M.I.M.; Awaisu, A.; Palaian, S.; Radoui, A.; Atwa, H. Do community pharmacists in Qatar manage acute respiratory conditions rationally? A simulated client study. *J. Pharm. Health Serv. Res.* **2018**, *9*, 33–39. [CrossRef]

76. Cohen, J. The effect size. In *Statistical Power Analysis for the Behavioral Sciences*, 2nd ed.; Lawrence Erlbaum Associates: New Jersey, NJ, USA, 1988.

77. Glasier, A.; Baraitser, P.; McDaid, L.; Norrie, J.; Radley, A.; Stephenson, J.M.; Battison, C.; Gilson, R.; Cameron, S. Emergency contraception from the pharmacy 20 years on: A mystery shopper study. *BMJ Sex. Reprod Health* **2020**. [CrossRef]

78. Bullock, H.; Steele, S.; Kurata, N.; Tschann, M.; Elia, J.; Kaneshiro, B.; Salcedo, J. Pharmacy access to ulipristal acetate in Hawaii: Is a prescription enough? *Contraception* **2016**, *93*, 452–454. [CrossRef]

79. Shigesato, M.; Elia, J.; Tschann, M.; Bullock, H.; Hurwitz, E.; Wu, Y.Y.; Salcedo, J. Pharmacy access to Ulipristal acetate in major cities throughout the United States. *Contraception* **2018**, *97*, 264–269. [CrossRef]

80. Kaur, G.; Fontanilla, T.; Bullock, H.; Tschann, M. 'The difference between plan b and ella®? They're basically the same thing': Results from a mystery client study. *Pharmacy* **2020**, *8*, 77. [CrossRef]

81. Edalat, A. Pro und Kontra: Pflichtfortbildung für Apotheker? DAZ.ONLINE vom 12.07.2018. Available online: https://www.deutsche-apotheker-zeitung.de/news/artikel/2018/07/12/pflichtfortbildung-fuer-apotheker/chapter:all (accessed on 22 April 2020).

82. Gabler Wirtschaftslexikon. Leistungskurve. Available online: https://wirtschaftslexikon.gabler.de/definition/leistungskurve-37379/version-260815 (accessed on 22 April 2020).

83. Watson, M.C.; Bond, C.M.; Grimshaw, J.; Johnston, M. Factors predicting the guideline compliant supply (or non-supply) of non-prescription medicines in the community pharmacy setting. *Qual. Saf. Health Care* **2006**, *15*, 53–57. [CrossRef] [PubMed]

84. Ogbo, P.U.; Aina, B.A.; Aderemi-Williams, R.I. Management of acute diarrhea in children by community pharmacists in Lagos, Nigeria. *Pharm. Prac.* **2014**, *12*, 376. [CrossRef] [PubMed]

85. Surur, A.S.; Getachew, E.; Teressa, E.; Hailemeskel, B.; Getaw, N.S.; Erku, D.A. Self-reported and actual involvement of community pharmacists in patient counseling: A cross-sectional and simulated patient study in Gondar, Ethiopia. *Pharm. Pract.* **2017**, *15*, 890. [CrossRef] [PubMed]

International Journal of
***Environmental Research
and Public Health***

Article

Advancing Prevention of STIs by Developing Specific Serodiagnostic Targets: *Trichomonas vaginalis* as a Model

John F. Alderete

School of Molecular Biosciences, College of Veterinary Medicine, Washington State University, Pullman, WA 99164, USA; alderete@wsu.edu

Received: 22 June 2020; Accepted: 6 August 2020; Published: 10 August 2020

Abstract: Point-of-Care (POC) serum antibody screening of large cohorts of women and men at risk for the sexually transmitted infection (STI) caused by *Trichomonas vaginalis* requires the availability of targets with high specificity. Such targets should comprise epitopes unique to *T. vaginalis* immunogenic proteins detected by sera of women and men patients with trichomonosis but not uninfected controls. Three enzymes to which patients make serum IgG antibody were identified as fructose-1,6-bisphosphate aldolase (A), α-enolase (E), and glyceraldehyde-3-phosphate dehydrogenase (G). Epitopes within these proteins were identified that had no sequence identity to enzymes of humans and other pathogens. Therefore, I constructed a chimeric recombinant String-Of-Epitopes (SOE) protein consisting of 15-mer peptides, within which are the epitopes of A, E, and G. This novel protein of ~36-kD is comprised of two epitopes of A, ten epitopes of E, and seven epitopes of G (AEG::SOE2). The AEG::SOE2 protein was detected both by immunoblot and by enzyme-linked immunosorbent assay (ELISA) using highly reactive sera of women and men but not negative serum unreactive to *T. vaginalis* proteins. Finally, AEG::SOE2 was found to be immunogenic, as evidenced by serum IgG from immunized mice. I discuss how this approach is important in relation to infectious disease diagnostic targets for detection of serum IgG antibody in exposed and/or infected individuals and how such novel targets may have potential as subunit vaccine candidates against microbial pathogens.

Keywords: diagnostic; diagnostic targets; ELISA-enzyme linked immunosorbent assay; epitopes; immunogens; sera; serodiagnosis; sexually transmitted infections; *Trichomonas vaginalis*

---

1. Introduction

Trichomonas vaginalis causes a non-viral sexually transmitted infection (STI) with adverse outcomes to infected women [1,2]. This STI is highly prevalent [3–5], and persistence within individuals may be due to the asymptomatic nature of infection. It is accepted that male partners of infected women with trichomonosis become infected. The organism and *T. vaginalis* DNA have been detected in hyperplastic prostate tissue [6,7], and there remains the possibility of a link between seropositivity to *T. vaginalis* in relation to prostate cancer (PCa) development [8–10]. More recently, a gene-expression model for *T. vaginalis*-mediated PCa was proposed [11], and other studies lend support to this hypothesis [6,7,12–15].

A rapid, inexpensive and specific serodiagnostic that could be used for screening large cohorts of at-risk individuals is desirable. A lateral flow, immunochromatographic Point-of-Care (POC) diagnostic (OSOM™ Trichomonas Rapid Test, Sekisui Diagnostics, Lexington, MA, USA) for rapid detection of active trichomonosis in women was invented in my laboratory [16]. Although the test meets criteria of being inexpensive, simple, rapid, and highly sensitive and specific, drawbacks of this test are that it is invasive for women, requiring a vaginal swab for obtaining sample, and the POC test fails to detect the specific parasite protein in the urine of male patients. Although there are numerous reports of accurate nucleic acid amplification-based tests [17–19], these tests are neither compatible for large scale screening in non-sterile settings nor are suitable for use in community-based clinics and at under-developed countries.

It is acknowledged that advancing the prevention of STIs in general will require specific and sensitive POC tests [20]. Such POC tests should be rapid, inexpensive, and highly dependable for serum IgG antibody detection that can be employed for broad screening of populations regardless of geographic setting. POC diagnostics are needed for surveillance of the global burden of STIs in both developed and undeveloped countries. In the case of *T. vaginalis*, surveillance is necessary among sexually active populations [20], reinforcing the view that development of a serum-antibody POC test would advance the reproductive health of at-risk women and men. Control and even elimination of *T. vaginalis* and other STIs requires an approach and method for the development of highly specific serodiagnostic targets. In this report, I provide an approach for the identification and development of serodiagnostic targets using *Trichomonas vaginalis* as a model.

As infection by *T. vaginalis* results in an IgG response [8–11,21]; I hypothesize that an approach can be developed that will lead to the synthesis of a protein for detection of serum IgG to *T. vaginalis*. Using *T. vaginalis* as a model, I present the concept that a novel, chimeric protein comprised of a String-Of-Epitopes (SOE) can be synthesized as a serodiagnostic target. My laboratory has previously determined that women and men patients make serum IgG antibody to numerous *T. vaginalis* immunogenic proteins, including the enzymes fructose-1,6-bisphosphate aldolase (referred to as A), α-enolase (E), and glyceraldehyde-3-phosphate dehydrogenase (G) [21–23]. Epitope mapping of these proteins with women and men patient sera identified epitopes unique to the trichomonad proteins [21]. This earlier report showed a proof-of-principle for the construction of a novel recombinant chimeric protein, called AEG::SOE, with two each of the A, E, and G epitopes of the three enzymes. This earlier construct, however, failed to detect some positive sera when compared with the gold standard immunogenic truncated version of α-actinin called ACTP2 [8–10,24,25]. In this report I test this hypothesis and develop a stepwise approach to show that a new recombinant protein, two epitopes of A, ten epitopes of E, and seven epitopes of G (AEG::SOE2), is a serodiagnostic target equal to ACTP2. I discuss how the approach used here may advance the development of serodiagnostic targets for this and other STIs. Finally, I show that AEG::SOE2 is immunogenic in immunized mice.

2. Materials and Methods

2.1. Epitopes Unique to the T. vaginalis A, E and G Proteins

The identification of immunogenic epitopes reactive to women and men patient sera was done using oligopeptides (Custom Peptide Arrays) immobilized on membranes (SPOTs system; Sigma-Aldrich Corp, St Louis, MO, USA) as recently detailed [21,24]. As before, oligopeptides of fructose-1,6-bisphosphate aldolase (A), α-enolase (E), and glyceraldehyde-3-phosphate dehydrogenase (G) were derived from GenBank® accession numbers AAW78351 (A), AAK73099 (E), and AAA30325 (G). The protocols for probing of SPOTs membranes with sera were as detailed before [21]. The epitopes of A, E, and G reactive with women and men positive control sera were presented in an earlier publication [21]. Further, sequence identity analysis was performed with the enzyme homologs of human and other eukaryote and bacterial pathogens [21]. Finally, each 15-mer peptide within which the epitopes resided (Figure 1) was analyzed using the Immune Epitope Database and Analysis Resource (www.iedb.org) to show the linear nature of the peptide–epitope sequence.

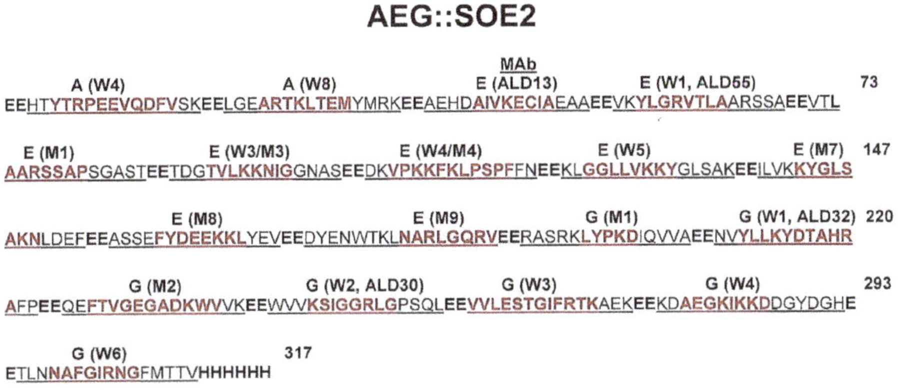

AEG::SOE2

```
                                                    MAb
   A (W4)                A (W8)              E (ALD13)        E (W1, ALD55)            73
EEHTYTRPEEVQDFVSKEELGEARTKLTEMYMRKEEAEHDAIVKECIAEAAEEVKYLGRVTLAARSSAEEVTL

   E (M1)              E (W3/M3)           E (W4/M4)            E (W5)          E (M7)  147
AARSSAPSGASTEETDGTVLKKNIGGNASEEDKVPKKFKLPSPFFNEEKLGGLLVKKYGLSAKEEILVKKYGLS

            E (M8)                    E (M9)              G (M1)        G (W1, ALD32) 220
AKNLDEFEEASSEFYDEEKKLYEVEEDYENWTKLNARLGQRVEERASRKLYPKDIQVVAEENVYLLKYDTAHR

          G (M2)          G (W2, ALD30)       G (W3)              G (W4)              293
AFPEEQEFTVGEGADKWVVKEEWVVKSIGGRLGPSQLEEVVLESTGIFRTKAEKEEKDAEGKIKKDDGYDGHE

          G (W6)                317
ETLNNAFGIRNGFMTTVHHHHHH
```

Figure 1. Linear amino acid sequence of the chimeric, recombinant AEG::SOE2 protein. The peptide-epitope sequences linked by EE amino acids are underlined. Epitopes within the protein–epitope sequences are colored red. The letters A, E, and G refer to amino acid sequences derived from respective proteins. The epitopes were detected by either or both pooled women (W) or men (M) sera, as indicated, and the number next to W and M represents the order the epitope was identified within the protein during epitope mapping on the SPOTs system.

2.2. ACT-P2, a Truncated α-Actinin Protein Used for Screening of Patient Sera

The α-actinin protein is one of the most immunogenic proteins of *T. vaginalis* [26,27] and has been shown to be a serodiagnostic target because it has no sequence identity to other known proteins in databases [24]. The truncated α-actinin protein called ACT-P2 of 558-amino acids (64.1-kDa) has been previously described [25]. ACT-P2 has been used to screen both male and female patient sera [24] and more recently to examine the relation of serum antibody in men and prostate cancer [8–10]. Thirteen epitopes were identified as reactive with women sera. In men, 5 of the 13 reactive epitopes were detected [24].

2.3. Plasmid Encoding AEG::SOE2 and ACT-P2

The DNA coding sequence for the chimeric, recombinant AEG::SOE2 protein was synthesized by GenWay Biotech, Inc (San Diego, CA, USA). The pET-23a (+) plasmid that was prepared was encoded for ampicillin (Amp $^+$) and chloramphenicol (Cam $^+$) genes. The recombinant *E. coli* BL21DE3 cells were used to synthesize the protein. The description of ACT-P2 in *E. coli* has been described [24]. Both recombinant proteins have hexa-histidine at the carboxy terminus for purification [21,25].

2.4. Recombinant Proteins

Detailed protocols for growing bacteria on Luria Broth (LB) agar plates with either 25 µg/mL Kanamycin (ACT-P2) or 100 µg/mL Ampicillin (AEG::SOE2) have been published [25]. A starter culture of LB medium with antibiotics was inoculated with recombinant *E. coli* and grown as before [21–24] prior to inoculation of medium and induction for expression with isopropyl β-D-1-thiogalactopyranoside (IPTG) for synthesis of recombinant proteins. Purification of the recombinant proteins was as before [21,24,25]. Purified proteins were obtained by Ni^{2+}-NTA superflow affinity column chromatography (Qiagen Inc., Valencia, CA, USA), and soluble AEG::SOE2 was recovered from *E. coli* lysates using the Ni^{2+}-NTA column, as evidenced by Figure 2. It is noteworthy that protein was also detected within inclusion bodies, and GenWay, Inc. indicates it is able to purify AEG::SOE2 from these inclusion bodies.

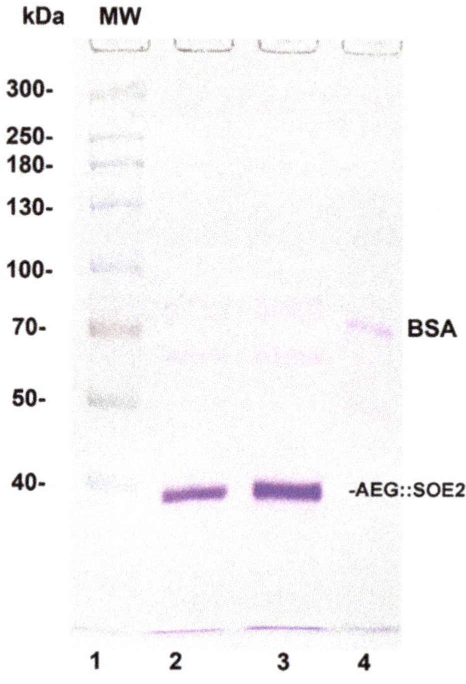

Figure 2. Sodium dodecyl sulfate-polyacrylamide gel electrophoresis (SDS-PAGE) in 8% acrylamide of AEG::SOE2 from two different experiments (lanes 2 and 3) after purification by Ni^{2+} NTA affinity chromatography. Lane 1 is of molecular weight (MW) standards, and numbers refer to daltons (×1000). Lane 4 is of 1 µg bovine serum albumin (BSA) electrophoresed for comparative purposes.

2.5. Sera for Enzyme-Linked Immunosorbent Assay (ELISA) and Derivation of Positive/Negative (P/N) Scores

Numerous reports have described the sera from women and men used for detection of ACTP2 [21,24]. Sera were also derived from research conducted at Washington University as reported earlier [8–10]. Institutional Review Board (IRB) approvals for collecting and using the sera were obtained at Washington University-Saint Louis, MO as well as by the IRB at Washington State University (approval number 01058) [8–10]. Equally importantly, the use of sera of patients has been reported before [28–30], and patients were diagnosed by microscopy and positive cultures.

My laboratory has published detailed descriptions for the use of sera by ELISA for detection of proteins, and identical materials and methods published earlier were used in this study to compare reactivities with ACTP2 and AEG::SOE2 [8–10,21,25]. Briefly, wells of microtiter plates prepared as before were stored at 4 °C prior to use. The processing of plates has been detailed before [25]. All washes of wells used phosphate-buffered saline (PBS), pH 7.4 containing 0.05% Tween-20 (PBS-Tween). Importantly, blocking of wells prior to addition of sera and dilutions of sera were done using a solution of 2% ELISA-grade bovine serum albumin (eBSA) (Sigma Chemical Co., St. Louis, MO, USA) prepared in PBS (eBSA-PBS). Where indicated, a 50 µL volume of undiluted hybridoma supernatant of a monoclonal antibody (MAb) or a cocktail of 25 µL for each of 4 MAbs (Figure 1) were used. Unless where indicated, all sera used for ELISA were diluted 1:25 (v/v) with eBSA-PBS.

In order to perform comparative analyses of ELISA using different targets, it was necessary to have serum standards as negative (scored as 0, 1+ and 2+) and positive (scored as 3+ and 4+) controls derived from testing ACTP2 as the gold standard for screening [8–10,24,25]. My laboratory previously determined the range of ELISA values for the different scores. Positive 3+ and 4+ sera detected trichomonad proteins by immunoblot, and 0, 1+ and 2+ sera did not detect proteins under the same conditions [24]. For these experiments obtaining P/N values during ELISAs to provide scores 0 to 4+, the absorbance mean ± standard deviations were as follows: the blank control with eBSA-PBS was 0.050 ± 0.002, the 0 (zero) score was 0.131 ± 0.012, the 1+ score was 0.187 ± 0.010, and the 2+ score was 0.233 ± 0.023. The scores for 3+ and 4+ for obtaining absorbance values were 0.311 ± 0.025 and 0.441 ± 0.20, respectively. All scores were derived by subtracting the average blank optical density (OD) reading.

2.6. Mouse Anti-AEG::SOE2 Serum and Anti-T. vaginalis Serum and ELISA for Detecting Antibody to T. vaginalis

Mouse anti-*T. vaginalis* serum was obtained by immunizing BALB/c mice as previously described [31]. Mouse anti-AEG::SOE2 serum (IMS) was derived by immunizing mice using the Washington State University Antibody Core Facility of the College of Veterinary Medicine. In this case, mice were immunized subcutaneously twice with 50 µg of purified protein (Figure 2) at 2-week intervals followed by the last booster injected into the tail vein. Prebleed normal mouse serum (NMS) was obtained prior to immunization and used as a negative control. All animals were treated humanely as governed by the Institutional Animal Use and Care Committee (IACUC number 6317) and National Institutes of Health protocols.

We also performed a whole cell ELISA for detecting antibody using microtiter wells coated with trichomonads. Parasites at logarithmic growth were washed three times with PBS, and a 50 µL suspension containing 1.25×10^5 organisms was added to individual wells of microtiter plates. After drying at 37 °C, 50 µL of 95% ethanol was added as fixative and wells were allowed to dry. Both protein-and trichomonad-coated wells were washed three times with PBS-Tween followed by blocking with 200 µL of eBSA-PBS. The remaining standard protocol is as previously reported elsewhere [8–10,21,24,25].

2.7. Reproducibility

All experiments were performed at least three times under identical conditions. ELISAs on microtiter plates coated with proteins or trichomonads were done in quadruplicate unless otherwise indicated, and means and standard deviations were calculated. All statistical analyses were conducted with RStudio (Version 1.2.5033; RStudio, Inc.: Vienna, Austria), and figures were made in Prism (Version 8.4.3; GraphPad, LLC: San Diego, CA, USA). T-tests were used to examine differences in absorbance levels for ACTP2 and AEG::SOE2 for mouse sera, negative and positive human sera, and monoclonal antibodies. Statistical significance was defined as a *p*-value less than 0.05.

3. Results

3.1. Epitopes of A, E, and G Unique to T. vaginalis and the AEG::SOE2 Protein Sequence

Epitope mapping revealed that pooled women and men positive serum recognized a total of 12 epitopes of A, 18 epitopes of E, and 19 epitopes of G, for a total of 49 epitopes for the three proteins [21]. Table 1 shows there were only 2 of 12, 9 of 18, and 7 of 19 epitopes unique to the *T. vaginalis* A, E, and G proteins, respectively, that had no sequence identity with other bacterial, fungal, parasite, and human sequences [21]. The red underlined amino acid sequences are the epitopes detected by IgG antibody in the SPOTS system used for screening. Not surprisingly, the epitopes that were not unique to *T. vaginalis* proteins had identity, albeit to different degrees, with protein sequences of enzymes of other bacterial, fungal, parasite, and human proteins (data not shown).

Figure 1 presents the recombinant protein amino acid sequence representing the peptides containing the unique epitopes shown above in Table 1, and each peptide sequence was linked with two glutamic acid residues. The reactivity with women (W) and/or men (M) sera is shown above the red epitope amino acid sequences. The protein has a M_r of 35,896.31 daltons and pI of 5.05. Overall, there are 8 and 11 epitopes detected by positive women and men sera, respectively. The four monoclonal antibodies (MAbs), labeled ALD 13, ALD 55, ALD 32, and ALD 30, are reactive with the epitopes, as indicated, and these MAbs have been previously characterized [21].

Table 1. The amino acid sequences containing the immunogenic epitopes (red underlined) of two epitopes of A, ten epitopes of E, and seven epitopes of G (AEG::SOE2) protein [‡].

		Proteins	
No.	**A**	**E**	**G**
1	HTYTRPEEVODFVSK	AEHDAIVKECIAEEAA	RACRKLYPKDIQVVA
2	LGEARTKLTEMYMRK	VKYLGRVTLAARSSA	NVYLLKYDTAHRAFP
3		VTLAARSSAPSGAST	QEFTVGEGADKWVVK
4		TDGTVLKKNIGGNAC	WVVKSIGGRLGPSQL
5		DKVPKKFKLPSPFFN	VVLESTGIFRTKAEK
6		KLGGLLVKKYGLSAK	KDAEGKIKKDDGYDGH
7		ILVKKYGLSAKNLDEF	TLNNAFGIRNGFMTTV
8		ASSEFYDEEKKLYEV	
9		DYENWTKLNARLGQRV	

[‡] AEG::SOE2 refers to the chimeric protein derived from epitopes unique to the *T. vaginalis* proteins fructose-1,6-bisphosphate aldolase (A), α-enolase (E), and glyceraldehyde-3-phosphate dehydrogenase (G), as described in Figure 1.

3.2. Purification of Recombinant AEG::SOE2

Figure 2 shows the Coomassie-blue-stained gel after sodium dodecyl sulfate-polyacrylamide gel electrophoresis (SDS-PAGE) of purified AEG::SOE2 from two different experiments (lanes 2 and 3). The relative mobility of the protein was compared with molecular weight standards (lane 1) and is consistent with the expected size of ~35.9-kDa. Lane 4 shows the stained band of 1 μg of BSA for comparison.

3.3. ELISA Using Different Amounts of AEG::SOE2-Coated Wells Probed with Different Antibodies

ELISA was then performed to determine the amount of AEG::SOE2 immobilized onto microtiter wells detected by pooled positive patient sera, mouse anti-*T. vaginalis* serum [31], and a cocktail of MAbs reactive to the protein (Figure 1). As seen in Figure 3, 1μg (**A**) and 5 μg (**B**) of protein on wells were detected by the mouse antiserum at a 1:100 dilution (wells numbered 2), the cocktail of IgG$_1$ monoclonal antibodies (MAbs) (wells numbered 3, 25 μL of hybridoma supernatant of each MAb), and pooled patient sera at a 1:25 dilution (wells numbered 4). Not surprisingly, the protein (Figure 2) was also detected by immunoblot with these antibody reagents. There was no detection of AEG::SOE2 even at 10 μg amounts using negative pooled sera of women and men (wells number 1). An irrelevant MAb to the actinin protein called HA423 [24], as shown also in Table 2, was unreactive to AEG::SOE2. One microgram amounts of AEG::SOE2 were chosen as the standard amount for ELISA as less than 1μg coated onto wells gave mixed results with patient sera and MAbs. The 1:25 dilution for patient sera was shown previously to be ideal for ELISA with 1μg of protein [8–10,21,24,25].

3.4. ELISA Comparing ACT-P2 and AEG::SOE2 with Different Antibodies

We now compared AEG::SOE2 by ELISA with the gold standard ACTP2 for detection of IgG. Figure 4 compares ELISA values for negative human sera (NHS) versus positive women and men sera (PHS) characterized previously [21,24,25]. The pooled PHS known to have antibody to trichomonad proteins were five each of women sera (PHS-1), men sera (PHS-2), or combined women and men sera (PHS-3). ELISA results show statistically significant higher absorbance values for PHS to both ACTP2 and AEG::SOE2 when compared with NHS of five different women and five different men sera that lacked reactivity to trichomonad proteins by immunoblot. Additionally, sera of mice immunized with *T. vaginalis* (IMS) gave statistically significant higher values to both ACTP2 and AEG::SOE2 when compared with both prebleed, normal mouse sera (NMS), and secondary peroxidase-conjugated goat anti-human IgG alone (P-G anti-H IgG). Finally, and not unexpectedly, the MAbs to AEG::SOE2 (Figure 2) gave statistically significant higher values with AEG::SOE2 but not ACTP2, and the MAb HA423 to ACTP2 was unreactive with AEG::SOE2 compared to statistically significant values with ACTP2. MAb L64 is a trichomonad cytoplasmic protein and was unreactive to both ACTP2 and AEG::SOE2.

Table 2. Testing by ELISA of women and men sera of different positive to negative (P/N) [‡‡] 0 to 4+ scores.

Expt	Serum Reaction Scores Against ACTP2:													
	0[‡‡]	0	1+	1+	2+	2+	3+	3+	4+	4+	4+	4+	4+	4+
	W	M	W	M	W	M	W	M	W	M	W	M	W	M
expt 1	0	0	1	1	2	2	3	3	4	4	4	4	4	4
expt 2	0	0	1	1	2	2	3	3	4	4	4	4	4	4
expt 3	0	0	1	1	2	2	3	3	4	4	4	4	4	4
expt 4	0	0	1	1	2	2	3	3	4	4	4	4	4	4

Expt	Serum Reaction Scores Against AEG::SOE2:													
	0	0	1+	1+	2+	2+	3+	3+	4+	4+	4+	4+	4+	4+
	W	M	W	M	W	M	W	M	W	M	W	M	W	M
expt 1	0	0	1	1	2	2	3	3	4	4	4	4	4	4
expt 2	0	0	1	1	2	2	3	3	4	4	4	4	4	4
expt 3	0	0	1	1	2	2	3	3	4	4	4	4	4	4
expt 4	0	0	1	1	2	2	3	3	4	4	4	4	4	4

[‡‡] Serum designations refer to the 0–4+ scoring of the sera following ELISA using standards as controls as described above. This scoring enables the examination of the relative reactivities against the ACT-P2 and ACT::SOE2 proteins. The red scores represent sera with reactivities to trichomonad proteins by immunoblot.

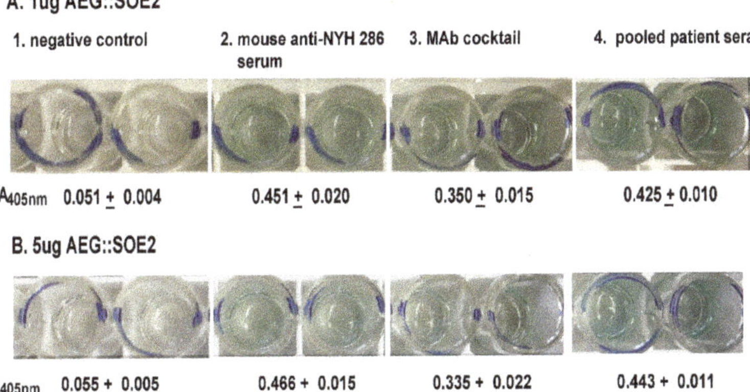

Figure 3. ELISA for detection of AEG::SOE2 at 1µg (**A**) and 5µg (**B**) AEG::SOE2 immobilized onto individual wells of 96-well microtiter plates. The negative controls were wells incubated with five different pooled negative sera each from women and men. This sera were shown previously to have no reactivity with any *T. vaginalis* proteins by immunoblot [24] (wells numbered 1). Wells with the different concentrations of AEG::SOE2 protein were incubated with mouse anti-*T. vaginalis* serum [31] (wells numbered 2), a cocktail of MAbs reactive with AEG::SOE2 epitopes as shown in Figure 1 (wells numbered 3), and with five different pooled positive women and men sera (wells numbered 4). The negative and positive pooled women and men sera have been previously reported [26]. Values were obtained by absorbance at 405 nm. As expected, a negative control irrelevant MAb to α-actinin called HA423 [24] was unreactive to AEG::SOE2 by ELISA, as shown in Figure 4. The ELISA was repeated on four different times with similar results.

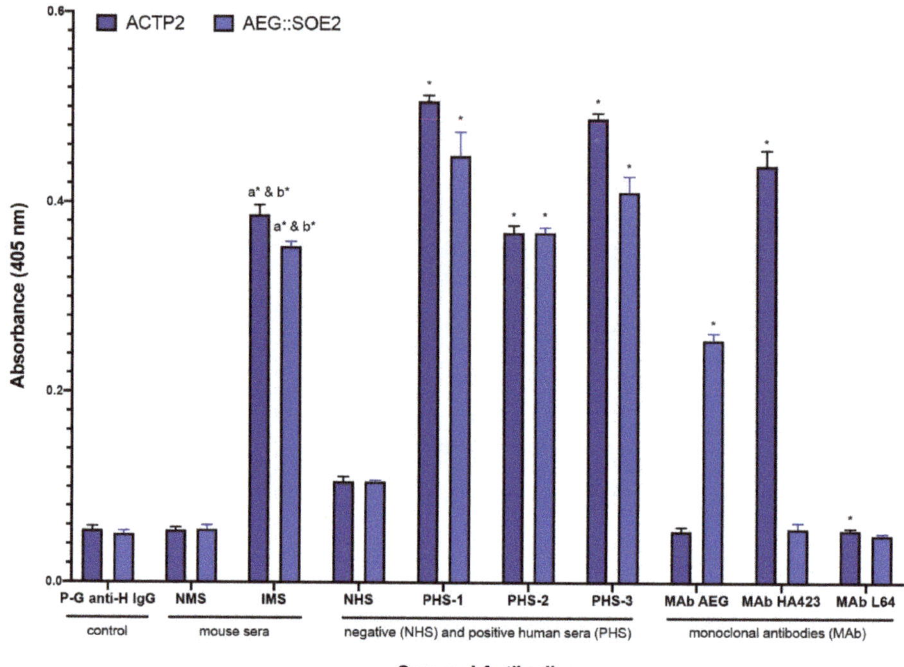

Figure 4. ELISA comparing negative (NHS) and positive women and men sera (PHS) and prebleed, normal mouse serum (NMS), and immunized mouse anti-*T. vaginalis* serum (IMS), for detection of IgG antibody to ACT-P2 (dark blue) and AEG::SOE2 (light blue). Bars represent means and standard deviations that were calculated for the average of all ELISA performed (n = 8). The secondary peroxidase-conjugated goat anti-human IgG (Fc fraction; labeled P-G anti-H IgG) is the secondary antibody used for ELISA for detecting human antibody and gave values equal to the use of 2% eBSA-PBS alone as a negative control. NMS sera gave values equivalent to secondary antibody alone. The pooled negative human sera (NHS) lacking reactivity to trichomonad proteins [24,25] represented five different women and five different men sera. The three pooled positive human sera (PHS) that detect trichomonad proteins represented either five women sera (number 1), five men sera (number 2), or five combined women and men sera (number 3). Independent t-tests were used to compare the mean absorbance levels for each target protein to its corresponding control. This included IMS vs. P-G anti-H IgG (a), IMS vs NMS (b), PHS-1 vs NHS, PHS-2 vs NHS, and PHS-3 vs NHS. The MAb cocktail mix is comprised of equal volumes (25 µL) of hybridoma supernatants of the four MAbs to the epitopes of the recombinant AEG::SOE2 (Figure 2). HA423 is an MAb directed to ACTP2 and is reactive with ACTP2 but not AEG::SOE2 [21,24,25]. MAb L64 is an irrelevant control antibody that reacts with a cytoplasmic protein of *T. vaginalis* [22]. All monoclonal antibodies are of the IgG$_1$ isotype. Lastly, differences in mean absorbance levels for each monoclonal antibody (i.e., cocktail MAb AEG, MAb HA423, and MAb L64) by protein type (i.e., ACTP2, AEG::SOE2) were examined. Absorbance values were obtained at 405nm. The means and standard deviations statistical significance is denoted as asterisks (*) and found to be $p < 0.001$.

3.5. Different ELISA Experiments Testing Women and Men Sera of Different 0 to 4+ Scores

Table 2 shows results of quadruplicate testing using up to ten individual women and men sera scored from 0 to 4+ based on reactivity to ACT-P2. Only 3+ and 4+ sera have been shown to have IgG antibody to trichomonad proteins [21,24,25]. Not unexpectedly, the negative women (W) and men (M) sera remained 0 to 2+ for ACT-P2 and AEG::SOE2. In contrast, and as shown within the red boxes, all of the 3+ and 4+ positive sera gave reproducible high reactions for both proteins.

For clarification of what the scores signify in terms of the mean and standard deviation (SD) absorbance readings for Table 2 and for Figure 5 and Table 3 below, the blank control with eBSA-PBS only was 0.050 ± 0.002. The mean and SD for sera giving different scores were as follows: sera with a 0 (zero) score was 0.131 ± 0.012, sera with a 1+ score was 0.187 ± 0.010, and the sera with a 2+ score was 0.233 ± 0.023. The sera with scores of 3+ and 4+ were 0.311 ± 0.025 and 0.441 ± 0.20, respectively. All scores were derived by subtracting from the average blank OD reading of 0.050.

3.6. ELISA of Women and Men Sera of 3+ and 4+ Scores at Different Dilutions

Figure 5 compares the positive 3+ and 4+ sera reactivities toward ACTP2 (dark blue) and AEG::SOE2 (light blue) using women and men sera at different dilutions. Not shown is that the 0 to 2+ sera remained negative at all dilutions for both target proteins. For both ACTP2 and AEG::SOE2, the 3+ women (part A) and men (part B) sera became negative (2+ or lower) at dilutions of 1:50. Part A shows that the three replicates of the 4+ women sera were positive for ACTP2 at 1:100 dilutions. Only the first and second replicates were positive for AEG::SOE2 at 1:100 dilutions, but the third replicate sera were positive only up to the 1:50 dilution. Interestingly, the first replicate 4+ sera remained positive for AEG::SOE2 at 1:200 dilution. Part B shows that the first and second replicates of 4+ men sera were positive for ACTP2 at the 1:50 dilution, and only the first replicate sera was positive at the 1:100 dilution. For AEG::SOE2, only the first and second replicates remained positive at the 1:50 dilution. Overall, the data suggest that the 4+ sera of both women and men have higher titers of IgG than the 3+ sera for both target proteins.

Table 3. Reactivities by ELISA using ACTP2 comparing with AEG::SOE2 as targets.

No.‡	Scores #	Scores	No.	Scores	Scores	No.	Scores	Scores	No.	Scores	Scores
	AEG::SOE2	ACTP2		AEG::SOE2	ACTP2		AEG::SOE2	ACTP2		AEG::SOE2	ACTP2
1	0	0	22	4+	3++	43	4+	4+	64	3+	2
2	4+	4+	23	3+	3+	44	3+	3+	65	4+	3++
3	3+	2++	24	4+	4+	45	3+	3+	66	2+	2+
4	1+	2+	25	4+	4+	46	4+	3+	67	3+	3++
5	2+	3+	26	3++	3+	47	2+	2+	68	2+	2+
6	2+	2+	27	4+	3++	48	3++	3+	69	2+	2+
7	3+	4+	28	4+	3++	49	3+	3+	70	3+	3++
8	2+	2+	29	2+	2+	50	2+	3+	71	1+	1+
9	2+	2+	30	1+	1+	51	3+	2+	72	0	0
10	4+	4+	31	1+	1+	52	4+	3++	73	1+	1+
11	4+	3++	32	1+	2+	53	3+	2+	74	0	0
12	2+	3+	33	4+	3++	54	3+	1+	75	2+	3+
13	1+		34	4+	3+	55	3++	3+	76	0	1+
14	4+	4+	35	3++	4+	56	2+	2+	77	4+	4+
15	0	0	36	3+	2+	57	3+	3++	78	3++	3++
16	2+	2+	37	3+	3+	58	4+	4+	79	3+	3+
17	4+	4+	38	3+	2+	59	3+	3+	80	3+	3+
18	3+	3+	39	3++	3+	60	4+	4+	81	4+	4+
19	4+	3++	40	1+	1+	61	4+	3++	82	3+	3+
20	3++	4+	41	2+	1+	62	3+	3+	83	4+	4+
21	4+	4+	42	3+	3+	63	3+	3+	84	4+	4+

‡ Refers to the number of the serum sample used for the ELISA comparisons of AEG::SOE2 and ACTP2. # All experiments were performed using quadruplicate wells. Agreement for 3+ and 4+ sera for both ACTP2 and AEG::SOE2 is denoted in yellow. The four 3+ sera that were reactive with ACTP2 but not AEG::SOE2 are denoted in green, and the four 3+ sera positive for AEG::SOE2 but not ACTP2 are denoted in blue.

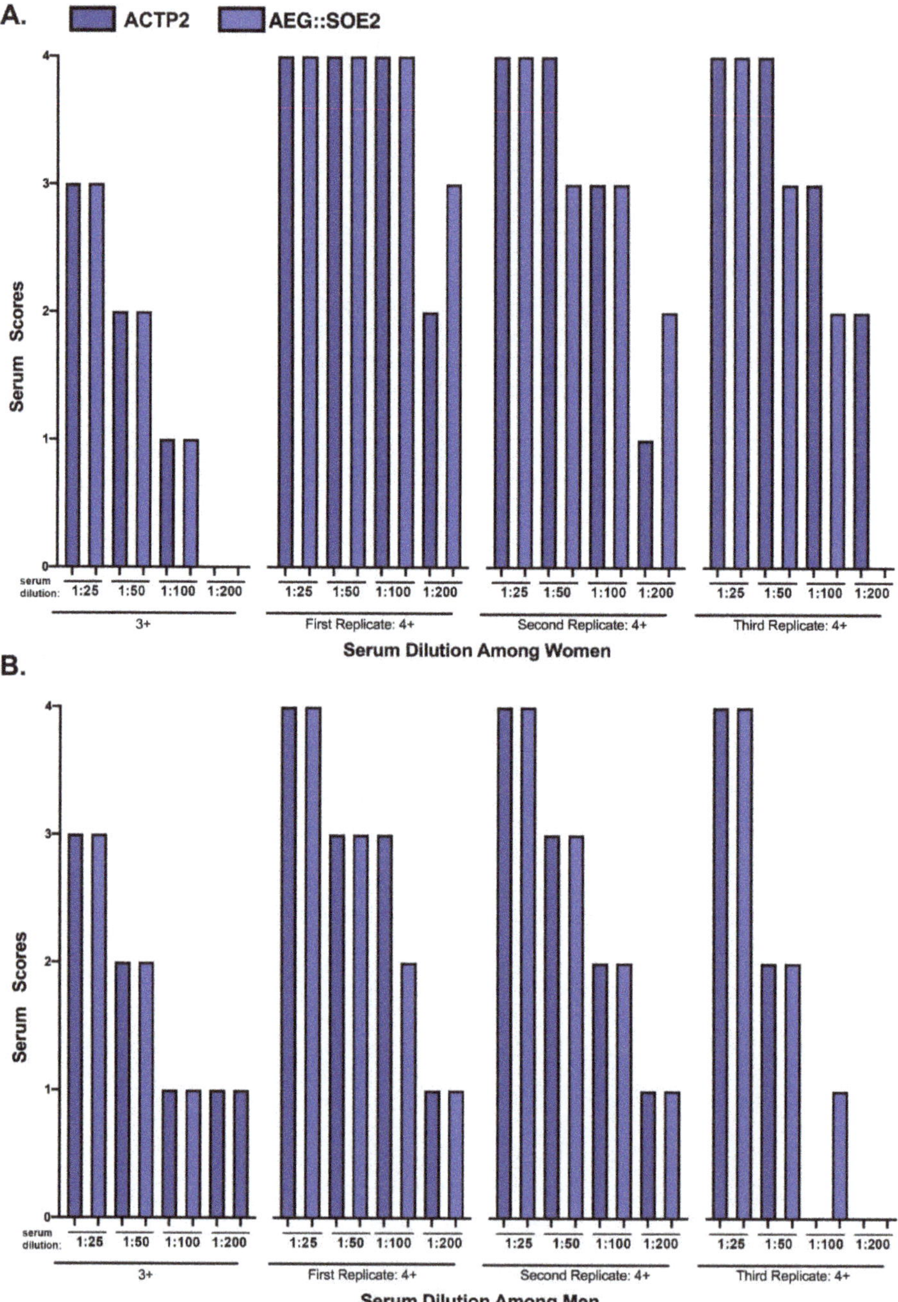

Figure 5. ELISA of women and men sera of 3+ and 4+ scores at different dilutions. The sera designations for 3+ and 4+ scores following ELISA using standards as controls is as described above for Table 2. This scoring enables the comparative examination of dilutions with eBSA-PBS on the relative reactivities against the target ACT-P2 and ACT::SOE2 proteins. Each replicate consists of pooled different positive sera (n = 10) of women and men.

3.7. ELISA Using ACT-P2 and AEG::SOE2 as Targets with 0 to 4+ Individual Sera

Finally, I randomly selected 42 each of sera of women and men with different scores for side-by-side evaluation. Table 4 presents results showing that, except for a few samples, there was almost 100% agreement for 3+ and 4+ positive sera for both ACT-P2 and AEG::SOE2. The score of 3++ had average and standard deviations much higher than the value for 3+ given above for Table 2. Likewise, there was almost 100% agreement for the 0 to 2+ negative sera. Interestingly, there were four 3+ sera that were reactive with ACT-P2 but not AEG::SOE2. Similarly, there were four 3+ sera that detected AEG::SOE2 but not ACT-P2. One possible explanation for these latter results is that the 3+ sera may be borderline negative, and this will require testing to determine whether or not there is IgG antibody that detects trichomonad proteins, such as by immunoblot, as before [21,24].

Table 4. Whole cell ELISA for detection of anti-*T. vaginalis* mouse serum IgG antibody and monoclonal antibody (MAb).

Antibodies *	Experiment No. (Mean ± SD)		
	1	**2**	**3**
goat anti-mouse IgG alone	0.075 ± 0.001 ‡	0.123 ± 0.005	0.130 ± 0.015
NMS (1:10)	0.160 ± 0.001	0.157 ± 0.002	0.140 ± 0.010
anti-*T. vaginalis* serum (1:10)	ND ‡‡	0.364 ± 0.005	0.455 ± 0.005
anti-*T. vaginalis* serum (1:100)	0.345 ± 0.005	ND	0.432 ± 0.001
anti-AEG::SOE2 serum (1:10)	0.225 ± 0.002	ND	ND
anti-AEG::SOE2 serum (1:10)	0.207 ±0.002	0.305 ± 0.005	0.190 ± 0.005
MAb cocktail (1:1:1:1, v/v)	0.275 ± 0.005	0.218 ± 0.005	0.228 ± 0.005
MAb ALD30A	0.335 ± 0.010	0.220 ± 0.010	0.253 ± 0.008
MAb L64	0.055, 0.055	0.051, 0.052	0.051, 0.050

* The peroxidase-conjugated goat anti-mouse IgG (Fc fraction) was used as a control and is the secondary antibody used for ELISA detection of mouse IgG antibody. It gave values equal to the use of 2% eBSA-PBS alone. Anti-*T. vaginalis* mouse serum has been described previously [29]. The normal mouse serum (NMS) and anti-AEG::SOE2 mouse serum are as described in Figure 4. MAb ALD30A and the MAb cocktail are to the epitopes of the recombinant AEG::SOE2 (Figure 1). MAb L64 is an irrelevant control antibody that reacts with a cytoplasmic protein of *T. vaginalis* and has been used previously [22,26]. All MAbs are the same IgG$_1$ isotype. ‡ Absorbance values were obtained at 405nm. The mean and standard deviation (SD) were calculated for the average of ELISA readings, and all experiments were performed using quadruplicate wells. ‡‡ ND, not done.

3.8. AEG::SOE2 is Immunogenic, and Anti-AEG::SOE2 Serum IgG Antibody Detects T. vaginalis Organisms Immobilized onto Microtiter Wells by ELISA

Finally, we wanted to examine whether mice immunized with AEG::SOE2 produced IgG antibodies. We compared mouse anti-AEG::SOE2 serum with mouse anti-*T vaginalis* serum [29] in individual wells of microtiter plates coated with fixed trichomonads. As shown in Table 4, both antisera gave ELISA readings greater than secondary peroxidase-conjugated goat anti-mouse IgG alone and prebleed normal mouse serum (NMS). Likewise, a cocktail of hybridoma supernatants of the MAbs reactive with AEG::SOE2 epitopes and MAb ALD30A alone (Figure 1) gave higher values compared to controls. The MAb L64 of the same isotype as the MAbs to AEG::SOE2 was used as another negative control. Furthermore, that MAbs to E and G react with trichomonads in this whole cell ELISA further supports earlier work that these metabolic enzymes are on the surface of *T. vaginalis* [23,24].

4. Discussion

In this study, I use an approach to extend an earlier published work [21] and show the synthesis of a larger novel, chimeric String-Of-Epitopes (SOE) protein called AEG::SOE2 with additional epitopes of A, E, and G. This AEG::SOE2 protein possesses the same high specificity and sensitivity as ACT-P2, the gold-standard target for *T. vaginalis* seropositivity (Figure 4 and Table 2; Table 3). These data indicate that AEG::SOE2 may be a target for a rapid, accurate, and cost-effective POC test. Such a test would allow for screening of individuals with active *T. vaginalis* infection or permit identification of those previously exposed to the organism. Another reason for moving toward a serum-based diagnostic is the demonstration of positive IgG seroconversion in relation to *T. vaginalis* and PCa development and progression [8–11]. The next step now appears to be development of a platform incorporating AEG::SOE2 in order to demonstrate a POC diagnostic for broad application of *T. vaginalis* surveillance.

As discussed recently [25], little is known of the temporal nature and duration of the serum IgG antibody responses among patients after infection with *T. vaginalis* and after diagnosis and cure. Such a POC test would also permit the medical community to understand the specific IgG response to the parasite in relation to active and/or past infections. My laboratory showed the short-lived nature of both serum and vaginal IgG to trichomonad cysteine proteinases after treatment of patients [28,29]. My laboratory also reported that IgG to a 230-kDa trichomonad protein was still evident in vaginal washes of patients at 4-weeks post cure [30]. I believe that these earlier findings support the view that a serodiagnostic test is necessary in order to understand the antibody responses of infected individuals. The availability of proteins unique to *T. vaginalis*, like α-actinin [24] and AEG::SOE2, provides the opportunity to elucidate the extent and nature of the antibody response issues in the future. The data (Figure 4 and Table 2) presented here show that AEG::SOE2 is equivalent to ACTP2 in serum IgG immunoreactivity. The presence of serum IgG antibody to epitopes unique to *T. vaginalis* proteins further reinforces the legitimacy of the approach taken here for identifying a specific and novel target for *T. vaginalis*, and although speculative, it may be possible to develop tests for other STI microbial pathogens and additional infectious diseases using this approach.

These novel, chimeric SOE proteins are comprised of immunogenic epitopes unique to the microbial pathogen of interest and, in this case, *T. vaginalis*. It is intriguing to consider that these SOE proteins may have efficacy as vaccines. At present, there is no evidence of immune protection against *T. vaginalis* despite the presence of both serum and vaginal antibody responses among patients with trichomonosis [28–30]. The many trichomonad proteases that degrade immunoglobulins may be a reason for immune evasion [29]. Reports have proposed that whole *T. vaginalis* organisms or lysates may be used as vaccines for *T. vaginalis* [32,33]. I argue against using *T. vaginalis* organisms or lysate as vaccines. One reason is that serum IgG antibody is made to epitopes of these enzymes, which in fact have amino acid sequence identities with human enzymes (referred to as trichomonad non-unique epitopes) [21]. In other words, as mentioned for Table 1, 10 of 12 epitopes of A, 9 of 18 epitopes of E, and 12 of 19 epitopes of G had sequence identity to human, bacterial, parasite, and fungal enzyme proteins. This is an important finding in and of itself that should be considered when studying host antibody responses to microbial pathogens. Whether antibodies to these trichomonad, non-unique epitopes common to human proteins mediate auto-immune reactions and, therefore, possible tissue damage is presently unknown. I believe that this issue must be considered within the framework of pathogenesis of trichomonosis and also for other infectious diseases. Indeed, it has been shown that human serum antibody to α–enolase of group A streptococcus cross-reacts with host tissues [34]. Thus, the approach described here may circumvent potential immune-crossreactive problems posed by using whole cell or lysate vaccines.

Another reason against whole organisms and lysates as vaccines is that we now know that *T. vaginalis* acquires onto its surface numerous host serum proteins [22,23,35–38]. The coating of the parasite surface with host proteins [36,38] may represent yet another mechanism for parasite evasion of immune-antibody responses. Further, the *T. vaginalis* surface-associated E and G metabolic enzymes are ligands that bind host proteins, such as plasminogen, fibronectin, collagen, and laminin [22,23]. As these host proteins may play a role in pathogenesis, the proteins are referred to as host-pathogenicity factors [22,23,35–38]. It is possible that host proteins on *T. vaginalis* may have altered structures exposing epitopes to host antibody responses, creating possible auto-antibodies and adverse reactions with tissues. This possibility deserves more attention in host–parasite interactions.

Of interest is that trichomonad lysates were recently compared with α-actinin by ELISA for IgG reactivity [39], and both were found to be equivalent in serum IgG detection. Here, too, I argue that lysates are inappropriate diagnostic targets for the same reasons mentioned above. Seropositive reactions may be due, in part, to IgG antibody responses to metabolic enzyme epitopes common to other pathogens. In this scenario, infections by other bacterial, parasite, and fungal pathogens may induce IgG not only to the epitopes of the enzymes A, E, and G used here but also to epitopes of other proteins that are immuno-crossreactive with *T. vaginalis* proteins. This conclusion has merit based on the findings presented here. This, then, would lead to false-positive reactions for this STI. Therefore, future serodiagnostic targets for this and other infectious diseases must have specific epitopes that are unique to the pathogen causing the disease. These concerns are relevant to the development of effective infectious disease diagnostics and vaccines and are also important considerations for surveillance and interventions of STIs and infectious diseases [40].

Finally, I have shown that the chimeric AEG::SOE2 protein is immunogenic, as evidenced by IgG antibody made by immunized mice (Table 4). Importantly, that anti-AEG::SOE2 serum detects whole organisms reaffirms the surface location of these proteins on *T. vaginalis* [22,23]. More importantly, multi-epitope constructs like AEG::SOE2 may be tested as a vaccine candidate for *T. vaginalis*, and this may be verified as was recently shown by others for α-actinin [41]. I believe that the approach described here may lead to future specific diagnostic targets and that such targets, comprised of immunogenic epitopes unique to the pathogen of interest, can be possible effective subunit vaccine candidates.

5. Conclusions

A stepwise approach is presented that may have applicability for infectious diseases by producing a unique and specific serodiagnostic target to an infectious agent, and such SOE proteins can be tested for specificity as a diagnostic target. *Trichomonas vaginalis* was used as a model to test the viability of the approach. This approach includes i) identification of immunogenic surface proteins; ii) epitope mapping; iii) selection of epitopes with amino acid sequences unique to the trichomonad proteins; and iv) construction of a hybrid String-Of-Epitopes AEG::SOE2 protein comprised of A, E, and G epitopes. Finally, because of the highly immunogenic nature of the epitopes, as evidenced by patients' serum IgG reactivities (Figure 3; Figure 4 and Table 3), the fact that AEG::SOE2 itself elicited IgG antibodies via immunization (Table 4), and the fact that the epitopes have no identity to other known proteins in databanks, this novel SOE protein of *T. vaginalis* and SOE proteins of infectious agents in general may have potential vaccine applicability.

6. Patent

J.F. Alderete. Strings of Epitopes Useful in Diagnosing and Eliciting Immune Responses to Sexually Transmitted Infections. No. 9910042, 5 March, 2018.
J.F. Alderete. Strings of Epitopes Useful in Diagnosing and Eliciting Immune Responses to Sexually Transmitted Infections. No. 10386369, 20 August, 2019.

Funding: This research received no external funding.

Acknowledgments: I thank a past collaborator at Washington University at St. Louis for providing sera that was highly seropositive for ACT-P2. I also thank past collaborators of the School of Medicine at The University of Texas Health Science Center at San Antonio for obtaining women and men patient sera used in prior studies by my laboratory there and used here at Washington State University. I want to acknowledge Grace Alderete for her voluntary assistance and laboratory maintenance throughout this work.

Conflicts of Interest: The author declares no conflict of interest. I alone designed the study and was responsible in the collection, analyses, or interpretation of data; in the writing of the manuscript; and in the decision to publish the results.

References

1. Hobbs, M.M.; Sena, A.C.; Swygard, H.; Schwebke, J.R. Trichomonas Vaginalis and Trichomoniasis. In *Sexually Transmitted Diseases*; Holmes, K.K., Sparling, P.F., Stamm, W.E., Piot, P., Wasserheit, J.N., Corey, L., Cohen, M.S., Watts, D.H., Eds.; McGraw-Hill MedicalHolmes: New York, NY, USA, 2008.

2. Swygard, H.; Seña, A.C.; Hobbs, M.M.; Cohen, M.S. Trichomoniasis: Clinical manifestations, diagnosis and management. *Sex. Transm. Infect.* **2004**, *80*, 91–95. [CrossRef] [PubMed]

3. Bachmann, L.H.; Hobbs, M.M.; Seña, A.C.; Sobel, J.D.; Schwebke, J.R.; Krieger, J.N.; McClelland, R.S.; Workowski, K.A. Trichomonas vaginalis genital infections: Progress and challenges. *Clin. Infect. Dis.* **2011**, *53*, S160–S172. [CrossRef] [PubMed]

4. Krashin, J.W.; Koumans, E.H.; Bradshaw-Sydnor, A.C.; Braxton, J.R.; Secor, W.E.; Sawyer, M.K.; Markowitz, L.E. Trichomonas vaginalis prevalence, incidence, risk factors and antibiotic-resistance in an adolescent population. *Sex. Transm. Dis.* **2010**, *37*, 440–444. [CrossRef] [PubMed]

5. Huppert, J.S. Trichomoniasis in teens: An update. *Curr. Opin. Obstet. Gynecol.* **2009**, *21*, 371–378. [CrossRef]

6. Gardner, W.A., Jr.; Culberson, D.E.; Bennett, B.D. Trichomonas vaginalis in the prostate gland. *Arch. Pathol. Lab. Med.* **1986**, *110*, 430–432.

7. Mitteregger, D.; Aberle, S.W.; Makristathis, A.; Walochnik, J.; Brozek, W.; Marberger, M.; Kramer, G. High detection rate of Trichomonas vaginalis in benign hyperplastic prostatic tissue. *Med. Microbiol. Immunol.* **2011**, *201*, 113–116. [CrossRef]

8. Stark, J.R.; Judson, G.; Alderete, J.F.; Mundodi, V.; Kucknoor, A.S.; Giovannucci, E.L.; Platz, E.A.; Sutcliffe, S.; Fall, K.; Kurth, T.; et al. Prospective study of Trichomonas vaginalis infection and prostate cancer incidence and mortality: Physicians' health study. *J. Natl. Cancer Inst.* **2009**, *101*, 1406–1411. [CrossRef]

9. Sutcliffe, S.; Alderete, J.F.; Till, C.; Goodman, P.J.; Hsing, A.W.; Zenilman, J.M.; De Marzo, A.M.; Platz, E.A. Trichomonosis and subsequent risk of prostate cancer in the prostate cancer prevention trial. *Int. J. Cancer* **2009**, *124*, 2082–2087. [CrossRef]

10. Sutcliffe, S.; Giovannucci, E.; Alderete, J.F.; Chang, T.-H.; Gaydos, C.A.; Zenilman, J.M.; De Marzo, A.M.; Willett, W.C.; Platz, E.A. Plasma antibodies against Trichomonas vaginalis and subsequent risk of prostate cancer. *Cancer Epidemiol. Biomarkers Prev.* **2006**, *15*, 939–945. [CrossRef]

11. Sutcliffe, S.; Neace, C.; Magnuson, N.S.; Reeves, R.; Alderete, J.F. Trichomonosis, a common curable STI, and prostate carcinogenesis—A proposed molecular mechanism. *PLOS Pathog.* **2012**, *8*, e1002801. [CrossRef]

12. Han, I.H.; Kim, J.H.; Kim, S.S.; Ahn, M.H.; Ryu, J.-S. Signalling pathways associated with IL-6 production and epithelial-mesenchymal transition induction in prostate epithelial cells stimulated with Trichomonas vaginalis. *Parasite Immunol.* **2016**, *38*, 678–687. [CrossRef] [PubMed]

13. Kim, J.-H.; Kim, S.-S.; Han, I.-H.; Sim, S.; Ahn, M.-H.; Ryu, J.-S. Proliferation of prostate stromal cell induced by benign prostatic hyperplasia epithelial cell stimulated with Trichomonas vaginalisvia crosstalk with mast cell. *Prostate* **2016**, *76*, 1431–1444. [CrossRef]

14. Kim, J.-H.; Han, I.-H.; Kim, S.-S.; Park, S.-J.; Min, D.-Y.; Ahn, M.-H.; Ryu, J.-S. Interaction between Trichomonas vaginalis and the prostate epithelium. *Korean J. Parasitol.* **2017**, *55*, 213–218. [CrossRef] [PubMed]

15. Twu, O.; Dessi', D.; Vu, A.; Mercer, F.; Stevens, G.C.; De Miguel, N.; Rappelli, P.; Cocco, A.R.; Clubb, R.T.; Fiori, P.L.; et al. Trichomonas vaginalis homolog of macrophage migration inhibitory factor induces prostate cell growth, invasiveness, and inflammatory responses. *Proc. Natl. Acad. Sci. USA* **2014**, *111*, 8179–8184. [CrossRef] [PubMed]

16. Pillay, A.; Lewis, J.; Ballard, R.C. Evaluation of Xenostrip-Tv, a rapid diagnostic test for Trichomonas vaginalis infection. *J. Clin. Microbiol.* **2004**, *42*, 3853–3856. [CrossRef]

17. Lee, J.J.; Moon, H.S.; Lee, T.Y.; Hwang, H.S.; Ahn, M.H.; Ryu, J.S. PCR for diagnosis of male Trichomonas vaginalis infection with chronic prostatis and urethritis. *Korean J. Parasitol.* **2012**, *50*, 157–159. [CrossRef] [PubMed]

18. Ginocchio, C.C.; Chapin, K.; Smith, J.S.; Aslanzadeh, J.; Snook, J.; Hill, C.S.; Gaydos, C.A. Prevalence of Trichomonas vaginalis and coinfection with chlamydia trachomatis and Neisseria gonorrhoeae in the United States as determined by the aptima Trichomonas vaginalis Nucleic acid amplification assay. *J. Clin. Microbiol.* **2012**, *50*, 2601–2608. [CrossRef]

19. Yar, T.M.; Karakus, M.; Toz, S.; Karabulut, A.B.; Ozbel, Y.; Atambay, M. Diagnosis of trichomonosis in male patients on performing nested polymerase chain reaction. *Turkiye Parzitol. Derg.* **2017**, *41*, 130–134. [CrossRef]

20. Toskin, I.; Murtagh, M.; Peeling, R.W.; Blondeel, K.; Cordero, J.P.; Kiarie, J. Advancing prevention of sexually transmitted infections through point-of-care testing: Target product profiles and landscape analysis. *Sex. Transm. Infect.* **2017**, *93*, S69–S80. [CrossRef]

21. Alderete, J.F.; Neace, C.J. Identification, characterization, and synthesis of peptide epitopes and a recombinant six-epitope protein for Trichomonas vaginalis serodiagnosis. *Immuno. Targets Ther.* **2013**, *2*, 91–103. [CrossRef]

22. Mundodi, V.; Kucknoor, A.S.; Alderete, J.F. α-Enolase is a surface-associated plasminogen-binding protein of Trichomonas vaginalis. *Infect. Immun.* **2008**, *76*, 523–531. [CrossRef] [PubMed]

23. Lama, A.; Kucknoor, A.; Mundodi, V.; Alderete, J. Glyceraldehyde-3-phosphate dehydrogenase is a surface-associated, fibronectin-binding protein of Trichomonas vaginalis. *Infect. Immun.* **2009**, *77*, 2703–2711. [CrossRef] [PubMed]

24. Neace, C.J.; Alderete, J.F. Epitopes of the highly immunogenic Trichomonas vaginalis—Actinin are serodiagnostic targets for both women and men. *J. Clin. Microbiol.* **2013**, *51*, 2483–2490. [CrossRef] [PubMed]

25. Alderete, J. Epitopes within recombinant α-actinin protein is serodiagnostic target for Trichomonas vaginalis sexually transmitted infections. *Heliyon* **2017**, *3*, e00237. [CrossRef]

26. Addis, M.F.; Rappelli, P.; De Andrade, A.M.P.; Rita, F.M.; Colombo, M.M.; Cappuccinelli, P.; Fiori, P.L. Identification of Trichomonas vaginalis? Actinin as the most common immunogen recognized by sera of women exposed to the parasite. *J. Infect. Dis.* **1999**, *180*, 1727–1730. [CrossRef]

27. Addis, M.F.; Rappelli, P.; Delogu, G.; Carta, F.; Cappuccinelli, P.; Fiori, P.L. Cloning and molecular characterization of a cDNA clone coding for Trichomonas vaginalis alpha-actinin and intracellular localization of the protein. *Infect. Immun.* **1998**, *66*, 4924–4931. [CrossRef]

28. Alderete, J.F.; Newton, E.; Dennis, C.; Neale, K.A. The vagina of women infected with Trichomonas vaginalis has numerous proteinases and antibody to trichomonad proteinases. *Sex. Transm. Infect.* **1991**, *67*, 469–474. [CrossRef]

29. Alderete, J.F.; Newton, E.; Dennis, C.; Neale, K.A. Antibody in sera of patients infected with Trichomonas vaginalis is to trichomonad proteinases. *Sex. Transm. Infect.* **1991**, *67*, 331–334. [CrossRef]

30. Alderete, J.F.; Newton, E.; Dennis, C.; Engbring, J.; Neale, K.A. Vaginal antibody of patients with trichomoniasis is to a prominent surface immunogen of Trichomonas vaginalis. *Sex. Transm. Infect.* **1991**, *67*, 220–225. [CrossRef]

31. Alderete, J.F. Antigen analysis of several pathogenic strains of Trichomonas vaginalis. *Infect. Immun.* **1983**, *39*, 1041–1047. [CrossRef]

32. Cudmore, S.L.; Garber, G.E. Prevention or treatment: The benefits of Trichomonas vaginalis vaccine. *J. Infect. Public Health.* **2010**, *3*, 47–53. [CrossRef] [PubMed]

33. Smith, J.; Garber, G.E. Current status and prospects for development of a vaccine against Trichomonas vaginalis infections. *Vaccine* **2014**, *32*, 1588–1594. [CrossRef] [PubMed]

34. Sunblad, V.; Bussmann, L.; Chiauzzi, V.A.; Pancholi, V.; Charreau, E.H. Alpha-enolase: A novel autoantigen in patients with premature ovarian failure. *Clin. Endocrinol.* **2006**, *65*, 745–751. [CrossRef]

35. Peterson, K.M.; Alderete, J.F. Host plasma proteins on the surface of pathogenic Trichomonas vaginalis. *Infect. Immun.* **1982**, *37*, 755–762. [CrossRef] [PubMed]

36. Crouch, M.-L.; Alderete, J.F. Trichomonas vaginalis interactions with fibronectin and laminin. *Microbiol.* **1999**, *145*, 2835–2843. [CrossRef] [PubMed]

37. Crouch, M.-L.; Benchimol, M.; Alderete, J. Binding of fibronectin by Trichomonas vaginalis is influenced by iron and calcium. *Microb. Pathog.* **2001**, *31*, 131–144. [CrossRef]

38. Ibáñez-Escribano, A.; Nogal-Ruiz, J.J.; Serrano, J.P.; Barrio, G.; Escario, J.A.; Alderete, J. Sequestration of host-CD59 as potential immune evasion strategy of Trichomonas vaginalis. *Acta. Trop.* **2015**, *149*, 1–7. [CrossRef]

39. Kim, S.R.; Kim, J.H.; Park, S.J.; Lee, H.Y.; Kim, Y.S.; Kim, Y.M.; Hong, Y.C.; Ryu, J.S. Comparison between mixed lysate antigen and α-actinin antigen in ELISA for serodiagnosis of trichomonosis. *Parasitol. Int'l.* **2015**, *64*, 405–407. [CrossRef]

40. Dodet, B. Current barriers, challenges and opportunities for the development of effective STI vaccines: Point of view of vaccine producers, biotech companies and funding agencies. *Vaccine* **2014**, *32*, 1624–1629. [CrossRef]

41. Xie, Y.-T.; Gao, J.-M.; Wu, Y.-P.; Tang, P.; Hide, G.; Lai, D.-H.; Lun, Z.-R. Recombinant α-actinin subunit antigens of Trichomonas vaginalis as potential vaccine candidates in protecting against trichomoniasis. *Parasite. Vector.* **2017**, *10*, 83. [CrossRef]

International Journal of
**Environmental Research
and Public Health**

MDPI

Article

Validation of the Women's Views of Birth Labor Satisfaction Questionnaire (WOMBLSQ4) in the Spanish Population

María Dolores Pozo-Cano [1], Adelina Martín-Salvador [2,*], María Ángeles Pérez-Morente [3,*], Encarnación Martínez-García [1], Juan de Dios Luna del Castillo [4], María Gázquez-López [5], Rafael Fernández-Castillo [1] and Inmaculada García-García [1]

[1]   Faculty of Health Sciences, University of Granada, 18071 Granada, Spain; pozocano@ugr.es (M.D.P.-C.); emartinez@ugr.es (E.M.-G.); rafaelfernandez@ugr.es (R.F.-C.); igarcia@ugr.es (I.G.-G.)
[2]   Faculty of Health Sciences, University of Granada, 52005 Melilla, Spain
[3]   Faculty of Health Sciences, University of Jaen, 23071 Jaen, Spain
[4]   Faculty of Medicine, University of Granada, 18071 Granada, Spain; jdluna@ugr.es
[5]   Faculty of Health Sciences, University of Granada, 51001 Ceuta, Spain; mgazquez@ugr.es
*   Correspondence: ademartin@ugr.es (A.M.-S.); mmorente@ujaen.es (M.Á.P.-M.)

Received: 4 June 2020; Accepted: 31 July 2020; Published: 2 August 2020

Abstract: The satisfaction of women with the birth experience has implications for the health and wellness of the women themselves and also of their newborn baby. The objectives of this study were to determine the factor structure of the Women's Views of Birth Labor Satisfaction Questionnaire (WOMBLSQ4) questionnaire on satisfaction with the attention received during birth delivery in Spanish women and to compare the level of satisfaction of pregnant women during the birth process with that in other studies that validated this instrument. A cross-sectional study using a self-completed questionnaire of 385 Spanish-speaking puerperal women who gave birth in the Public University Hospitals of Granada (Spain) was conducted. An exploratory factor analysis of the WOMBLSQ4 questionnaire was performed to identify the best fit model. Those items that showed commonalities higher than 0.50 were kept in the questionnaire. Using the principal components method, nine factors with eigenvalues greater than one were extracted after merging pain-related factors into a single item. These factors explain 90% of the global variance, indicating the high internal consistency of the full scale. In the model resulting from the WOMBLSQ4 questionnaire, its nine dimensions measure the levels of satisfaction of puerperal women with childbirth care. Average scores somewhat higher than those of the original questionnaire and close to those achieved in the study carried out in Madrid (Spain) were obtained. In clinical practice, this scale may be relevant for measuring the levels of satisfaction during childbirth of Spanish-speaking women.

Keywords: validation study; satisfaction questionnaire; birth attention; patient satisfaction

---

1. Introduction

The birth of a child is one of the most significant events in the lives of women and their families. Knowing the level of satisfaction regarding the care received during the birth and postpartum periods is of special interest as it may help to improve the quality of health systems [1]. In Western countries, these experiences are becoming less frequent due to the drop in birth rates observed in many of them, especially in southern European countries, and it is hoped that the birth experience can become as rewarding as possible, despite not being free of serious consequences for the health of women and their newborns [1–3].

Many authors have explained the importance of women's satisfaction with the birth process, because it influences such important aspects as the maintenance of breastfeeding [4], which is crucial for the health of mothers and newborns [5,6].

When women experience unsatisfactory or traumatic births, their memories will be of pain, anger, fear, or sadness, and they may even suffer from post-traumatic stress disorders or may not remember anything about the delivery process [7–9]. Furthermore, a bad experience in a previous delivery increases the anxiety and fear in subsequent deliveries [10,11]. The proximity of childbirth activates memories of previous traumatic experiences and abuse as well as psychiatric disorders in women that can trigger a fear of vaginal childbirth and increase the demand for caesarean births, thus increasing the risks to maternal and perinatal health [12].

This is why, at present, the perceived satisfaction regarding care received during the birth is considered an essential indicator to measure quality of care [13]. Hodnett describes the personal expectations of pregnant women, the support and quality of the relationship with health professionals, especially midwives, and the participation of women in decision making as the most influential elements [14].

There are various instruments that measure the satisfaction of women with childbirth [2,15–22]. The Women's Views of Birth Labor Satisfaction Questionnaire (WOMBLSQ4) [23] has been used extensively in the recent literature and identifies women's satisfaction with their birth labor and delivery experiences, as well as the pain relief received during and after. It was developed in the United Kingdom by Smith and has been translated into French and validated to be applied to French-speaking women in University Hospitals in Geneva (Switzerland), and into Spanish, where Marín-Morales et al. did the same with women who gave birth in hospitals in Madrid (Spain) [23–25].

In the Autonomous Community of Andalusia (Spain), the Public Health System is committed to achieving excellence in healthcare. This is understood as a comprehensive concept involving multiple variables, among which citizen satisfaction is an inalienable element [26]. Birth care is focused on women, providing them with personalized care and promoting their autonomy and their role in decision-making [27].

The version translated into Spanish also presents discrepancies in the number of factors with respect to the original version and its translation into French due to significant convergence problems that make it necessary to eliminate the "control" factor, thus leaving the scale in nine dimensions. In order to assess the satisfaction of Andalusian mothers in the process of birth labor and to check the structure of the instrument, the WOMBLSQ4 scale was translated into Spanish and validated.

The objectives of this study are to determine the factor structure of the WOMBLSQ4 questionnaire on satisfaction with the care received during birth in Spanish women and to compare the level of satisfaction of pregnant women during the birth process with other studies that validated this instrument.

2. Materials and Methods

2.1. Sample and Data Collection

A cross-sectional study was carried out between January and March 2019 in puerperal women who had given birth in the Public University Hospitals of the city of Granada (Spain). In the year prior, an average of 5000 deliveries had taken place at both hospitals. Through intentional sampling, 385 Spanish-speaking puerperal women aged 18 years old or older were selected by collaborating with midwives in the studio. The included women voluntarily agreed to participate and signed an informed consent self-completed questionnaire that was delivered in a sealed envelope and later collected by the principal investigator. Those who did not understand Spanish and had elective caesarean births were excluded.

Postpartum surveys were administered to 450 women, of whom 15 refused to complete them, 40 did not deliver babies, and 10 did not provide informed consent. The questionnaires that were complete for all items were considered valid. The final sample consisted of 385 women, which constitutes a response rate of 85.5%.

2.2. Materials

To evaluate women's satisfaction with care received during delivery, the final version of the WOMBLSQ4 scale was used, which consists of 32 questions with Likert-type responses and 10 dimensions: professional support during the birth (5 questions), expectations of delivery (4 questions), assessment at home at the beginning of birth labor (3 questions), first contact with the newborn (3 questions), support of the husband/partner during labor (3 questions), pain relief during labor (3 questions), pain relief immediately after delivery (3 questions), continuity (2 questions), environment during delivery (2 questions), and control (2 questions). The measure of general satisfaction involved two questions [23]. The factorial validity of the scale was confirmed, as well as an adequate global reliability (Cronbach's alpha 0.89), and the validity of the subscales was also shown (Cronbach's alpha values ranged between 0.62 and 0.91). The score for each dimension was obtained by adding the values obtained in each question (some of them with an inverse score), and later on, the result was transformed so that the minimum possible score was 0 and the maximum possible one was 100 (total satisfaction in the dimension) [23–25]. Higher scores indicated greater satisfaction on the part of the women.

For this research, the scale was translated into Spanish by two English language translators and its final content was agreed upon by three midwives with extensive experience in childbirth assistance. The translated questionnaire was piloted to 50 women, and it was demonstrated that the instrument presented an excellent level of comprehension and an adequate completion time since, when collecting it, the participants were asked if they had difficulty completing it, if they understood all the questions, and if it seemed too long.

In addition, the following sociodemographic variables were incorporated: age, marital status, educational level, and employment situation.

2.3. Data Analysis

A descriptive analysis was performed in which means and standard deviations were calculated for the quantitative variables and frequencies and percentages for the qualitative ones. The factorial structure of the scale was explored by extraction of the main components followed by a Varimax rotation. In the first analysis, the Kaiser–Meyer–Olkin (KMO) sample adequacy measure was calculated, accepting values greater than 0.70 as optimal measures. Subsequently, the Bartlett sphericity test was applied to show significant differences between the items in the correlation and the unit matrix.

Next, the communality of each of the items on the scale was studied, and those that showed values less than 0.30 were eliminated, as they were poorly represented in the factorial set obtained. Those factors with eigenvalues greater than 1 were considered, and the percentage of variance explained with the said factors was determined to assess the weight of each one. After the rotation and analysis of the item saturation table, these were assigned to the dimension in which their saturation was highest. Once the items were eliminated, the previous steps were repeated in order to obtain the final factor structure. The internal consistency of each of the subscales was measured using Cronbach's alpha. Data analysis was performed with the SPSS v. Statistical package. 26.0 (International Busines Machines Corporation (IBM), Armonk, NY, USA) for Windows.

2.4. Ethical Considerations

The study complies with the standards of good clinical practice, explicit in the European Directive 2001/20/EC and Law 14/2007 (of 3 July) on biomedical research. The treatment of personal data in health research is governed by the provisions of the Organic Law 3/2018, 5 December, Protection of Personal Data and Guarantee of Digital Rights in Spain. The research protocol obtained a favorable resolution from the Ethics and Research Committee of Health Institutions.

3. Results

The sociodemographic characteristics of the analyzed sample are reflected in Table 1. The mean age of the participants was 31.62 years (SD 5.32), with a range of 18 to 46 years old. Regarding the level of education, almost half (175, 46.2%) had a university-level education. In relation to marital status, the majority were married or had a partner (359, 94.0%).

Table 1. Sociodemographic characteristics of the sample.

Socio-Demographic Variables		
	x ± SD	**Range**
Age	31.6 ± 5.32	18–46
	n	%
Level of Education (*n* = 379)		
University	175	46.2
Vocational training	87	23.0
Secondary education	66	17.4
Primary/Elementary/Basic education	51	13.5
Marital Status (*n* =382)		
Married or with a partner	359	94.0
Single	23	6.0
Labor Situation (*n* = 384)		
Employed workers	184	47.9
Housewives	68	17.7
Busines women	26	6.8
Unemployed	92	24.0
Other work circumstances (studying, retired, etc.)	14	3.6

Regarding the labor situation, 184 (47.9%) were employed workers.

3.1. Exploratory Facial Analysis

To carry out the factor analysis, firstly, all items on the scale were considered, and a mean KMO sample adequacy of 0.80 was obtained, with the result of the Bartlett sphericity test being statistically significant ($p < 0.001$).

Of the 32 items in the original questionnaire, only three showed communalities below 0.50: 25 (I am satisfied with just one or two things about the labor care that I received: 0.441), 31 (I didn't need a lot of pain relief after the birth: 0.475) and 12 (The way my labor care was provided could not have been improved: 0.476). However, they have not yet been removed from the questionnaire.

Using the main components method, nine factors that showed self-values greater than one, explaining 68.0% of the global variance, were extracted. A new dimension (3) was designed—pain during and after delivery—after merging dimensions six (pain during delivery) and seven (pain after delivery) from Smith's original questionnaire [23].

Table 2 shows the Cronbach's alpha and variance explained by each factor, as well as the saturations of each item, once the Varimax rotation had been performed.

Table 2. Analysis of each dimension and items on the scale.

	1 Professional Support (Cronbach'sAlpha = 0.867, % Variance Explained = 14.02)	Coefficient *
Q19	During labor there was always a carer to explain things so that I could understand.	0.843
Q7	All my labor carers were very supportive.	0.834
Q13	Carers always listened very, very carefully to everything that I had to say.	0.808
Q27	All my carers treated me in the most friendly and courteous manner possible.	0.778
Q32	My carers couldn't have been more helpful.	0.733
Q12	The way my labor care was provided could not have been improved.	0.556
	2 Expectations (Cronbach's Alpha = 0.861, % Variance Explained = 9.19)	
Q17	The delivery went almost completely as I had hoped that it would	0.809
Q11	The labor went nearly exactly as I had hoped that it would.	0.794
Q22	My labor was just about the right length.	0.719
Q1	My labor went totally normally.	0.710
	3 Pain During and After the Birth (Cronbach's Alpha = 0.781, % Variance Explained = 8.75)	
Q26	More pain relief would have made my labor easier. (−)	0.719
Q6	I should have been offered something more to relieve the pain I had after my baby was born. (−)	0.702
Q16	I was in a fair bit of pain immediately after the birth. (−)	0.669
Q9	I should have been offered something more to relieve my labor pains. (−)	0.668
Q20	I got excellent pain relief in labor.	0.586
Q31	I didn't need a lot of pain relief after the birth.	0.399
	4 Home Assessment (Cronbach's Alpha = 0.843, % Variance Explained = 7.52)	
Q15	When I thought that my labor had started, I would have liked a carer to come and see me at home to confirm that I had. (−)	0.914
Q28	Early home assessment of me in labor would have been very helpful. (−)	0.904
Q8	I should have had a home assessment in early labor. (−)	0.761
	5 Support from Husband (Cronbach's Alpha = 0.750, % Variance Explained = 6.80)	
Q2	My birth partner/husband helped me to understand what was going on when I was in labor.	0.937
Q23	My birth partner/husband couldn't have supported me any better.	0.920
Q29	I could have had a bit more help from my birth partner/husband. (−)	0.511

Table 2. *Cont.*

	6 Holding Baby (Cronbach's Alpha = 0.675, % Variance Explained = 6.74)	
Q18	I needed to hold my baby a little earlier than I did. (−)	0.842
Q10	After my baby was born, I was not given him/her quite as soon as I wanted. (−)	0.786
Q3	I got to see my baby at exactly the right time after she/he was born.	0.577
	7 Knowledge of Women about Professionals During Childbirth Assistance (Cronbach's Alpha = 0.797, % Variance Explained = 5.21)	
Q24	I knew the carer(s) present at the birth of my baby.	0.855
Q5	At the start of my labor I knew my carers very well.	0.844
	8 Environment (Cronbach's Alpha = 0.711, % Variance Explained = 4.97)	
Q4	My birth room was a little impersonal and clinical. (−)	0.810
Q14	The area where I gave birth was very pleasant and relaxing.	0.764
	9 Control (Cronbach's Alpha = 0.436, % Variance Explained = 4.76)	
Q21	Everyone seemed to tell me what to do in labor. (−)	0.753
Q30	Labor was just a matter of doing what I was told by my carers. (−)	0.729
Q25	I am satisfied with just one or two things about the labor care that I received. (−)	0.460

* Correlation coefficients of each item with its subscale.

Item 12 (the way my labor care was provided could not have been improved) showed a saturation of 0.55 and the generalization of its statement could be confusing. Item 25 did not saturate well with respect to the other two (0.46), and due to its statement (I am satisfied with just one or two things about labor care that I received), it did not seem to correspond to the being analyzed. Finally, both items were removed from the questionnaire.

Later on, a second analysis was performed with the remaining items, obtaining a sample adequacy of KMO of 0.86 and maintaining statistical significance in the Bartlett sphericity test ($p < 0.001$). This time, only items 3 and 31 showed communalities of less than 0.50, (0.44 and 0.47, respectively), although we decided to keep them in the model. The number of factors extracted by the principal component method with eigenvalues greater than 1 was also nine, which explained 70.0% of the global variance.

Table 3 represents the saturation level in the rotated components, the corresponding Cronbach's alphas, and the variance explained by each factor.

In Table 3, the dimension of pain again appears to be merged. Item 31 (I didn't need a lot of pain relief after the birth) has a saturation level close to 0.50 and continues to remain on the scale, although it is poorly associated with the other items, because its contents belong to this dimension.

Table 4 shows the Cronbach's alpha values from the validation carried out in this study as well as those from the English version, the French adaptation, and the puerperal period in Madrid (Spain). It can be seen that the Cronbach's alpha values of this study are in the range of previous studies or, in some cases, even higher.

3.2. Level of Satisfaction in the Different Versions

Table 5 shows the mean scores in each of the dimensions for the different versions. It can be seen that the three best valued dimensions in the four versions were professional support, support of the husband, and first contact with the newborn.

Table 3. Analysis of each dimension and items on the scale after the removal of items Q12 and Q25.

1 Professional Support (Cronbach's Alpha = 0.869, % Variance Explained = 13.403)		Coefficient *
Q19	During labor there was always a carer to explain things so that I could understand.	0.836
Q7	All my labor carers were very supportive.	0.830
Q13	Carers always listened very, very carefully to everything that I had to say.	0.801
Q27	All my carers treated me in the most friendly and courteous manner posible.	0.772
Q32	My carers couldn't have been more helpful.	0.722
2 Expectations (Cronbach's Alpha = 0.861, % Variance Explained = 9.817)		
Q17	The delivery went almost completely as I had hoped that it would.	0.819
Q11	The labor went nearly exactly as I had hoped that it would.	0.808
Q22	My labor was just about the right length.	0.725
Q1	My labor went totally normally.	0.719
3 Pain during and after the Birth (Cronbach's Alpha = 0.749, % Variance Explained = 9.085)		
Q6	I should have been offered something more to relieve the pains I had after my baby was born. (−)	0.717
Q26	More pain relief would have made my labor easier. (−)	0.716
Q16	I was in a fair bit of pain immediately after the birth. (−)	0.682
Q9	I should have been offered something more to relieve my labor pains. (−)	0.660
Q20	I got excellent pain relief in labor.	0.575
Q31	I didn't need a lot of pain relief after the birth.	0.418
4 Home Assessment (Cronbach's Alpha = 0.843, % Variance Explained = 8.026)		
Q15	When I thought that my labor had started, I would have liked a carer to come and see me at home to confirm that I had. (−)	0.912
Q28	Early home assessment of me in labor would have been very helpful. (−)	0.903
Q8	I should have had a home assessment in early labor. (−)	0.762
5 Support from Husband (Cronbach's Alpha = 0.750, % Variance Explained = 7.209)		
Q2	My birth partner/husband helped me to understand what was going on when I was in labor.	0.940
Q23	My birth partner/husband couldn't have supported me any better.	0.927
Q29	I could have had a bit more help from my birth partner/husband. (−)	0.498

Table 3. *Cont.*

6 Holding Baby (Cronbach's Alpha = 0.675, % Variance Explained = 7.042)		
Q18	I needed to hold my baby a little earlier than I did. (−)	0.842
Q10	After my baby was born, I was not given him/her quite as soon as I wanted. (−)	0.784
Q3	I got to see my baby at exactly the right time after she/he was born.	0.579
7 Knowledge of Women about Professionals during Childbirth Assistance (Cronbach's Alpha = 0.797, % Variance Explained = 5.532)		
Q24	I knew the carer(s) present at the birth of my baby.	0.855
Q5	At the start of my labor I knew my carers very well.	0.847
8 Environment (Cronbach's Alpha = 0.711, % Variance Explained = 5.297)		
Q4	My birth room was a little impersonal and clinical. (−)	0.834
Q14	The area where I gave birth was very pleasant and relaxing.	0.771
9 Control (Cronbach's Alpha = 0.481, % Variance Explained = 4.646)		
Q30	Labor was just a matter of doing what I was told by my carers. (−)	0.789
Q21	Everyone seemed to tell me what to do in labor. (−)	0.778

* Correlation coefficients of each item with its subscale.

Table 4. Cronbach's alpha values from the different studies analyzed.

	V. Granada (Spanish)	V. Original (English)	V. Geneva (French)	V. Madrid (Spanish)
	C.'s Alpha	C.'s Alpha	C.'s Alpha	C.'s Alpha
1 Professional support	0.869	0.91	0.84	0.74
2 Expectations	0.861	0.90	0.86	0.80
3 Pain in labor and Pain after the birth	0.749	0.83 & 0.65	0.79 & 0.59	0.68
4 Home assessment	0.843	0.90	0.87	0.83
5 Support from husband	0.750	0.83	0.56	0.61
6 Holding baby	0.675	0.87	0.78	0.51
7 Knowledge of women about their caretakers during the birth	0.797	0.82	0.84	0.36
8 Environment	0.711	0.80	0.67	0.43
9 Control	0.481	0.62	0.53	——
10 General satisfaction	0.421	0.75	0.85	——

Table 5. Average scores in the different versions.

Dimensions	V. Granada (Spanish)	V. Original (English)	V. Geneva (French)	V. Madrid (Spanish)
	Mean ± SD	Mean ± SD	Mean ± SD	Mean ± SD
1 Professional support	83.71 ± 12.4	72.3 ± 20.1	80.9 ± 19.5	91.35 ± 12.9
2 Expectations	64.16 ± 20.8	59.0 ± 27.7	64.2 ± 29.7	60.88 ± 29.8
3 Pain in labor and Pain after delivery	65.59 ± 27.9	60.8 ± 23.5	65.15 ± 27.0	64.98 ± 22.9
4 Home assessment	60.16 ± 17.9	54.3 ± 20.5	64.6 ± 24.7	69.31 ± 30.0
5 Support from husband	80.42 ± 25.5	72.7 ± 21.2	75.2 ± 21.0	90.48 ± 15.9
6 Holding baby	78.09 ± 25.5	74.2 ± 21.8	78.1 ± 25.7	82.61 ± 25.8
7 Knowledge of women about their caretakers during the birth	51.75 ± 23.2	38.8 ± 21.0	38.0 ± 30.2	67.49 ± 27.7
8 Environment	56.74 ± 19.8	61.6 ± 28.2	59.6 ± 27.4	40.18 ± 22.2
9 Control	49.31 ± 21.2	53 ± 23.7	46.5 ± 27.1	——
10 General satisfaction	70.11 ± 18.3	53.1 ± 22.2	66.5 ± 13.5	83.33 ± 22.2

4. Discussion

The response rate was 85.0%, similar to that obtained in the study by Floris et al. [24] in Geneva, Switzerland, and somewhat higher than that obtained by Marín-Morales et al. [25] in Madrid (Spain).

The scale designed in this study showed a high validity and some good psychometric characteristics for measuring childbirth satisfaction in women based on their sociocultural environment.

It can be used in primiparous and/or multiparous women; pregnant women of low, medium, and high risk; puerperal women who have had vaginal births, whether spontaneous or instrumental; and even those who have had unscheduled caesarean sections. However, it is not applicable for women who have had scheduled (elective) caesarean births since, in most cases, these women would not be able to complete some items. In these circumstances, other dimensions not considered in the original scale should be considered.

The obtained percentage of women's satisfaction with the care received during their births by factor analysis was somewhat lower than that shown by the original scale [23]; however, we understand that it is adequate for identifying those aspects that can be improved.

In relation to the first dimension, "Professional support", the psychometric characteristics of our study showed slightly lower values than those of Smith [23], but higher than those achieved by Floris et al. [24] and Marín-Morales et al. [25]. This is the first factor identified in all of these studies and the one that best explains women's satisfaction. It is made up of five items, all of them stated in a positive way, and results similar to those of previous studies were obtained [28–31]. All of them indicate that the kind and correct treatment of professionals and good communication favor the satisfaction of women, especially highlighting the role of the midwife as the professional who provides the most support during the delivery process, describing her as "competent", "inspiring confidence", or "wonderful" [32].

In the second dimension, "Expectation", made up of four questions expressed in a positive way, the parameters obtained are similar to those obtained in the previous dimension. In their research, Melender et al. postulated that if the expectations of the pregnant women are in accordance with their lived experiences during childbirth, their evaluation of childbirth will be satisfactory [33]. Many women look for information on the sensations that they may experience during the labor process and idealize how it should go. In preparation for childbirth sessions the expectations created, information from other mothers, previous experiences, and the signals of their own bodies influence the elaboration of a mental image of delivery [34]. In other cases, despite experiences of severe pain or complications during a previous delivery that are different from their expectations, women feel motivated and encouraged to have another child due to having received good support from the midwife during the process [32].

The third dimension, "Pain during and after childbirth", integrates two dimensions of the original version, which also appears in the French version. Our results coincide with a study carried out on Spanish women from Madrid (Spain). In both cases, the items were related to pain. In this new dimension, there are 2 items stated positively and 4 negatively, which coincides with the original and French versions. The reliability of the original version is superior to that of the other studies, while the results of this investigation are superior to the version carried out with women from the center of Madrid (Spain) and partially superior to the French version.

Item 31 "I didn't need a lot of pain relief after the birth" is the only one of these new dimensions that has an adjusted value in our study. This is probably explained by the fact that most of the women in our sample received epidural analgesia for childbirth, and the effects of this analgesia remain in the immediate postpartum period [35–37]. This fact means that puerperal women do not need many analgesics in this period.

In the early puerperium period (from 3 h after birth to 10–15 days later), women may experience pain due to uterine involution, the presence of hemorrhoids, and even breast pain [38,39]. There may also be perineal pain following or without an episiotomy [39,40], but in most cases, pain relief is necessary. On the other hand, this item is not relevant for those women who wanted a natural childbirth and therefore would consider it unnecessary. Therefore, it could be a potentially upgradeable question.

The fourth dimension, "Home assessment", presents a very similar Cronbach's alpha value in the four versions, the highest of which corresponds to the original version. In all versions, with the exception of the version from Madrid (Spain), this dimension consists of three items written in negative form. The coefficients obtained in our study were slightly lower than those obtained by Smith [23], but higher than those found by Marín-Morales et al. [25].

In Spain, as in most countries of the Organization for Economic Cooperation and Development (OCEDE) [41], except in the Netherlands and the United Kingdom, there is no culture of maternal care at home [42,43]. In the Spanish Health System, both public and private, most pregnant women go to hospitals when they have their first contractions to be evaluated at the beginning of labor and births are completed in them. Anecdotally, births at home are attended by professionals from the private sector.

In relation to the fifth dimension, "Support from husband", the psychometric parameters obtained are similar to those in the original version. Item 29 "I would have preferred to have more support from my partner/husband" is negative, and its value is the lowest on our scale. This factor identified in the analysis is a component of satisfaction that, in our environment, is favored by legislation [44]. Along the line of humanization of perinatal care, in 1995, in the Autonomous Community of Andalusia, the right of pregnant women to be accompanied by a person they trust during the prepartum, delivery, and postpartum periods was legislated [45]. This legal support for the figure of the companion has been highly valued in numerous studies [30,32,46].

The sixth dimension, "Holding baby", consists of two negative and one positive item, and the psychometric characteristics found in our study have somewhat higher values than those obtained in Madrid (Spain) [25]. Early contact with the newborn, in addition to being an indicator of women's satisfaction, favors the establishment of an emotional bond between mother and child [47]. The search for greater prominence, that is granted to women in our Public Health System through the implementation of the Childbirth and Birth Plan of the Autonomous Community of Andalusia, provides mothers with the possibility of expressing their preferences during birth as the right to have their son or daughter by their side during the hospital stay [46,48]. Whenever the state of the newborn and the mother allow it, skin-to-skin contact between the two should be promoted, since it provides benefits to both the mother and the newborn: maintenance of a good body temperature, an increase in blood glucose levels, and helps to maintain breastfeeding and weight [49–52].

The seventh dimension, "Knowledge of women by professionals during birth assistance", showed an adequate Cronbach's alpha value. The title of this dimension is formulated in the same sense as the French version and both differently from the original version. The content of the questions is focused on the continuity raised by Australian authors [53,54] and adapted to the title formulated in our research.

The values of the parameters obtained in the eighth dimension, "Environment", are adequate and most of them are superior to those collected the study of the Center of Spain [25]. This dimension consists of two questions and one of them is negative. An environment that facilitates intimacy, silence, environmental warmth, and the absence of medicalized furniture contributes to the satisfaction of women [55,56]. However, in the qualitative study carried out by Jenkins et al. in the state of New South Wales (Australia), most women did not highlight the environment as one of the three most important aspects in their care [53].

"Control" was the ninth dimension and showed a great relationship with the satisfaction of women and their experience in childbirth. Various authors have pointed out that perceived control over the situation increases satisfaction; this is a dimension that has been widely incorporated in Anglo-Saxon research [53,55,57–60]. However, in our study, similarly to that of Smith and Floris et al., this dimension was the last to be shown as a factor and also the one that explained the smallest percentage of satisfaction [23,24]. Probably for this reason, this dimension was excluded in the version carried out by Marín-Morales et al. with women from Centre of Spain [25]. This result shows that, despite the fact that, in recent years, it has been an incentive by the State and Health Institutions for women and families to take control and responsibility over their health through the inclusion of the Autonomy Law of the patient (2002), the change in the healthcare model based on paternalism has prevailed for so many years in our healthcare system that it is not yet something that the population considers to be of special importance [61].

The two questions about General Satisfaction, number 12 (the care during the delivery process could not have been better) and number 25, (I am satisfied with only one or two things about the care I received during the delivery process) were not analyzed as in the original study. Both questions, especially the last one, are so general that they suggest a certain ambiguity, since it is difficult to assess satisfaction with one or two of these aspects. However, the resulting mean scores are mostly somewhat higher than those obtained by Smith [23] in the original version, although they are similar to those obtained by Marín-Morales et al. [25] in their study carried out in Madrid (Spain).

We understand that, at present, childbirth satisfaction questionnaires should incorporate the control dimension, since, in the current context, the empowerment of women and decision-making during childbirth is a priority in healthcare [62].

4.1. Limitations

Given the good understanding of the scale items and the good adaptation of the scale, it would have been interesting to have expanded the sample to other Spanish-speaking areas.

4.2. Recommendations for Future Research

It would be advisable for future research to merge dimensions 6 "pain relief during childbirth" and 7 "pain relief immediately after childbirth" into one dimension, as shown in the original version. Both dimensions are related to pain relief.

In this dimension, there is item 31, which asks about the need to relieve pain immediately after delivery. We believe that this question should be changed, as many analgesics are not necessary immediately after delivery, only if in the postpartum period.

5. Conclusions

In the resulting model from the WOMBLSQ4 questionnaire, the nine dimensions measure the level of satisfaction with childbirth care in puerperal women. The resulting average scores are mostly somewhat higher than those obtained in the original version, and close to those achieved in the study carried out in Madrid (Spain). In clinical practice, this scale is relevant for measuring satisfaction levels in Spanish-speaking women.

Author Contributions: Conceptualization, M.D.P.-C. and A.M.-S.; methodology, M.D.P.-C., A.M.-S., M.Á.P.-M., J.d.D.L.d.C. and I.G.-G.; software, J.d.D.L.d.C. and I.G.-G.; investigation, M.D.P.-C.; data curation, M.D.P.-C., E.M.-G., J.d.D.L.d.C. and I.G.-G.; writing—original draft preparation, M.D.P.-C., A.M.-S., M.Á.P.-M., M.G.-L., R.F.-C., I.G.-G.; writing—review and editing, M.D.P.-C., A.M.-S., M.ÁP.-M., E.M.-G., M.G.-L., R.F.-C. and I.G.-G; supervision, M.D.P.-C and I.G.-G. All authors have read and agreed to the published version of the manuscript.

Funding: This research received no external funding.

Acknowledgments: We gratefully acknowledge the women who participated in this study.

Conflicts of Interest: The authors declare no conflict of interest.

References

1. Peristat, E. Better Statistics for Better Health for Pregnant Women and Their Babies. Available online: https://www.europeristat.com/ (accessed on 22 April 2020).
2. Goodman, P.; Mackey, M.C.; Tavakoli, A.S. Factors related to childbirth satisfaction. *J. Adv. Nurs.* **2004**, *46*, 212–219. [CrossRef] [PubMed]
3. Redshaw, M.; Martin, C.R.; Savage-McGlynn, E.; Harrison, S. Women's experiences of maternity care in England: Preliminary development of a standard measure. *BMC Pregnancy Childbirth* **2019**, *19*, 167. [CrossRef] [PubMed]
4. Aguilar-Cordero, M.J.; Sáez-Martín, I.; Menor-Rodríguez, M.J.; Mur-Villar, N.; Expósito- Ruiz, M.; Hervás-Pérez, A.; González Mendoza, J.L. Valoración del nivel de satisfacción en un grupo de mujeres de Granada sobre atención al parto, acompañamiento y duración de la lactancia. *Nutr. Hosp.* **2013**, *28*, 920–926. [PubMed]
5. Ip, S.; Chung, M.; Raman, G.; Chew, P.; Magula, N.; DeVine, D.; Trikalinos, T.; Lau, J. Breastfeeding and maternal and infant health outcomes in developed countries. *Evid. Rep. Technol. Assess.* **2007**, *153*, 1–186.
6. Deoni, S.C.L.; Dean, D.C.; Piryatinsky, I.; O'Muircheartaigh, J.; Waskiewicz, N.; Lehman, K.; Han, M.; Dirks, H. Breastfeeding and early white matter development: A cross-sectional study. *Neuroimage* **2013**, *82*, 77–86. [CrossRef]
7. Reynolds, J.L. Post-traumatic stress disorder after childbirth: The phenomenon of traumatic birth. *Can. Med. Assoc. J.* **1997**, *156*, 831–835.
8. Beck, C.T. Post-Traumatic Stress Disorder Due to Childbirth: The Aftermath. *Nurs. Res.* **2004**, *53*, 216–224. [CrossRef]

9.   Modarres, M.; Afrasiabi, S.; Rahnama, P.; Montazeri, A. Prevalence and risk factors of childbirth-related post-traumatic stress symptoms. *BMC Pregnancy Childbirth* **2012**, *12*, 88. [CrossRef]

10. Hosseini, V.M.; Nazarzadeh, M.; Jahanfar, S. Interventions for reducing fear of childbirth: A systematic review and meta-analysis of clinical trials. *Women Birth* **2018**, *31*, 254–262. [CrossRef]

11. Ternström, E.; Hildingsson, I.; Haines, H.; Rubertsson, C. Higher prevalence of childbirth related fear in foreign born pregnant women—Findings from a community sample in Sweden. *Midwifery* **2015**, *31*, 445–450. [CrossRef]

12. Nerum, H.; Halvorsen, L.; Sørlie, T.; Øian, P. Maternal Request for Cesarean Section due to Fear of Birth: Can It Be Changed Through Crisis-Oriented Counseling? *Birth* **2006**, *33*, 221–228. [CrossRef] [PubMed]

13. Caminal, J. La medida de la satisfacción: Un instrumento de participación de la población en la mejora de la calidad de los servicios sanitarios. *Rev Calid. Asist.* **2001**, *16*, 276–279. [CrossRef]

14. Hodnett, E.D. Pain and women's satisfaction with the experience of childbirth: A systematic review. *Am. J. Obstet. Gynecol.* **2002**, *186*, S160–S172. [PubMed]

15. Perriman, N.; Davis, D. Measuring Maternal Satisfaction with Maternity Care: A Systematic Integrative Review: What Is the Most Appropriate, Reliable and Valid Tool That Can Be Used to Measure Maternal Satisfaction with Continuity of Maternity Care? *Women Birth* **2016**, *29*, 293–299. [CrossRef] [PubMed]

16. Ängeby, K.; Sandin-Bojö, A.K.; Persenius, M.; Wilde-Larsson, B. Early Labor Experience Questionnaire: Psychometric Testing and Women's Experiences in a Swedish Setting. *Midwifery* **2018**, *64*, 77–84. [CrossRef] [PubMed]

17. Oweis, A. Jordanian mother's report of their childbirth experience: Findings from a questionnaire survey. *Int. J. Nurs. Pract.* **2009**, *15*, 525–533. [CrossRef]

18. Dencker, A.; Taft, C.; Bergqvist, L.; Lilja, H.; Berg, M. Childbirth experience questionnaire (CEQ): Development and evaluation of a multidimensional instrument. *BMC Pregnancy Childbirth* **2010**, *10*, 81. [CrossRef]

19. Christiaens, W.; Bracke, P. Place of birth and satisfaction with childbirth in Belgium and the Netherlands. *Midwifery* **2009**, *25*, 11–19. [CrossRef]

20. Mas-Ponsa, R.; Barona-Vilara, C.; Carreguí-Vilar, S.; Ibáñez-Gil, N.; Margaix-Fontestad, L.; Escribà-Agüir, V. Satisfacción de las mujeres con la experiencia del parto: Validación de la Mackey Satisfaction Childbirth Rating Scale. *Gac. Sanit.* **2012**, *26*, 236–242. [CrossRef]

21. Janssen, P.A.; Dennis, C.L.; Reime, B. Development and Psychometric Testing of the Care in Obstetrics: Measure for Testing Satisfaction (COMFORTS) Scale. *Res. Nurs. Health* **2006**, *29*, 51–60. [CrossRef]

22. Antoniotti, S.; Baumstarck-Barrau, K.; Siméoni, M.C.; Sapin, C.; Labarère, J.; Gerbaud, L.; Boyer, L.; Colin, C.; François, P.; Auquier, P. Validation of a French hospitalized patients' satisfaction questionnaire: The QSH-45. *Int. J. Qual. Health Care* **2009**, *21*, 243–252. [CrossRef] [PubMed]

23. Smith, L.F. Development of a multidimensional labor satisfaction questionnaire: Dimensions, validity, and internal reliability. *Qual. Health Care* **2001**, *10*, 17–22. [CrossRef] [PubMed]

24. Floris, L.; Mermillod, B.; Chastonay, P. Translation and validation in French of a multidimensional scale to evaluate the degree of satisfaction during childbirth. *Rev. Epidemiol. Sante Publique* **2010**, *58*, 13–22. [CrossRef] [PubMed]

25. Marín-Morales, D.; Carmona-Monge, F.J.; Peñacoba-Puente, C.; Olmos Albacete, R.; Toro Molina, S. Factor structure, validity, and reliability of the Spanish version of the Women's Views of Birth Labor Satisfaction Questionnaire. *Midwifery* **2013**, *29*, 1339–1345. [CrossRef] [PubMed]

26. Junta de Andalucía. Proceso Asistencia integrado al embarazo parto y puerperio. In *Consejería de Salud*; Junta de Andalucía: Sevilla, Spain, 2014.

27. Ministerio de Sanidad, Servicios Sociales e Igualdad. *Guía de Práctica Clínica de Atención en el Embarazo y Puerperio*; Ministerio de Sanidad, Servicios Sociales e Igualdad: Madrid, Spain, 2014.

28. Díaz-Sáez, J.; Catalán-Matamoros, D.; Fernández-Martínez, M.M.; Granados-Gámez, G. La comunicación y la satisfacción de las primíparas en un servicio público de salud. *Gac. Sanit.* **2011**, *25*, 483–489. [CrossRef]

29. CY Chan, Z.; Wong, K.S.; Lam, W.M.; Wong, K.Y.; Kwok, Y.C. An exploration of postpartum women's perspective on desired obstetric nursing qualities. *J. Clin. Nurs.* **2014**, *23*, 103–112. [CrossRef]

30. Fernández, M.A.; Fernández, R.; Pavón, I.; López, L. La calidad percibida por el usuario, su relación con la información y la presencia de acompañante en una unidad de paritorios. *Matronas Prof.* **2003**, *4*, 29–34.

31. Van Stenus, C.M.B.; Boere-Boonekamp, M.M.; Kerkhof, E.F.G.M.; Ariana, N. Client Satisfaction and Transfers Across Care Levels of Women With Uncomplicated Pregnancies at the Onset of Labor. *Midwifery* **2017**, *48*, 11–17. [CrossRef]

32. Rilby, L.; Jansson, S.; Lindblom, B.; Martensson, L.B.A. Qualitative Study of Women's Feelings about Future Childbirth: Dread and Delight. *J. Midwifery Women's Health* **2012**, *57*, 120–125. [CrossRef]

33. Melender, H.L. What Constitutes a Good Childbirth? A Qualitative Study of Pregnant Finnish Women. *J. Midwifery Women's Health* **2006**, *51*, 331–339. [CrossRef]

34. Beebe, K.R.; Humphreys, J. Expectations, Perceptions, and Management of Labor in Nulliparas Prior to Hospitalization. *J. Midwifery Women's Health* **2006**, *51*, 347–353. [CrossRef] [PubMed]

35. Martínez, G.J.M.; Delgado, R.M. Nivel de dolor y elección de analgesia en el parto determinada por la realización de educación maternal. *Rev. Chil. Obstet. Ginecol.* **2013**, *78*, 293–297. [CrossRef]

36. Fernández-Guisasola, J.; Rodríguez Caravaca, G.; Serrano Rodríguez, M.L.; Delgado González, T.; García del Valle, S.; Gómez-Arnau, J.I. Analgesia epidural obstétrica: Relación con diversas variables obstétricas y con la evolución del parto. *Rev. Esp. Anestesiol. Reanim.* **2004**, *51*, 121–127. [PubMed]

37. Biedma Velázquez, L.; García de Diego, J.M.; Serrano del Rosal, R. Análisis de la no elección de la analgesia epidural durante el trabajo de parto en las mujeres andaluzas: "la buena sufridora". *Rev. Soc. Esp. Dolor.* **2010**, *17*, 3–15. [CrossRef]

38. Aviño, J. Tema 10. Puerperio fisiológico y lactancia. In *Reproducción y Ginecología Básicas*, 1st ed.; Musoles, F.B., Pellicer, A., Obstetricia, A., Eds.; Panamericana: Madrid, Spain, 2008; pp. 197–207.

39. Eshkevari, L.; Trout, K.K.; Damore, J. Management of Postpartum Pain. *J. Midwifery Women's Health* **2013**, *58*, 622–631. [CrossRef]

40. Karbanova, J.; Rusavy, Z.; Betincova, L.; Jansova, M.; Necesalova, P.; Kalis, V. Clinical evaluation of early postpartum pain and healing outcomes after mediolateral versus lateral episiotomy. *Int. J. Gynecol. Obstet.* **2014**, *127*, 152–156. [CrossRef]

41. Escuriet, R.; Pueyo, M.; Biescas, H.; Espiga, I.; Colls, C.; Sanders, M.; Kinnear, A. La atención al parto en diferentes países de la Organización para la Cooperación y el Desarrollo Económico (OCDE). *Matronas Prof.* **2014**, *15*, 62–70.

42. Van der Kooy, J.; Poeran, J.; de Graaf, J.P.; Birnie, E.; Denktasş, S.; Steegers, E.A.; Bonsel, G.J. Planned home compared with planned hospital births in the Netherlands: Intrapartum and early neonatal death in low-risk pregnancies. *Obstet. Gynecol.* **2011**, *118*, 1037–1046. [CrossRef]

43. Walker, J.J. Planned home birth. *Best Pract. Res.Clin. Obstet. Gynaecol.* **2017**, *43*, 76–86. [CrossRef]

44. Ministerio de Sanidad y Consumo. *Low Risk Labor Care Strategy in the National Health System*; Ministerio de Sanidad y Consumo: Madrid, Spain, 2008.

45. Junta de Andalucía. *Decreto 101/1995, de 18 de Abril, por el que se Determinan los Derechos de los Padres y de los Niños Durante el Proceso de Nacimiento. BOJA 72*; Junta de Andalucía: Sevilla, Spain, 1995.

46. Porrett, L.; Barkla, S.; Knights, J.; de Costa, C.; Harmen, S. An exploration of the perceptions of male partners involved in the birthing experience at a regional Australian hospital. *J. Midwifery Women's Health* **2013**, *58*, 92–97. [CrossRef]

47. Souza, L.H.; Soler, Z.A.S.G.; Santos, M.L.S.G.; Sasaki, N.S.G.M.S. Puerperae bonding with their children and labor experiences. *Invest. Educ. Enferm.* **2017**, *35*, 364–370. [CrossRef] [PubMed]

48. Junta de Andalucía Consejería de Salud. Plan de Parto y Nacimiento Sistema Sanitario Público de Andalucía. 2017. Available online: https://www.juntadeandalucia.es/export/drupaljda/PLANPARTOS2017JUNIO_completo.pdf (accessed on 22 April 2020).

49. De Alba-Romero, C.; Camaño-Gutiérrez, I.; López-Hernández, P.; Castro-Fernández, J.; Barbero-Casado, P.; Salcedo-Vázquez, M.L.; Sánchez-López, D.; Cantero-Arribas, P.; Moral-Pumarega, M.T.; Pallás-Alonso, C.R. Postcesarean Section Skin-to-Skin Contact of Mother and Child. *J. Hum. Lact.* **2014**, *30*, 283–286. [CrossRef] [PubMed]

50. Srivastava, S.; Gupta, A.; Bhatnagar, A.; Dutta, S. Effect of very early skin to skin contact on success at breastfeeding and preventing early hypothermia in neonates. *Indian J. Public Health* **2014**, *58*, 22–26. [CrossRef] [PubMed]

51. Raies, C.L.; Doren, F.M.; Torres, C.U. Efectos del Contacto Piel Con Piel del Recién Nacido Con su Madre. *Index Enferm.* **2012**. Available online: http://www.index-f.com/index-enfermeria/v21n4/7863.php (accessed on 22 April 2020).

52. Moore, E.R.; Anderson, G.C.; Bergman, N.; Dowswell, T. Early skin-to-skin contact for mothers and their healthy new-born infants (Review). *Cochrane Libr.* **2012**, *5*, 1–107.

53. Jenkins, M.G.; Ford, J.B.; Morris, J.M.; Roberts, C.L. Women's expectations and experiences of maternity care in NSW—What women highlight as most important. *Women Birth* **2014**, *27*, 214–219. [CrossRef]

54. Jenkins, M.G.; Ford, J.B.; Todd, A.L.; Forsyth, R.; Morris, J.M.; Roberts, C.L. Women's views about maternity care: How do women conceptualise the process of continuity? *Midwifery* **2015**, *31*, 25–30. [CrossRef]

55. Guittier, M.J.; Cedraschi, C.; Jamei, N.; Boulvain, M.; Guillemin, F. Impact of mode of delivery on the birth experience in first-time mothers: A qualitative study. *BMC Pregnancy Childbirth* **2014**, *14*, 254. [CrossRef]

56. Hodnett, E.D.; Stremler, R.; Weston, J.A.; McKeever, P. Re-Conceptualizing the Hospital Labor Room: The PLACE (Pregnant and Laboring in an Ambient Clinical Environment). *Pilot Trial. Birth* **2009**, *36*, 159–166. [CrossRef]

57. Lankin, P.; Begley, C.M.; Devane, D. Women's experiences of labor and birth: An evolutionary concept analysis. *Midwifery* **2009**, *25*, 49–59.

58. Lally, J.E.; Thomson, R.G.; MacPhail, S.; Exley, C. Pain relief in labor: A qualitative study to determine how to support women to make decisions about pain relief in labor. *BMC Pregnancy Childbirth* **2014**, *8*, 14–16.

59. Snowden, A.; Martin, C.; Jomeen, J.; Hollins Martin, C. Concurrent analysis of choice and control in childbirth. *BMC Pregnancy Childbirth* **2011**, *1*, 11–40. [CrossRef] [PubMed]

60. Namey, E.E.; Lyerly, A.D. The meaning of "control" for childbearing women in the US. *Soc. Sci. Med.* **2010**, *71*, 769–776. [CrossRef] [PubMed]

61. Ley 41/2002, de 14 de Noviembre Básica Reguladora de la Autonomía del Paciente y de Derechos y Obligaciones en Materia de Información y Documentación Clínica. BOE núm. 274. Available online: https://www.boe.es/buscar/act.php?id=BOE-A-2002-22188 (accessed on 23 May 2020).

62. Martínez-García, E.; Pozo-Cano, M.D.; Martín-Salvador, A.; Pérez-Morente, M.A.; Gázquez-López, M.; Medina-Casado, M. Cambio de paradigma en la atención al parto en España ¿Realidad o ficción? *Rev. Paraninfo Digit.* **2019**, *XIII*, e091.

International Journal of
***Environmental Research
and Public Health***

Article

Contribution of Chronic Fatigue to Psychosocial Status and Quality of Life in Spanish Women Diagnosed with Endometriosis

Antonio Mundo-López [1,2], Olga Ocón-Hernández [3,4], Ainhoa P. San-Sebastián [1], Noelia Galiano-Castillo [3,5,6], Olga Rodríguez-Pérez [1], María S. Arroyo-Luque [1], Manuel Arroyo-Morales [3,5,6], Irene Cantarero-Villanueva [3,5,6], Carolina Fernández-Lao [3,5,6,*] and Francisco Artacho-Cordón [1,3,6,7,*]

[1] Department of Radiology and Physical Medicine, University of Granada, E-18016 Granada, Spain; antonio@alarconpsicologos.com (A.M.-L.); ainhoassp@correo.ugr.es (A.P.S.-S.); olgarope@correo.ugr.es (O.R.-P.); marisi_7@hotmail.com (M.S.A.-L.)
[2] Clinic Psychology Center Alarcón (CPCA), E-18004 Granada, Spain
[3] Biohealth Research Institute in Granada (ibs.GRANADA), E-18012 Granada, Spain; ooconh@ugr.es (O.O.-H.); noeliagaliano@ugr.es (N.G.-C.); marroyo@ugr.es (M.A.-M.); irenecantarero@ugr.es (I.C.-V.)
[4] Gynaecology and Obstetrics Unit, 'San Cecilio' University Hospital, E-18016 Granada, Spain
[5] Department of Physiotherapy, University of Granada, E-18016 Granada, Spain
[6] "Cuídate" Support Unit for Oncology Patients (UAPO), Sport and Health University Research Institute (iMUDS), E-18016 Granada, Spain
[7] CIBER Epidemiology and Public Health (CIBERESP), E-28029 Madrid, Spain
* Correspondence: carolinafl@ugr.es (C.F.-L.); fartacho@ugr.es (F.A.-C.)

Received: 15 April 2020; Accepted: 23 May 2020; Published: 28 May 2020

Abstract: Aim: To analyze the levels of chronic fatigue in Spanish women with endometriosis and its relationship with their psychosocial status and quality of life (QoL). Methods: A total of 230 Spanish women with a clinical diagnosis of endometriosis were recruited. Chronic fatigue (Piper Fatigue Scale) and pelvic pain (Numeric Rating Scale) were evaluated. An on-line battery of validated scales was used to assess psychosocial status [Hospital Anxiety and Depression Scale, Scale for Mood Assessment, Pain Catastrophizing Scale, Pittsburgh Sleep Quality Index, Gastrointestinal Quality of Life Index, Female Sexual Function Index and Medical Outcomes Study-Social Support Survey] and QoL [Endometriosis-Health Profile questionnaire-30]. Associations between fatigue and both psychosocial and QoL outcomes were explored through multivariate regression models. Results: One-third and one-half of women showed moderate and severe fatigue, respectively. Fatigue was associated with higher anxiety and depression, poorer sleep quality, poorer sexual functioning, worse gastrointestinal health, higher catastrophizing thoughts, higher anger/hostility scores and lower QoL (p-values < 0.050). Moreover, fatigue and catastrophizing thoughts showed a mediating effect on the association between pelvic pain and QoL. Conclusion: This work reveals the important role of fatigue in the association between pain, psychosocial status, and QoL of Spanish women with endometriosis.

Keywords: chronic fatigue; endometriosis; psychosocial status; quality of life

---

1. Introduction

Endometriosis, characterized by the ectopic development of endometrial-like tissue, is among the most commonly diagnosed benign diseases in women of reproductive age [1]. Diagnostic delay and the fact that diagnosis is often overlooked by primary care doctors make the prevalence of the disease difficult to establish. Nevertheless, prevalence estimates range from 1–2% when considering populations at low risk to 10% when high-risk populations are considered [2]. However, despite the benign nature of this disease, huge direct and indirect costs (raising up to more than $12,000 – $15,000 in some countries) have been evidenced to be associated with endometriosis [3].

Pain in the pelvic region is acknowledged to be the most characteristic symptom of women with endometriosis, which is intensified during the menstruation period (dysmenorrhea) and during the performance of daily activities such as defecation (dyschezia), urination (dysuria) or sexual relationships (dyspareunia) [4,5]. The contribution of pelvic pain (PP) to the psychosocial status and the quality of life (QoL) of women with endometriosis has been extensively addressed [6–10]. Additionally, chronic fatigue, i.e., the perception of physical tiredness and lack of energy distinct from sadness or weakness, is another endometriosis-related symptom, as recently identified in women with endometriosis [11]. However, the role of chronic fatigue on patients' lives has been poorly addressed, although a few qualitative studies have indicated that affected women ascribed social and work impairments to fatigue [12,13]. Contrary, there is consistent evidence of the relevant role of fatigue in different subsets of patients experiencing chronic pain, suggesting that fatigue hinders the completion of routines and significant activities, and therefore, severely reduces QoL [14,15].

Moreover, the relevant contribution of chronic fatigue to the presence of psychosocial impairments in patients with autoimmune diseases [16] or neurological problems [17] has been reported. However, contrary to the well-established relationship between chronic pain, psychosocial problems and QoL in women with endometriosis, there is a scarcity of published studies addressing the contribution of chronic fatigue to the symptomatic burden in women with endometriosis under medical treatment. Thus, the aim of this study was to explore the presence of chronic fatigue in Spanish women diagnosed with endometriosis and its contribution to the psychosocial status and QoL.

2. Material and Methods

2.1. Study Population

A total of 230 women with a clinical diagnosis of endometriosis, from different regions of Spain, were enrolled in this cross-sectional study from January to July 2019. Recruitment of women was carried out in combination with both gynecologists and Spanish associations of endometriosis patients, which advertised the study in their social networks. The inclusion criteria were: to have attended a gynecological visit with any participating gynecologist or to belong to any of the Spanish associations of endometriosis patients; to be diagnosed with endometriosis (either by laparoscopy, magnetic resonance or ultrasound imaging, or based on symptoms); and to have the ability and availability to use an electronic device with internet connection (computer, tablet or mobile phone). The exclusion criteria were: to live in another country or to be a non-Spanish speaker. The survey was designed to consider researchers', gynecologists' and patients' opinion about the most relevant aspects that should be addressed. Interested women received a link for the completion of an on-line questionnaire. Prior to this, women were informed about the nature and objectives of the study, and they were requested to read and sign the informed consent. No personal information was asked in the questionnaire, and data was extracted to create an anonymized database. This study was carried out following the principles of the Declaration of Helsinki and Biomedical Research Law 14/2007 and was approved by the Research Ethics Committee of Granada (1733-N-18).

2.2. Assessment of Self-Reported Intensity of Chronic Fatigue and Pain

Chronic fatigue was assessed through the Spanish version of the Piper Fatigue Scale (PFS) [18,19]. PFS is a validated 22-item tool for self-reported chronic fatigue in breast cancer survivors, but also in patients with gynecological disorders [20] or coronary diseases [21]. It includes four dimensions of subjective fatigue: "behavioral/severity", "affective meaning", "sensory" and "cognitive/mood". Scores range from 0 to 10, with higher scores indicating greater fatigue. It has demonstrated high reliability and validity (Cronbach's alpha 0.86). Participants were divided into three groups according to the clinically significant fatigue criteria: mild (≤ 4.0), moderate (4–7) or fatigued (≥ 7), according to the value obtained for the PFS total score [18,22].

Intensity of self-reported PP during the last week was assessed through a numeric rating scale (NRS). This 11-point Likert scale (0 = no pain; 10 = unbearable pain) is one of the best single-item methods available to estimate the intensity of pain [23–25]. Pain severity was categorized as mild (0–3), moderate (4–7) and severe (8–10), as reported elsewhere [26].

2.3. Psychosocial Assessment

Participants were asked to complete the Hospital Anxiety and Depression Scale (HADS), the Scale for Mood Assessment (EVEA), the Pain Catastrophizing Scale (PCS), the Pittsburgh Sleep Quality Index (PSQI), the Female Sexual Function Index (FSFI), the Gastrointestinal Quality of Life Index (GIQLI) and the Medical Outcomes Study-Social Support Survey (MOS-SSS).

Anxiety and depression were assessed through the Spanish version of the HADS [27], a self-assessment mood scale validated for its use in non-psychiatric hospital outpatients to determine their levels of anxiety and depression [28]. It has two subscales (anxiety and depression), each ranging from 0 to 21, showing adequate reliability (Cronbach's alpha 0.86) [29]. Higher scores on the subscale indicate higher degrees of anxiety and depression [30]. For both subscales, available cut-off scores allowed the identification of non-cases (≤7) mild (8–10), moderate (11–14) and severe cases (15–21) [28,31]. Additionally, the Spanish version of the EVEA scale was partially used to evaluate "anger/hostility" and "happiness" through the corresponding subscales. Item scores are evaluated with Likert scales ranging from 0 to 10, and the values per category are obtained from mean scores. EVEA subscales have shown good reliability (Cronbach's alpha range between 0.88 and 0.93) [32].

Catastrophic thoughts about pain were assessed through the Spanish version of the PCS, a 13-item, validated, self-report instrument with adequate reliability (Cronbach's alpha 0.79) [33]. This measure has a 5-point Likert-style response scale and the scoring range is 0–52, with higher scores indicating higher levels of catastrophic thoughts. Previous studies have shown a cut-off of more than 30 points to be associated with clinical relevance [34].

Sleep quality was assessed using the Spanish validated version of the PSQI [35]. The PSQI is a 19-item, validated, self-report scale used to measure quality and patterns of sleep, with adequate reliability (Cronbach's alpha 0.87). Scores range from 0 to 21, with higher scores representing poorer sleep quality [36]. It has been proposed that a total score ≤5 indicates good sleep quality while a total score > 5 indicates poor sleep quality [36].

Sexual function was assessed through the Spanish version of the FSFI [37]. This is a 19-item questionnaire, validated, multidimensional self-report instrument for assessing the major aspects of female sexual dysfunction and sexual satisfaction [37,38]. The FSFI score ranges from 0 to 36. Higher scores represent better sexual function, considering that patients with a FSFI total score below 26 are categorized as sexual dysfunctional, whereas those scoring at or above this cut-off score are categorized as sexually functional [39]. Adequate reliability has been reported (Cronbach's alpha >0.70 for all domains).

Digestive complaints were assessed through the Spanish version of the GIQLI [40], a self-administered 36-item questionnaire concerning digestive symptoms, physical status, emotions, social dysfunction and effects of medical treatment. Each item scores from 0 to 4 with the total score ranging from 0 to 144, higher scores representing better quality of life. GIQLI also measures physical well-being, mental well-being, digestion and defecation [41].

The Spanish version of the MOS-SSS scale was used to assess the extent to which the person has the support of others to face stressful situations [42]. It is comprised of 19 items with a 5-point Likert-style response, with higher scores representing better social support. This measure has shown good psychometric quality in different studies using diverse populations and clinical scenarios (Cronbach's alpha 0.94) [43].

2.4. Quality of Life

The Spanish version of the validated Endometriosis Health Profile-30 (EHP-30) questionnaire was used for the assessment of QoL in participating women [44]. This 30-item scale has five subscale scores: pain, control and powerlessness, social support, emotional well-being and self-image. Each subscale is standardized on a scale of 0–100, where 0 indicates the best health status and 100 the worst health status. Scale scores for each scale are calculated from the total of the raw scores of each item in the scale divided by the maximum possible raw score of all the items in the scale, multiplied by 100. This instrument has shown good internal consistency reliability, with Cronbach's alpha >0.88 for all subscales.

2.5. Statistical Analysis

The sociodemographic and gynecological characteristics of participants and scores for PP and chronic fatigue were expressed as geometric means (GMs) with geometric standard deviation (GSD), or as percentages, depending on the continuous or categorical nature of the variable. Scores for QoL, i.e., psychosocial outcomes, were expressed as GM with GSD, as minimum and maximum values, and as percentiles (25, 50, and 75). When clinical cut offs were available, variables were categorized and expressed as percentages.

To improve normality of the data, psychosocial outcomes were log-transformed and, therefore, β coefficients are also presented as exp(β). Associations between fatigue severity (mild/moderate/severe), psychosocial outcomes and QoL were assessed by using linear regression models adjusted for sociodemographic and gynecological characteristics, including age, schooling, civil status, severity of premenstrual syndrome (none, mild, moderate or severe), type of diagnosis, time since diagnosis and number of surgeries. Additional models adjusted for severity of last week PP are also presented. Moreover, the mediation effect of fatigue and pain catastrophizing thoughts on the relationship between last week PP intensity and QoL was assessed through the macro PROCESS for Statistical Package for the Social Sciences (SPSS) [45], and mediating effects were considered significant when zero was not located within the confidence intervals.

The statistical significance level was set at $p = 0.05$. Analyses were performed using SPSS v23.0 statistical software (IBM, Chicago, IL, USA), while figures were designed with Graphad Prism 5.0 software (San Diego, CA, USA). The post-hoc analysis to estimate the power (1-β) of the statistical analysis was conducted using G*Power 3.1.9.7 statistical software (Düsseldorf University, Düsseldorf, Germany). For the main analysis between chronic fatigue and QoL, it revealed that, for an R^2 of 0.28 assuming an α-error of 0.05, the power was >0.99.

3. Results

A total of 241 women were interested in the study. However, 11 (4.6%) women did not meet inclusion/exclusion criteria. Finally, 230 women agreed to participate. Baseline characteristics of the study population are summarized in Table 1. The mean (±SD) age of the study population was 36.7 ± 5.2 years, the majority of them hold a university degree (53.9%), are currently working (64.3%) and declared the presence of premenstrual syndrome at any level of severity (56.6%). A total of 155 out of 230 women had a laparoscopic confirmation of the presence of endometriosis lesions at the time of this survey, while in 62 (27.0%) the diagnosis was based on magnetic resonance imaging (MRI) and/or ultrasound (US) imaging techniques. Only 13 (5.7%) were diagnosed based on symptoms but not confirmed by MRI and/or US imaging. Finally, the mean time since diagnosis was 5.0 ± 5.3 years, and 68 (29.6%) had undergone at least two endometriosis surgeries.

3.1. Intensity of Chronic Fatigue and Pain in Spanish Women Diagnosed with Endometriosis

Self-reported severity of chronic fatigue and last week PP are summarized in Table 2. GM (±GSD) intensity of chronic fatigue was 5.9 ± 1.7 points, with almost half of the participating women reporting severe fatigue. Concerning last week PP intensity, GM (±GSD) was 5.0 ± 1.9. A total of 46.3% and 27.1% of the entire study population showed moderate and severe PP during the last week. Using multivariate linear regression modelling, a positive association was found between intensities of both chronic fatigue and last week PP scores after adjustment for potential confounders (Supplementary Table S1).

Table 1. Baseline characteristics of the study population (N = 230).

	N (%)		N (%)
Sociodemographic characteristics		*Gynecological characteristics*	
Age (years)*	36.4 ± 1.2	**N° children**	
Schooling		*None*	162 (70.4)
Primary/secondary studies	43 (18.7)	*1*	40 (17.4)
Vocational training	63 (27.4)	*2*	23 (10.0)
University studies	124 (53.9)	*3*	5 (2.2)
Civil status		**PMS severity**	
Single/Divorced	112 (48.7)	*None*	100 (43.5)
Married	118 (51.3)	*Mild*	14 (6.1)
Working outside home		*Moderate*	77 (33.5)
No	26 (11.3)	*Severe*	39 (17.0)
No, sick leave	30 (13.0)	**Endometriosis diagnosis**	
No, loss due to endometriosis	26 (11.3)	*Laparoscopy*	155 (67.4)
Yes	148 (64.3)	*MRI and/or US*	62 (27.0)
		Based on symptoms	13 (5.7)
		Time since endometriosis diagnosis (years) *	2.4 ± 4.6
		N° surgeries	
		None	75 (32.6)
		1 surgery	87 (37.8)
		2 surgeries	40 (17.4)
		3 or more surgeries	28 (12.2)

* Geometric mean ± geometric standard deviation; PMS: premenstrual symptoms; MRI: magnetic resonance imaging; US: ultrasound.

Table 2. Intensity of chronic fatigue and pelvic pain in women with endometriosis.

	N
Chronic fatigue	5.9 ± 1.7 *
Mild (<4)	34 (14.8)
Moderate (4–7)	82 (35.7)
Severe (>7)	114 (49.6)
Last week pelvic pain intensity	5.0 ± 1.9 *
Mild (<4)	64 (26.7)
Moderate (4–7)	111 (46.3)
Severe (>7)	65 (27.1)

* Geometric mean ± geometric standard deviation.

3.2. Psychosocial Impairments and Quality of Life in Spanish Women Diagnosed with Endometriosis

Table 3 summarizes the results from the descriptive analysis of each analyzed psychosocial dimension. Considering available cut-off scores (not shown in tables), anxiety was present in 169 out of 230 (73.4%) of the women, with 58 (25.2%) and 48 (20.9%) showing moderate and severe anxiety, respectively. Similarly, depression was detected in 111 (48.3%) of the women, with 45 (19.6%) and 19 (8.3%) showing moderate and severe depression, respectively. Additionally, pain catastrophizing thoughts were found in 108 (47.0%) of the participants, while poor sleep quality and sexual dysfunction were found in 187 (81.3%) and 174 (75.7%) of the participating women, respectively. Moreover, anger/hostility and happiness dimensions, assessed through the EVEA subscales, showed a GM (±GSD) of 12.2 ± 2.6 and 11.1 ± 2.3 points, respectively. GM (±GSD) MOS-SSS score was 73.7 ± 18.7 points, while for gastrointestinal problems, GM (±GSD) GIQLI score was 65.2 ± 1.4 points. Regarding QoL, GM (±GSD) EHP-30 score was 55.0 ± 1.7 points.

Table 3. Psychosocial status and quality of life in women with endometriosis.

	GM	**GSD**	**Min.**	**Percentiles**			**Max.**
				25	**50**	**75**	
Quality of life							
Total score	55.0	1.7	1.7	47.2	62.5	73.5	96.6
Gastrointestinal quality of life							
Total score	65.2	1.4	11.0	56.8	70.0	82.0	116.0
Sexual function							
Total score	14.1	2.3	2.0	9.8	19.7	25.8	36.0
Mental health							
Anxiety	9.4	1.7	1.0	7.0	10.0	13.3	20.0
Depression	6.6	1.9	0.0	4.0	7.0	11.0	21.0
Pain catastrophizing scale							
Total score	23.4	2.0	0.0	17.0	28.0	39.0	52.0
Sleep quality							
Total score	9.2	1.7	1.0	6.8	10.0	14.0	21.0
Scale for Mood Assessment							
Anger hostility	12.2	2.6	0.0	4.0	13.0	26.0	40.0
Happiness	11.1	2.3	0.0	5.0	13.0	21.0	38.0
Medical Outcomes Study-Social Support Survey							
Total score	70.8	1.4	23.0	59.8	76.5	90.0	95.0

GM: geometric mean; GSD: geometric standard deviation.

3.3. Contribution of Fatigue Intensity to Psychosocial Impairment in Spanish Women

Results from the multivariate analyses assessing associations between self-perceived fatigue severity and psychosocial impairments are depicted in Figure 1. Results from the bivariate and multivariate analyses are summarized in Supplementary Table S2. After adjustment for potential confounders (sociodemographic and gynecological characteristics and intensity of PP during the last week), moderate and severe fatigue was found to be related to anxiety and depression, poorer sleep quality, poorer sexual functioning and less gastrointestinal quality of life in an intensity-dependent manner, while higher PCS and EVEA-anger/hostility scores were associated with severe fatigue. Moreover, multivariate logistic regression analyses that were run in parallel when cut-off points were available showed similar results (data not shown in tables). Sensitivity analyses stratified by endometriosis diagnosis yielded similar results.

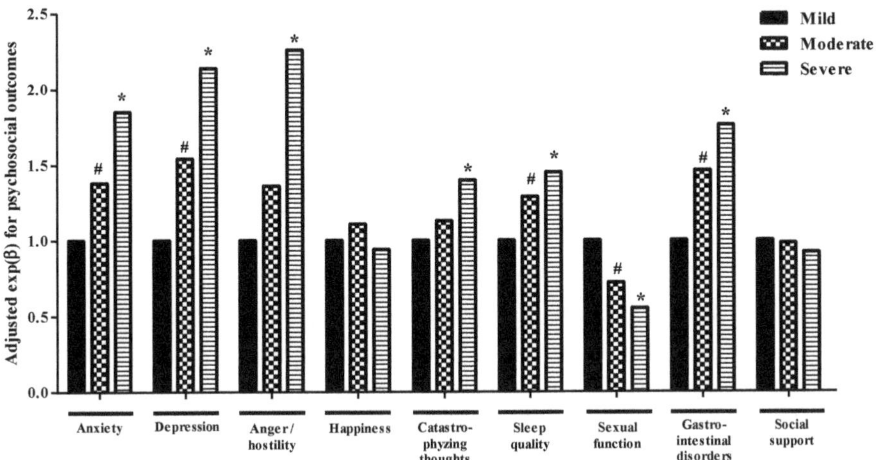

Figure 1. Influence of chronic fatigue on psychosocial status in Spanish women with endometriosis. Results from multivariate linear regression analyses adjusted for age (years), schooling, civil status, number of surgeries, type of diagnosis, time since diagnosis, number of children, premenstrual symptom severity and last week pelvic pain intensity. [#] *p*-value ≤ 0.05 between mild and moderate groups; * *p*-value ≤ 0.05 between mild and severe groups.

3.4. Contribution of Pain, Fatigue and Psychosocial Impairment to Quality of Life in Spanish Women

Results from the multivariate linear regression analyses are summarized in Table 4. Severity of chronic fatigue and last week PP were associated with poorer QoL in an intensity-dependent manner. Moreover, anxiety, depression, anger/hostility and catastrophizing thoughts were associated with poorer QoL. Similarly, poorer gastrointestinal health, sexual function and sleep quality were also related to poorer QoL, although the latter showed a close to statistically significant association with QoL when models were further adjusted for intensity of PP during the last week (*p*-value 0.059).

Mediation effects of fatigue and psychosocial impairments on QoL were also accomplished (Figure 2). All chronic fatigue, gastrointestinal complaints, sexual function, anxiety, depression, anger/hostility, sleep quality and catastrophizing thoughts showed a mediating effect on the association between last week PP and QoL when assessed on an individual level (data not shown). However, when the combined mediating effect was evaluated, only chronic fatigue and catastrophizing thoughts revealed a statistically significant mediating effect on the association between last week PP and QoL (1.128 and 0.863, respectively; *p*-value < 0.05).

4. Discussion

To our knowledge, this study constitutes the first attempt to objectively evaluate levels of endometriosis-related fatigue in a comprehensive population, and to address its relevant contribution to the psychosocial status and QoL in Spanish women diagnosed with endometriosis. Moreover, our results suggest that endometriosis-related fatigue and catastrophizing thoughts also exert a mediating effect on the association between intensity of PP and poorer QoL in affected women, evidencing that these factors also need to be addressed within appropriate treatment approaches in women with endometriosis.

Table 4. Relationship between psychosocial impairments and quality of life in women with endometriosis. Linear regression analyses.

	Bivariate Analysis			Adjusted Analysis *			Adjusted Analysis **		
	Exp (β)	95%CI	p-Value	Exp (β)	95%CI	p-Value	Exp (β)	95%CI	p-Value
Self-perceived pelvic pain severity (NRS)									
NRS = Moderate	1.39	1.20 1.62	<0.001	1.28	1.09 1.50	0.003	-	- -	-
NRS = Severe	1.50	1.26 1.78	<0.001	1.39	1.15 1.68	0.001	-	- -	-
Chronic fatigue (PFS)									
PFS = Moderate	1.41	1.17 1.71	<0.001	1.42	1.18 1.71	<0.001	1.37	1.13 1.65	0.001
PFS = Severe	1.74	1.45 2.08	<0.001	1.67	1.39 2.01	<0.001	1.56	1.28 1.91	<0.001
Gastrointestinal quality of life (GIQLI)									
Total score	1.96	1.66 2.32	<0.001	1.83	1.51 2.23	<0.001	1.83	1.51 2.23	<0.001
Sexual function (FSFI)									
Total score	0.87	0.80 0.94	<0.001	0.89	0.82 0.96	0.004	0.91	0.84 0.99	0.022
Mental health (HADS)									
Anxiety	1.43	1.27 1.61	<0.001	1.33	1.18 1.50	<0.001	1.29	1.15 1.46	<0.001
Depression	1.29	1.18 1.41	<0.001	1.25	1.13 1.37	<0.001	1.21	1.09 1.34	<0.001
Pain catastrophizing scale (PCS)									
Total score	1.52	1.41 1.63	<0.001	1.49	1.38 1.61	<0.001	1.47	1.35 1.59	<0.001
Sleep quality (PSQI)									
Total score	1.30	1.16 1.46	<0.001	1.21	1.06 1.38	0.005	1.14	0.99 1.31	0.059
Scale for Mood Assessment (EVEA)									
Anger hostility	1.13	1.06 1.22	<0.001	1.09	1.02 1.17	0.012	1.08	1.01 1.15	0.031
Happiness	1.02	0.94 1.10	0.674	1.01	0.94 1.09	0.782	1.04	0.96 1.12	0.346
Medical Outcomes Study-Social Support Survey (MOS-SSS)									
Total score	0.73	0.59 0.91	0.005	0.79	0.63 1.00	0.048	0.83	0.66 1.03	0.096

NRS: numeric rating scale; PFS: Piper Fatigue Scale; GIQLI: Gastrointestinal Quality of Life Index; FSFI: Female Sexual Function Scale; HADS: Hospital Anxiety and Depression Scale; PCS: Pain Catastrophizing Scale; PSQI: Pittsburgh Sleep Quality Index; EVEA: Scale for Mood Assessment; MOS-SSS: Medical Outcomes Study-Social Support Survey. * Adjusted for age (yrs), schooling, civil status, number of surgeries, type of diagnosis, time since diagnosis, number of children and PMS severity; ** Additionally adjusted for intensity of pelvic pain during last week.

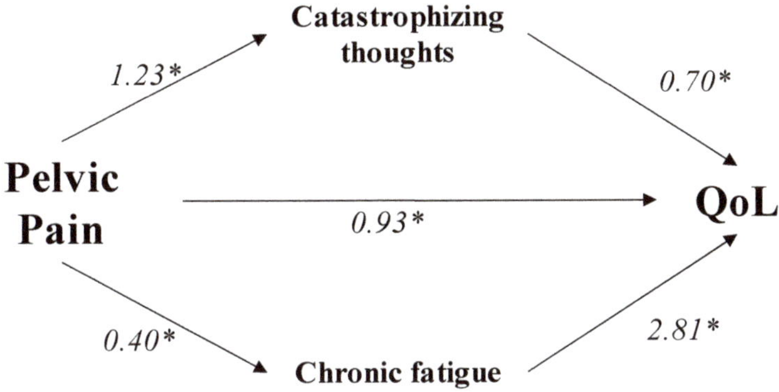

Total effect: 2.925 (2.118-3.732)
Direct effect: 0.934 (0.244-1.624)
Indirect effects
- Catastrophizing thoughts: 0.863 (0.397-1.379)
- Chronic fatigue: 1.128 (0.679-1.626)

Figure 2. Mediating effect of chronic fatigue and catastrophizing thoughts on the association between pelvic pain and quality of life (QoL) in Spanish women with endometriosis. * Mediating effects were considered significant when zero was not located within the confidence intervals. Analyses were conducted with the macro PROCESS for SPSS.

Previous studies have stated that women with endometriosis frequently report the presence of chronic fatigue [46], with authors defending the effect of endometriosis on its generation, independently from other symptoms of the disease [11]. Our study shows that 85.3% of the patients enrolled have moderate to severe fatigue. Our findings are in accordance with previous studies where affected women were asked if they felt fatigue. Hence, a total of 50.7% and 27.1% of women reported frequent and occasional fatigue, respectively [11]. Similarly, Surrey et al. [47] recently reported that 54–74% of affected women with moderate to severe pain reported experiencing fatigue. Although the underlying mechanisms are not still fully elucidated, it has been reported that the endometriotic lesions usually develop a complex and dynamic environment dominated by inflammatory, angiogenic, and endocrine signals [48]. Similarly, Suryawanshi et al. [49] reported that endometriotic lesions generate a specific immune microenvironment similar to a tumor-like inflammatory profile. Thus, in accordance with the positive correlation between inflammatory cytokines and fatigue shown in cancer patients [50], elevated cytokine levels found in endometriosis might be hypothesized to play a role in the development of fatigue symptomology in these affected women.

Regarding the relationship identified in this study between levels of fatigue experienced by women with endometriosis and severity of PP, it was not unexpected, as this association has been previously stated in different populations suffering different chronic conditions such as rheumatic diseases [51] or cancer [52]. In fact, both symptoms have been found to be related to an inflammatory microenvironment. Hence, studies from basic sciences evidenced that changes in immune surveillance and central sensitization were related to the pathophysiology of endometriosis [53]. Interestingly, a misbalance in estrogen levels, as widely reported in patients with endometriosis, may be the first responsible for the generation of an inflammatory microenvironment [48] that ultimately can lead to the development of not only PP [54] but also endometriosis-related fatigue. Moreover, the dysregulation of the hypothalamic–pituitary–adrenal (HPA) axis has been reported to contribute to the development of fatigue in chronic illness [55].

Our sample of patients show high levels of psychosocial impairments such as anxiety, depression, poor sleep quality or sexual dysfunction. In this respect, several studies have stated the association between endometriosis and psychological impairments, with depression and anxiety as the most common disorders related to endometriosis, deeply affecting the QoL of these women [6,56–59]. Interestingly, our results suggest an association of chronic fatigue and psychosocial factors, as reported in another multicenter study by Ramin-Wright et al. [11] that comprised 554 women with endometriosis, where fatigue was associated with insomnia and depression among other factors. In this respect, a previous review stated the influence of chronic fatigue on different inflammatory conditions and the possible association between inflammation, pain and depression [51,60]. Moreover, in addition to poorer sleep quality and depression, our study suggests for the first time that the presence of chronic fatigue is associated with higher levels of anxiety and anger/hostility, as well as poorer sexual function. In accordance with our results, it has been reported that fatigue was associated with poorer sexual functioning in women with chronic conditions such as breast cancer [61] or multiple sclerosis [62]. In this regard, fitness level, crucially related to the presence of chronic fatigue, has been recently identified as a strong predictor of sexual function in middle-aged adult women [63]. Although the molecular links between chronic fatigue and psychosocial impairments remain unclear, it has been suggested that the HPA axis might be behind this cluster of symptoms typically observed in cancer patients [64]. Interestingly, an aberrant HPA response has been reported in women with endometriosis [65]. Moreover, social factors might also contribute to the development of fatigue. Hence, in a different subset of patients, it has been reported that social support, through promoting self-confidence and rational thoughts, may have an impact on the reinforcement of the immunity system, and in turn, on the reduction of fatigue levels [66]. Care practitioners and clinicians' perception of women's experiences of endometriosis [67] and low-value healthcare [68] might also contribute to the endometriosis-related burden of symptoms. Additionally, diagnosis delay, infertility and worries related to low work productivity or job loss, in addition to depressive symptoms or disturbed sleep, might also negatively impact on energy and vitality [69–71], revealing the complex inter-relationships between physiological and psychosocial factors in women with endometriosis.

Regarding the interrelationship between intensity of PP, chronic fatigue and psychosocial impairments, our findings indicate that chronic fatigue and catastrophizing thoughts may mediate the association between last week PP intensity and QoL. A similar relationship was described in a previous work showing an association between pain and psychological stress with a worsening of the QoL in women living with endometriosis [72]. In this study, we have added for the first time the contribution of chronic fatigue, in addition to psychosocial distress, on QoL impairment in women with endometriosis. Taken together, these studies would support the idea that pain is associated with chronic fatigue and negative emotions [73,74], that in turn could affect QoL in women with endometriosis. Therefore, our data supports the necessity of multimodal treatments that address fatigue and psychosocial distress in addition to PP intensity in order to improve QoL in women with endometriosis, in line with previous suggestions [11,72]. Thus, besides medical therapy [47], physical and psychological interventions might be beneficial in endometriosis treatment, as evidenced for a variety of chronic illnesses in women [75–78]. More attention should be paid to non-pharmacological approaches to manage the symptom burden of this silenced female disease.

Regarding limitations, study population might not be fully representative of Spanish women with endometriosis. In this regard, although we have included affected women from all Spanish regions, the presence of a selection bias is plausible, given that participants might have a different symptom burden than non-participants. In this regard, it has been reported that outcomes related to QoL are influenced by recruitment strategy [79]. Secondly, this study has a cross-sectional design that does not allow the assessment of the causal relation between studied variables. The absence of a control group also limited understanding of the differential impact of this symptom on lives of women with and without endometriosis. Nevertheless, in a case-control study comprised of 25 women with endometriosis and 25 healthy controls, we have recently reported that mean fatigue score was 2.9 ± 2.0 among controls and

5.3 ± 2.3 among women with endometriosis [80]. In fact, a large majority of healthy women had mild fatigue and none of them had severe fatigue. Contrarily, the majority of women with endometriosis (72.0%) had moderate-severe fatigue. Moreover, the information retrieved in the present study was obtained from self-administered questionnaires and, therefore, a risk of information bias could also exist. However, the use of validated scales for this assessment may counteract this information bias. Finally, we have no information about medication taken by the women during the study, although all participants reported to be adhering to the prescribed medical treatment. However, it is possible that fatigue in endometriosis could be partially attributed to side effects from medication [46]. In addition, contraceptive hormonal therapy is usually prescribed to many women with endometriosis, and its use has been previously associated with different depression symptoms [81].

5. Conclusions

This work provides preliminary evidence of the relevance of chronic fatigue for the psychosocial status and the QoL of women living with endometriosis. We consider that it has important implications for the evaluation and treatment of this population, as the main goal of their management is usually to ameliorate symptoms and to improve general QoL. The habitual treatment of the disease is focused on classic symptoms such as pain or infertility [1,82], but our findings also support the importance of addressing fatigue when treating patients with endometriosis, highlighting the necessity for an interdisciplinary management of the disease. Thus, our results warrant future studies that assess the effectiveness of multidisciplinary approaches (i.e., physical and psychological rehabilitation interventions, in addition to medical therapy) for symptom management.

Supplementary Materials: The following are available online at http://www.mdpi.com/1660-4601/17/11/3831/s1, Table S1. Relationship between intensity of chronic pelvic pain and chronic fatigue; Table S2. Relationship between intensity of chronic fatigue, psychosocial status and quality of life in women with endometriosis.

Author Contributions: F.A.-C., C.F.-L., A.M.-L. and M.S.A.-L. conceived, designed, and implemented the study. O.O.-H. helped with the design and the implementation of the study and coordinated the patient recruitment. I.C.-V., N.G.-C., C.F.-L. and F.A.-C. analyzed and interpreted the data and A.M.-L. and A.P.S.-S. drafted the paper. M.A.-M., I.C.-V. and O.R.-P. contributed to the interpretation of the data and critically revised the article for important intellectual content. All authors provided a critical revision of the manuscript and helped with data interpretation and manuscript preparation. F.A.-C. and C.F.-L. contributed equally to this work. All authors have read and agreed to the published version of the manuscript.

Funding: This research was funded by Health Institute Carlos III (ISCIII)-FEDER (grant number PI17/01743) and donations from particular women with endometriosis that believed in this project from the first time. It was also partly supported by the PAIDI group CTS-206 (Oncología Básica y Clínica) funds. This study takes place thanks to the additional funding from the University of Granada, Plan Propio de Investigación 2016, Excellence actions: Units of Excellence; Unit of Excellence on Exercise and Health (UCEES), and by the Junta de Andalucía, Consejería de Conocimiento, Investigación y Universidades and European Regional Development Fund (ERDF), ref. SOMM17/6107/UGR.

Acknowledgments: The authors are indebted to all participants, without whom this work would not have been possible. We are grateful to Ana Yara Postigo-Fuentes for her assistance with the English language. This paper is part of the thesis developed by A. Mundo-López in the Official Doctoral Program in Clinical Medicine and Public Health of the University of Granada.

Conflicts of Interest: The authors declare no conflicts of interest.

References

1.   Vercellini, P.; Viganò, P.; Somigliana, E.; Fedele, L. Endometriosis: Pathogenesis and treatment. *Nat. Rev. Endocrinol.* **2014**, *10*, 261. [CrossRef] [PubMed]

2.   Eisenberg, V.H.; Weil, C.; Chodick, G.; Shalev, V. Epidemiology of endometriosis: A large population-based database study from a healthcare provider with 2 million members. *BJOG* **2018**, *125*, 55–62. [CrossRef] [PubMed]

3.   Soliman, A.M.; Yang, H.; Du, E.X.; Kelley, C.; Winkel, C. The direct and indirect costs associated with endometriosis: A systematic literature review. *Hum. Reprod.* **2016**, *31*, 712–722. [CrossRef] [PubMed]

4. Berkley, K.J.; Rapkin, A.J.; Papka, R.E. The pains of endometriosis. *Science* **2005**, *308*, 1587–1589. [CrossRef] [PubMed]

5. Loeser, J.D.; Treede, R.-D. The Kyoto protocol of IASP Basic Pain Terminology. *Pain* **2008**, *137*, 473–477. [CrossRef] [PubMed]

6. Soliman, A.M.; Coyne, K.S.; Zaiser, E.; Castelli-Haley, J.; Fuldeore, M.J. The burden of endometriosis symptoms on health-related quality of life in women in the United States: A cross-sectional study. *J. Psychosom. Obstet. Gynaecol.* **2017**, *38*, 238–248. [CrossRef] [PubMed]

7. Facchin, F.; Barbara, G.; Saita, E.; Mosconi, P.; Roberto, A.; Fedele, L.; Vercellini, P. Impact of endometriosis on quality of life and mental health: Pelvic pain makes the difference. *J. Psychosom. Obstet. Gynaecol.* **2015**, *36*, 135–141. [CrossRef]

8. Nnoaham, K.E.; Hummelshoj, L.; Webster, P.; d'Hooghe, T.; de Cicco Nardone, F.; de Cicco Nardone, C.; Jenkinson, C.; Kennedy, S.H.; Zondervan, K.T.; World Endometriosis Research Foundation Global Study of Women's Health consortium. Impact of endometriosis on quality of life and work productivity: A multicenter study across ten countries. *Fertil. Steril.* **2011**, *96*, 366–373.e8. [CrossRef]

9. Eriksen, H.-L.F.; Gunnersen, K.F.; Sørensen, J.-A.; Munk, T.; Nielsen, T.; Knudsen, U.B. Psychological aspects of endometriosis: Differences between patients with or without pain on four psychological variables. *Eur. J. Obstet. Gynecol. Reprod. Biol.* **2008**, *139*, 100–105. [CrossRef]

10. Nunes, F.R.; Ferreira, J.M.; Bahamondes, L. Pain threshold and sleep quality in women with endometriosis. *Eur. J. Pain* **2015**, *19*, 15–20. [CrossRef]

11. Ramin-Wright, A.; Schwartz, A.S.K.; Geraedts, K.; Rauchfuss, M.; Wölfler, M.M.; Haeberlin, F.; von Orelli, S.; Eberhard, M.; Imthurn, B.; Imesch, P. Fatigue–a symptom in endometriosis. *Hum. Reprod.* **2018**, *33*, 1459–1465. [CrossRef] [PubMed]

12. Gilmour, J.A.; Huntington, A.; Wilson, H.V. The impact of endometriosis on work and social participation. *Int. J. Nurs. Pract.* **2008**, *14*, 443–448. [CrossRef] [PubMed]

13. Jones, G.; Jenkinson, C.; Kennedy, S. The impact of endometriosis upon quality of life: A qualitative analysis. *J. Psychosom. Obstet. Gynaecol.* **2004**, *25*, 123–133. [CrossRef] [PubMed]

14. Pappalardo, A.; Reggio, E.; Patti, F.; Reggio, A. Management of fatigue in multiple sclerosis. *Eur. J. Phys. Rehabil. Med.* **2003**, *39*, 147.

15. Ahlberg, K.; Ekman, T.; Gaston-Johansson, F.; Mock, V. Assessment and management of cancer-related fatigue in adults. *Lancet* **2003**, *362*, 640–650. [CrossRef]

16. Omdal, R.; Koldingsnes, W.; Husby, G.; Mellgren, S.I. Fatigue in patients with systemic lupus erythematosus: The psychosocial aspects. *J. Rheumatol.* **2003**, *30*, 283–287.

17. Trojan, D.; Arnold, D.; Collet, J.; Shapiro, S.; Bar-Or, A.; Robinson, A.; Le Cruguel, J.; Ducruet, T.; Narayanan, S.; Arcelin, K. Fatigue in multiple sclerosis: Association with disease-related, behavioural and psychosocial factors. *Mult. Scler.* **2007**, *13*, 985–995. [CrossRef]

18. Piper, B.F.; Dibble, S.L.; Dodd, M.J.; Weiss, M.C.; Slaughter, R.E.; Paul, S.M. The revised Piper Fatigue Scale: Psychometric evaluation in women with breast cancer. *Oncol. Nurs. Forum.* **1998**, *25*, 677–684.

19. Cantarero-Villanueva, I.; Fernández-Lao, C.; Díaz-Rodríguez, L.; Cuesta-Vargas, A.I.; Fernández-de-las-Peñas, C.; Piper, B.F.; Arroyo-Morales, M. The Piper Fatigue Scale-Revised: Translation and psychometric evaluation in Spanish-speaking breast cancer survivors. *Qual. Life Res.* **2014**, *23*, 271–276. [CrossRef]

20. Zhang, Q.; Li, F.; Zhang, H.; Yu, X.; Cong, Y. Effects of nurse-led home-based exercise & cognitive behavioral therapy on reducing cancer-related fatigue in patients with ovarian cancer during and after chemotherapy: A randomized controlled trial. *Int. J. Nurs. Stud.* **2018**, *78*, 52–60.

21. Pozehl, B.; Duncan, K.; Hertzog, M. The effects of exercise training on fatigue and dyspnea in heart failure. *Eur. J. Cardiovasc. Nurs.* **2008**, *7*, 127–132. [CrossRef] [PubMed]

22. O'Regan, P.; Hegarty, J. The importance of self-care for fatigue amongst patients undergoing chemotherapy for primary cancer. *Eur. J. Oncol. Nurs.* **2017**, *28*, 47–55. [CrossRef] [PubMed]

23. Jensen, M.P.; Turner, J.A.; Romano, J.M.; Fisher, L.D. Comparative reliability and validity of chronic pain intensity measures. *Pain* **1999**, *83*, 157–162. [CrossRef]

24. Breivik, E.K.; Björnsson, G.A.; Skovlund, E. A comparison of pain rating scales by sampling from clinical trial data. *Clin. J. Pain* **2000**, *16*, 22–28. [CrossRef] [PubMed]

25. Boonstra, A.M.; Stewart, R.E.; Köke, A.J.; Oosterwijk, R.F.; Swaan, J.L.; Schreurs, K.M.; Schiphorst Preuper, H.R. Cut-off points for mild, moderate, and severe pain on the numeric rating scale for pain in patients with chronic musculoskeletal pain: Variability and influence of sex and catastrophizing. *Front. Psychol.* **2016**, *7*, 1466. [CrossRef] [PubMed]
26. Forchheimer, M.B.; Richards, J.S.; Chiodo, A.E.; Bryce, T.N.; Dyson-Hudson, T.A. Cut point determination in the measurement of pain and its relationship to psychosocial and functional measures after traumatic spinal cord injury: A retrospective model spinal cord injury system analysis. *Arch. Phys. Med. Rehabil.* **2011**, *92*, 419–424. [CrossRef]
27. Herrero, M.J.; Blanch, J.; Peri, J.M.; De Pablo, J.; Pintor, L.; Bulbena, A. A validation study of the hospital anxiety and depression scale (HADS) in a Spanish population. *Gen. Hosp. Psychiatry* **2003**, *25*, 277–283. [CrossRef]
28. Zigmond, A.S.; Snaith, R.P. The hospital anxiety and depression scale. *Acta Psychiatr Scand.* **1983**, *67*, 361–370. [CrossRef]
29. Quintana, J.; Padierna, A.; Esteban, C.; Arostegui, I.; Bilbao, A.; Ruiz, I. Evaluation of the psychometric characteristics of the Spanish version of the Hospital Anxiety and Depression Scale. *Acta Psychiatr. Scand.* **2003**, *107*, 216–221. [CrossRef]
30. Bjelland, I.; Dahl, A.A.; Haug, T.T.; Neckelmann, D. The validity of the Hospital Anxiety and Depression Scale: An updated literature review. *J. Psychosom. Res.* **2002**, *52*, 69–77. [CrossRef]
31. Olssøn, I.; Mykletun, A.; Dahl, A.A. The hospital anxiety and depression rating scale: A cross-sectional study of psychometrics and case finding abilities in general practice. *BMC Psychiatry* **2005**, *5*, 46. [CrossRef] [PubMed]
32. Fernández, J.S. Un instrumento para evaluar la eficacia de los procedimientos de inducción de estado de ánimo: La "Escala de Valoración del Estado de Ánimo" (EVEA). *Anal. Modif. Conduct.* **2001**, *27*, 71–110.
33. García, J.C.; Rodero, B.; Alda, M.; Sobradiel, N.; Montero, J.; Moreno, S. Validation of the Spanish version of the Pain Catastrophizing Scale in fibromyalgia. *Med. Clin. (Barc.)* **2008**, *131*, 487–492.
34. Sullivan, M.J.; Bishop, S.R.; Pivik, J. The pain catastrophizing scale: Development and validation. *Psychol. Assess.* **1995**, *7*, 524. [CrossRef]
35. Hita-Contreras, F.; Martínez-López, E.; Latorre-Román, P.A.; Garrido, F.; Santos, M.A.; Martínez-Amat, A. Reliability and validity of the Spanish version of the Pittsburgh Sleep Quality Index (PSQI) in patients with fibromyalgia. *Rheumatol. Int.* **2014**, *34*, 929–936. [CrossRef] [PubMed]
36. Buysse, D.J.; Reynolds, C.F., 3rd; Monk, T.H.; Berman, S.R.; Kupfer, D.J. The Pittsburgh Sleep Quality Index: A new instrument for psychiatric practice and research. *Psychiatry Res.* **1989**, *28*, 193–213. [CrossRef]
37. Blümel, J.E.; Binfa, L.; Cataldo, P.; Carrasco, A.; Izaguirre, H.; Sarrá, S. Índice de función sexual femenina: Un test para evaluar la sexualidad de la mujer. *Rev. Chil. Obstet. Ginecol.* **2004**, *69*, 118–125. [CrossRef]
38. Rosen, C.B.; Heiman, J.; Leiblum, S.; Meston, C.; Shabsigh, R.; Ferguson, D.; D'Agostino, R., Jr. The Female Sexual Function Index (FSFI): A multidimensional self-report instrument for the assessment of female sexual function. *J. Sex Marital Ther.* **2000**, *26*, 191–208. [CrossRef]
39. Wiegel, M.; Meston, C.; Rosen, R. The female sexual function index (FSFI): Cross-validation and development of clinical cutoff scores. *J. Sex Marital Ther.* **2005**, *31*, 1–20. [CrossRef]
40. Quintana, J.; Cabriada, J.; de Tejada López, I.; Varona, M.; Oribe, V.; Barrios, B.; Perdigo, L.; Bilbao, A. Translation and validation of the gastrointestinal Quality of Life Index (GIQLI). *Rev. Esp. Enferm. Dig.* **2001**, *93*, 693–706.
41. Nieveen Van Dijkum, E.J.; Terwee, C.B.; Oosterveld, P.; Van Der Meulen, J.H.; Gouma, D.J.; De Haes, J.C. Validation of the gastrointestinal quality of life index for patients with potentially operable periampullary carcinoma. *Br. J. Surg.* **2000**, *87*, 110–115. [CrossRef] [PubMed]
42. De la Revilla, L.; Luna, J.; Bailón, E.; Medina, I. Validación del cuestionario MOS de apoyo social en Atención Primaria. *Med. de Fam.* **2005**, *6*, 10–18.
43. Requena, G.C.; Salamero, M.; Gil, F. Validación del cuestionario MOS-SSS de apoyo social en pacientes con cáncer. *Med. Clin. (Barc.)* **2007**, *128*, 687–691. [CrossRef] [PubMed]
44. Jones, G.; Kennedy, S.; Barnard, A.; Wong, J.; Jenkinson, C. Development of an endometriosis quality-of-life instrument: The Endometriosis Health Profile-30. *Obstet. Gynecol.* **2001**, *98*, 258–264. [CrossRef] [PubMed]
45. Hayes, A.F. *Introduction to Mediation, Moderation, and Conditional Process Analysis: A Regression-Based Approach*; Guilford Press: New York, NY, USA, 2013; Volume 17, p. 507.

46. Grogan, S.; Turley, E.; Cole, J. 'So many women suffer in silence': A thematic analysis of women's written accounts of coping with endometriosis. *Psychol. Health* **2018**, *33*, 1364–1378. [CrossRef] [PubMed]

47. Surrey, E.S.; Soliman, A.M.; Agarwal, S.K.; Snabes, M.C.; Diamond, M.P. Impact of elagolix treatment on fatigue experienced by women with moderate to severe pain associated with endometriosis. *Fertil. Steril.* **2019**, *112*, 298–304.e3. [CrossRef] [PubMed]

48. Symons, L.K.; Miller, J.E.; Kay, V.R.; Marks, R.M.; Liblik, K.; Koti, M.; Tayade, C. The immunopathophysiology of endometriosis. *Trends Mol. Med.* **2018**, *24*, 748–762. [CrossRef]

49. Suryawanshi, S.; Huang, X.; Elishaev, E.; Budiu, R.A.; Zhang, L.; Kim, S.; Donnellan, N.; Mantia-Smaldone, G.; Ma, T.; Tseng, G. Complement pathway is frequently altered in endometriosis and endometriosis-associated ovarian cancer. *Clin. Cancer Res.* **2014**, *20*, 6163–6174. [CrossRef]

50. Bower, J.E. Cancer-related fatigue: Links with inflammation in cancer patients and survivors. *Brain Behav. Immun.* **2007**, *21*, 863–871. [CrossRef]

51. Louati, K.; Berenbaum, F. Fatigue in chronic inflammation-a link to pain pathways. *Arthritis Res. Ther.* **2015**, *17*, 254. [CrossRef]

52. So, W.K.W.; Marsh, G.; Ling, W.M.; Leung, F.Y.; Lo, J.C.K.; Yeung, M.; Li, G.K.H. The symptom cluster of fatigue, pain, anxiety, and depression and the effect on the quality of life of women receiving treatment for breast cancer: A multicenter study. *Oncol. Nurs. Forum.* **2009**, *36*, E205–E214. [CrossRef] [PubMed]

53. Matarese, G.; De Placido, G.; Nikas, Y.; Alviggi, C. Pathogenesis of endometriosis: Natural immunity dysfunction or autoimmune disease? *Trends Mol. Med.* **2003**, *9*, 223–228. [CrossRef]

54. Marquardt, R.M.; Kim, T.H.; Shin, J.H.; Jeong, J.W. Progesterone and Estrogen Signaling in the Endometrium: What Goes Wrong in Endometriosis? *Int. J. Mol. Sci.* **2019**, *20*, 3822. [CrossRef] [PubMed]

55. Matura, L.A.; Malone, S.; Jaime-Lara, R.; Riegel, B. A Systematic Review of Biological Mechanisms of Fatigue in Chronic Illness. *Biol. Res. Nurs.* **2018**, *20*, 410–421. [CrossRef] [PubMed]

56. Sepulcri Rde, P.; do Amaral, V.F. Depressive symptoms, anxiety, and quality of life in women with pelvic endometriosis. *Eur. J. Obstet. Gynecol. Reprod. Biol.* **2009**, *142*, 53–56. [CrossRef]

57. Pope, C.J.; Sharma, V.; Sharma, S.; Mazmanian, D. A Systematic Review of the Association Between Psychiatric Disturbances and Endometriosis. *JOGC* **2015**, *37*, 1006–1015. [CrossRef]

58. Chen, L.C.; Hsu, J.W.; Huang, K.L.; Bai, Y.M.; Su, T.P.; Li, C.T.; Yang, A.C.; Chang, W.H.; Chen, T.J.; Tsai, S.J.; et al. Risk of developing major depression and anxiety disorders among women with endometriosis: A longitudinal follow-up study. *J. Affect. Disord.* **2016**, *190*, 282–285. [CrossRef]

59. De Graaff, A.A.; Van Lankveld, J.; Smits, L.J.; Van Beek, J.J.; Dunselman, G.A. Dyspareunia and depressive symptoms are associated with impaired sexual functioning in women with endometriosis, whereas sexual functioning in their male partners is not affected. *Hum. Reprod. (Oxford, England)* **2016**, *31*, 2577–2586. [CrossRef]

60. Santos, J.C.; Pyter, L.M. Neuroimmunology of behavioral comorbidities associated with cancer and cancer treatments. *Front. Immunol.* **2018**, *9*, 1195. [CrossRef]

61. Webber, K.; Mok, K.; Bennett, B.; Lloyd, A.R.; Friedlander, M.; Juraskova, I.; Goldstein, D.; FolCan study group. If I am in the mood, I enjoy it: An exploration of cancer-related fatigue and sexual functioning in women with breast cancer. *Oncologist* **2011**, *16*, 1333–1344. [CrossRef]

62. Ghasemi, V.; Simbar, M.; Ozgoli, G.; Nabavi, S.M.; Alavi Majd, H. Prevalence, dimensions, and predictor factors of sexual dysfunction in women of Iran Multiple Sclerosis Society: A cross-sectional study. *Neurol. Sci.* **2020**. [CrossRef] [PubMed]

63. Dote-Montero, M.; De-la-O, A.; Castillo, M.J.; Amaro-Gahete, F.J. Predictors of Sexual Desire and Sexual Function in Sedentary Middle-Aged Adults: The Role of Lean Mass Index and S-Klotho Plasma Levels. The FIT-AGEING Study. *J. Sex Med.* **2020**, *17*, 665–677. [CrossRef] [PubMed]

64. Li, H.; Marsland, A.L.; Conley, Y.P.; Sereika, S.M.; Bender, C.M. Genes Involved in the HPA Axis and the Symptom Cluster of Fatigue, Depressive Symptoms, and Anxiety in Women With Breast Cancer During 18 Months of Adjuvant Therapy. *Biol. Res. Nurs.* **2020**, *22*, 277–286. [CrossRef] [PubMed]

65. Quinones, M.; Urrutia, R.; Torres-Reveron, A.; Vincent, K.; Flores, I. Anxiety, coping skills and hypothalamus-pituitary-adrenal (HPA) axis in patients with endometriosis. *J. Reprod. Biol. Health* **2015**, *3*, 2. [CrossRef] [PubMed]

66. Aghaei, N.; Karbandi, S.; Gorji, M.A.; Golkhatmi, M.B.; Alizadeh, B. Social Support in Relation to Fatigue Symptoms Among Patients with Multiple Sclerosis. *Indian J. Palliat. Care* **2016**, *22*, 163–167. [PubMed]

67. Young, K.; Fisher, J.; Kirkman, M. "Do mad people get endo or does endo make you mad?": Clinicians' discursive constructions of Medicine and women with endometriosis. *Fem. Psychol.* **2018**, *29*, 337–356. [CrossRef]

68. Vercellini, P.; Giudice, L.C.; Evers, J.L.; Abrao, M.S. Reducing low-value care in endometriosis between limited evidence and unresolved issues: A proposal. *Hum. Reprod. (Oxford, England)* **2015**, *30*, 1996–2004. [CrossRef]

69. Culley, L.; Law, C.; Hudson, N.; Denny, E.; Mitchell, H.; Baumgarten, M.; Raine-Fenning, N. The social and psychological impact of endometriosis on women's lives: A critical narrative review. *Hum. Reprod. Update* **2013**, *19*, 625–639. [CrossRef]

70. Wichniak, A.; Wierzbicka, A.; Walęcka, M.; Jernajczyk, W. Effects of Antidepressants on Sleep. *Curr. Psychiatry Rep.* **2017**, *19*, 63. [CrossRef]

71. Angst, J.; Gamma, A.; Gastpar, M.; Lépine, J.P.; Mendlewicz, J.; Tylee, A. Gender differences in depression. Epidemiological findings from the European DEPRES I and II studies. *Eur. Arch. Psychiatry Clin. Neurosci.* **2002**, *252*, 201–209. [CrossRef]

72. Márki, G.; Bokor, A.; Rigó, J.; Rigó, A. Physical pain and emotion regulation as the main predictive factors of health-related quality of life in women living with endometriosis. *Hum. Reprod.* **2017**, *32*, 1432–1438. [CrossRef] [PubMed]

73. Petrelluzzi, K.; Garcia, M.; Petta, C.; Grassi-Kassisse, D.; Spadari-Bratfisch, R.C. Salivary cortisol concentrations, stress and quality of life in women with endometriosis and chronic pelvic pain. *Stress* **2008**, *11*, 390–397. [CrossRef] [PubMed]

74. Carey, E.T.; Martin, C.E.; Siedhoff, M.T.; Bair, E.D.; As-Sanie, S. Biopsychosocial correlates of persistent postsurgical pain in women with endometriosis. *Int. J. Gynaecol. Obstet.* **2014**, *124*, 169–173. [CrossRef] [PubMed]

75. Montgomery, G.H.; Kangas, M.; David, D.; Hallquist, M.N.; Green, S.; Bovbjerg, D.H.; Schnur, J.B. Fatigue during breast cancer radiotherapy: An initial randomized study of cognitive-behavioral therapy plus hypnosis. *Health Psychol.* **2009**, *28*, 317–322. [CrossRef] [PubMed]

76. Anderson, F.J.; Winkler, A.E. An Integrated Model of Group Psychotherapy for Patients with Fibromyalgia. *Int. J. Group Psychother.* **2007**, *57*, 451–474. [CrossRef] [PubMed]

77. Doerr, J.M.; Fischer, S.; Nater, U.M.; Strahler, J. Influence of stress systems and physical activity on different dimensions of fatigue in female fibromyalgia patients. *J. Psychosom. Res.* **2017**, *93*, 55–61. [CrossRef]

78. Galiano-Castillo, N.; Cantarero-Villanueva, I.; Fernández-Lao, C.; Ariza-García, A.; Díaz-Rodríguez, L.; Del-Moral-Ávila, R.; Arroyo-Morales, M. Telehealth system: A randomized controlled trial evaluating the impact of an internet-based exercise intervention on quality of life, pain, muscle strength, and fatigue in breast cancer survivors. *Cancer* **2016**, *122*, 3166–3174. [CrossRef]

79. De Graaff, A.A.; Dirksen, C.D.; Simoens, S.; De Bie, B.; Hummelshoj, L.; D'Hooghe, T.M.; Dunselman, G.A. Quality of life outcomes in women with endometriosis are highly influenced by recruitment strategies. *Hum. Reprod.* **2015**, *30*, 1331–1341. [CrossRef]

80. Álvarez-Salvago, F.; Lara-Ramos, A.; Cantarero-Villanueva, I.; Mazheika, M.; Mundo-López, A.; Galiano-Castillo, N.; Fernández-Lao, C.; Arroyo-Morales, M.; Ocón-Hernández, O.; Artacho-Cordón, F. Chronic Fatigue, Physical Impairments and Quality of Life in Women with Endometriosis: A Case-Control Study. *Int. J. Environ. Res. Public Health* **2020**, *17*, 3610. [CrossRef]

81. Skovlund, C.W.; Mørch, L.S.; Kessing, L.V.; Lidegaard, Ø. Association of Hormonal Contraception With Depression. *JAMA Psychiatry* **2016**, *73*, 1154–1162. [CrossRef]

82. Dunselman, G.; Vermeulen, N.; Becker, C.; Calhaz-Jorge, C.; D'hooghe, T.; De Bie, B.; Heikinheimo, O.; Horne, A.; Kiesel, L.; Nap, A.; et al. ESHRE guideline: Management of women with endometriosis. *Hum. Reprod.* **2014**, *29*, 400–412. [CrossRef] [PubMed]

International Journal of
*Environmental Research
and Public Health*

Article

Laws Restricting Access to Abortion Services and Infant Mortality Risk in the United States

Roman Pabayo [1,*], **Amy Ehntholt** [2], **Daniel M. Cook** [2], **Megan Reynolds** [3], **Peter Muennig** [4] and **Sze Y. Liu** [5]

1   School of Public Health, University of Alberta, 11405-87, Edmonton, AB T6G 1C9, Canada
2   School of Community Health Sciences, University of Nevada, Reno, NV 89557, USA;
     ehntholt@gmail.com (A.E.); dmcook@unr.edu (D.M.C.)
3   Sociology Department, University of Utah, Salt Lake City, UT 84117, USA; megan.reynolds@soc.utah.edu
4   Mailman School of Public Health, Columbia University, New York, NY 10032, USA;
     m124@cumc.columbia.edu
5   Public Health Department, Montclair State University, Montclair, NJ 07043, USA; lius@montclair.edu
*   Correspondence: pabayo@ualberta.ca; Tel.: +1-780-492-9954

Received: 28 April 2020; Accepted: 20 May 2020; Published: 26 May 2020

Abstract: *Objectives:* Since the US Supreme Court's 1973 *Roe v. Wade* decision legalizing abortion, states have enacted laws restricting access to abortion services. Previous studies suggest that restricting access to abortion is a risk factor for adverse maternal and infant health. The objective of this investigation is to study the relationship between the type and the number of state-level restrictive abortion laws and infant mortality risk. *Methods:* We used data on 11,972,629 infants and mothers from the US Cohort Linked Birth/Infant Death Data Files 2008–2010. State-level abortion laws included Medicaid funding restrictions, mandatory parental involvement, mandatory counseling, mandatory waiting period, and two-visit laws. Multilevel logistic regression was used to determine whether type or number of state-level restrictive abortion laws during year of birth were associated with odds of infant mortality. *Results:* Compared to infants living in states with no restrictive laws, infants living in states with one or two restrictive laws (adjusted odds ratio (AOR) = 1.08; 95% confidence interval [CI] = 0.99–1.18) and those living in states with 3 to 5 restrictive laws (AOR = 1.10; 95% CI = 1.01–1.20) were more likely to die. Separate analyses examining the relationship between parental involvement laws and infant mortality risk, stratified by maternal age, indicated that significant associations were observed among mothers aged ≤19 years (AOR = 1.09, 95% CI = 1.00–1.19), and 20 to 25 years (AOR = 1.10, 95% CI = 1.03–1.17). No significant association was observed among infants born to older mothers. *Conclusion:* Restricting access to abortion services may increase the risk for infant mortality.

Keywords: US state laws; abortion; infant mortality

---

1. Introduction

The infant mortality rate (IMR), the number per 1000 live births of infant deaths before the age of 1, is one of the best predictors of a nation's life expectancy and widely used as an indicator of population health [1]. In 2011, around 24,000 infants died in the United States (US), resulting in an IMR of 6.1, nearly twice the Organization for Economic Co-operation and Development (OECD) average of 3.4/1000 [2]. Known risk factors for infant mortality include individual-level risk factors (e.g., socio-economic status, such as mother's education or household income) and larger contextual factors (e.g., neighborhood environment or state-level policies that affect access to health care) [3].

Within the US, IMRs vary greatly across socio-demographic groups. Infants born to Black mothers, single mothers, and low-income mothers have higher IMRs compared to infants born to White mothers, married mothers, and moderate or high-income mothers [4]. For example, in 2016, the IMR among non-Hispanic Blacks was 11.4 per 1000, whereas the IMR for non-Hispanic Whites was 4.9 [5]. These socio-demographic groups may have limited access to resources, such as psychosocial supports, material resources, and vital health and reproductive services.

Abortion, a medical procedure terminating a pregnancy, is recognized as a key component of reproductive health services [6], which the United Nations recognizes as an important predictor of national well-being and population health, generally, and maternal and infant mortality, specifically [7–9]. Beyond the abortion laws examined in this study, states continue to create policy restricting access to safe abortions. Currently, the Supreme Court is hearing one such case affecting a provider's rights to provide abortion services. Beyond the immediate constitutional value of any restrictive abortion law, courts should consider the health impact of such laws on women and their future unborn children.

Restrictive abortion policies may have a detrimental effect on both maternal and infant health via several mechanisms. First, restrictive abortion policies may jeopardize patient health by undermining providers' medical counsel. For example, some US states require counseling that provides inaccurate information about negative mental health consequences of abortion or a link between abortion and increased risk of breast cancer [10]. Second, restrictions may increase psychological distress, which is a known risk factor for poor birth outcomes. Both human and animal studies show that psychological stress produces a cascade of neuroendocrine changes that increase the risk of serious birth complications (e.g., pre-eclampsia) and thereby increase infant mortality [11–14].

Third, these policies may increase the risk for postpartum depression by inducing dramatic, but unwanted life changes. Postpartum depression can contribute to infant death in cases where the mother is unable to properly care for her infant [15,16]. Previous research indicated that women in states that prohibit the Medicaid funding of abortions have significantly higher rates of postpartum depression than those in states that fund Medicaid abortions [17]. Mental illness among pregnant women has been associated with increased risk for infant mortality [18]. Postnatal depression has also been observed as a risk factor for sudden infant death (SIDS) [19].

Fourth, restrictive policies may encourage the delivery of infants whom the mother is unable to economically or emotionally support. For example, women from low socioeconomic backgrounds who have limited resources and income may be less able to pay to terminate their pregnancies out of pocket and more likely to carry their infant to full term [20]. Moreover, women from lower socioeconomic backgrounds may also experience limited access to prenatal care and health care, and sufficient living expenses for optimal health and growth, all of which may have an adverse effect on maternal and infant health [21]. Findings from a previous study suggests denying women abortions may be associated with greater poverty than enabling women to postpone childbearing [22].

As shown in ecological studies conducted in the US, increased state funding for family planning and abortion services is associated with lower infant mortality rates, especially for low-income women of color [7]. Conversely, restrictions on abortion services have been associated with increased infant and maternal mortality risk [23,24]. However, because these previous studies are based on aggregated data, they limit our ability to make inferences at the individual level. Moreover, few of these studies have examined interactions between restrictive abortion policy and individual-level characteristics. Therefore, the overall objective of this study is to identify the association between the number and type of state-level restrictions to abortion services during year of birth and infant mortality risk while controlling for both individual and state-level confounders. Additionally, we examine whether observed relationships are heterogeneous across socio-demographic groups, such as mother's age, race, and educational attainment. By studying these heterogeneous associations, we aim to determine who is at greatest risk of infant mortality when such state-level abortion policies are enacted.

2. Methods

We obtained data from the US Cohort Linked Birth/Infant Death (LBID) Data Files on infants born 2008–2010, which is provided by the National Center for Health Statistics (NCHS). Infants were followed until their first birthdays. US state laws require birth certificates to be completed for all births. Of all deaths that occurred, around 98% were linked to the corresponding birth certificate. Instructed by US state laws, each birth requires a birth certificate to be completed, which contains information on maternal and socio-demographic characteristics and place of birth. Federal law mandates the national collection and publication of births, deaths, and other vital statistics. These data are then compiled by the National Vital Statistics System. Those with missing socio-demographic data, foreign residents, and those with records that were a mismatch between state of birth and mother's state of residence were excluded from this analysis. Ethical approval was obtained from the University of Nevada, Reno, Institutional Review Board (code 791378-1).

3. Measures

3.1. Outcome

Infant mortality (death within 365 days of birth) was our outcome of interest.

3.2. Main Exposure

Our main measure of exposure was the category of restrictive state abortion law. Since the US Supreme Court's 1973 *Roe v. Wade* decision legalizing abortion nationwide, individual states have enacted laws regulating the access and availability of abortion services in the hopes of moderating the effects of federal decision. Data on restrictive abortion laws for each US state were obtained in 2008, 2009, and 2010 from the Guttmacher Institute [25–27]. From 2008–2010, there were five types of state-level restrictive abortion laws directly impacting patients that the Supreme Court regarded as constitutionally permissible. These types of restrictive laws consist of: (1) Medicaid funding restrictions—prohibitions against use of state public funds to pay for abortions for indigent women [25–27]; (2) parental involvement laws—requirements that a parent be notified or give consent for an unmarried teen minor to obtain an abortion; (3) mandatory counselling—requirements that an abortion provider give or offer their patients information about abortion (usually written in a way to dissuade women from completing the abortion) by providing information on fetal development, fetal pain, fetal age, possible future health risks [e.g., substance abuse, breast cancer, suicide, or infertility], adoption options, and available public assistance to the birth mother); (4) mandatory waiting period—requirements that a specified time period (usually 24 h) elapse before the procedure can be performed; (5) two-visit laws—requirements that women make two separate trips to the abortion provider prior to the procedure, elevating the monetary and time burdens for prospective patients.

Table 1 shows a complete list of the status of the five restrictive abortion laws for all 50 states and the District of Columbia in 2008. Over the next two years, several states enacted a few more abortion restrictive laws. For example, in 2008, the only restrictive law in Arizona was mandatory parental involvement [25–27]. By 2010, Arizona had added mandatory counselling and mandatory wait times. We looked at the relationship between each type of restriction and the number of restrictions during the year of birth of each infants and risk for infant mortality.

Table 1. Abortion restrictions by State, 2008.

| | Medicaid | | | Mandatory | | |
State	Funding Restriction	Parental Involvement	Mandatory Counseling	Waiting Period	Two-Visit Law	Total Laws
Alabama	Yes	Yes	Yes	Yes	No	4
Alaska	No	No	Yes	No	No	1
Arizona	No	Yes	No	No	No	1
Arkansas	Yes	Yes	Yes	Yes	No	4
California	No	No	Yes	No	No	1
Colorado	Yes	Yes	No	No	No	2
Connecticut	No	No	Yes	No	No	1
Delaware	Yes	Yes	Yes	No	No	3
Washington DC	Yes	No	No	No	No	1
Florida	Yes	Yes	Yes	No	No	3
Georgia	Yes	Yes	Yes	Yes	No	4
Hawaii	No	No	No	No	No	0
Idaho	Yes	Yes	Yes	Yes	No	4
Illinois	No	No	No	No	No	0
Indiana	Yes	Yes	Yes	Yes	Yes	5
Iowa	Yes	Yes	No	No	No	2
Kansas	Yes	Yes	Yes	Yes	No	4
Kentucky	Yes	Yes	Yes	Yes	No	4
Louisiana	Yes	Yes	Yes	Yes	Yes	5
Maine	Yes	No	Yes	No	No	2
Maryland	No	Yes	No	No	No	1
Massachusetts	No	Yes	No	No	No	1
Michigan	Yes	Yes	Yes	Yes	No	4
Minnesota	No	Yes	Yes	Yes	No	3
Mississippi	Yes	Yes	Yes	Yes	Yes	5
Missouri	Yes	Yes	Yes	Yes	Yes	5
Montana	No	No	No	No	No	0
Nebraska	Yes	Yes	Yes	Yes	No	4
Nevada	Yes	No	Yes	No	No	2
New Hampshire	Yes	No	No	No	No	1
New Jersey	No	No	No	No	No	0
New Mexico	No	No	No	No	No	0
New York	No	No	No	No	No	0
North Carolina	Yes	Yes	No	No	No	2
North Dakota	Yes	Yes	Yes	Yes	No	4
Ohio	Yes	Yes	Yes	Yes	Yes	5
Oklahoma	Yes	Yes	Yes	Yes	No	4
Oregon	No	No	No	No	No	0
Pennsylvania	Yes	Yes	Yes	Yes	No	4
Rhode Island	Yes	Yes	Yes	No	No	3
South Carolina	Yes	Yes	Yes	Yes	No	4
South Dakota	Yes	Yes	Yes	Yes	No	4
Tennessee	Yes	Yes	Yes	No	No	3
Texas	Yes	Yes	Yes	Yes	No	4
Utah	Yes	Yes	Yes	Yes	Yes	5
Vermont	No	No	No	No	No	0
Virginia	Yes	Yes	Yes	Yes	No	4
Washington	No	No	No	No	No	0
West Virginia	No	Yes	Yes	Yes	No	3
Wisconsin	Yes	Yes	Yes	Yes	Yes	5
Wyomimg	Yes	Yes	No	No	No	2

Source. Guttmacher Institute (2008) [25].

For this study, we included several additional state-level and individual-level covariates that could act as confounders of the relationship between restrictions to abortion services and infant mortality risk. State-level covariates include median income, proportion of population that is African-American, population size, and US census division. Individual-level maternal covariates include mother's age, race/ethnicity, education, marital status, and nativity (US vs. foreign-born).

4. Statistical Analysis

Infants with missing data on their mother's education were excluded from the analyses. We used multilevel logistic modeling (mothers and their infant nested within states) to determine the association between abortion-restriction laws during the year of birth and infant mortality risk. We tested the relationship between each of the five types of restrictive abortion laws separately and looked at the total number of restrictions (no restrictions, one or two restrictions, and three or more restrictions). The internal consistency of restrictive laws was high (Cronbach's alpha = 0.82).

To investigate the potential effect of restrictive laws and risk of infant mortality, we adopted a step-up approach, where we systematically add variables to the models. The null model was first estimated to compute the overall predicted probability, which indicates the average probability of infant mortality across all states. Additionally, the 95% plausible value range, which is an indication of the degree of variability of the likelihood of infant mortality. For example, the plausible value range allows us to compute the range of plausible proportion (i.e., the maximum and minimum values) experiencing infant mortality across the US states. Second, we identified the crude relationship between each type of restrictive law at the year of birth (2008, 2009, or 2010) and infant mortality risk. Third, we fit our logistic regression model including state- and individual-level characteristics. Finally, we stratified the analyses by age ($\leq$19 years, 20–25 years, 25–30 years, 30–34 years, and $\geq$35 years of age), race (Black vs. White), and education (less than high school vs. high school or more) to determine if relationships were heterogeneous across socio-demographic groups (results not shown).

5. Results

Characteristics

Socio-demographic characteristics of infants born 2008 to 2010 are presented in Table 2. Over half of the mothers were White (53.5%), around a quarter were Hispanic (24.4%), and around 15% were Black. Just over a half of the mothers (52.5%) had some post-secondary education, almost two-thirds were married (59.0%) and roughly three-quarters were US-born (76.4%). Characteristics of the fifty states and District of Columbia also appear in Table 1. The median state-level income in 2010 was \$49,973.65 (SD = 8130.60) and the average proportion of African-Americans in a state's population was 12% (SD = 11.8).

By 2010, the most common state-level restrictive abortion law was mandatory parental involvement ($n = 35$) and the least common restrictive abortion law was the two-visit law ($n = 7$). Nine states (17.7%) had no restrictive laws, while seven (13.7%) had all five laws enacted.

From 2008 to 2010, there were 71,528 infant deaths corresponding to an infant mortality rate of 6.0 deaths/1000 births. The overall predicted probability by year was 0.61%, 0.60%, and 0.57% in 2008, 2009, and 2010, respectively. The plausible value range indicates that there is considerable variability in the cumulative incidence of infant death across US states. The plausible value range for infant mortality was 0.44–0.86%, 0.43–0.84%, and 0.40–0.80%, in 2008, 2009, and 2010, respectively.

When we tested the crude relationship between each of the five types of abortion restrictive laws, each was significantly associated with an increased odds for infant mortality (Table 3): the presence of Medicaid restrictions (OR = 1.22, 95% CI = 1.10–1.36); parental involvement (OR = 1.26, 95% CI = 1.13–1.40); mandatory counselling (OR = 1.08, 95%CI = 1.00–1.17); mandatory wait period (OR = 1.09, 95% CI = 1.01–1.18); and two-visit laws (OR = 1.18, 95% CI = 1.01–1.38). However, after adjusting for individual- and state-level characteristics, infants born in states with parental involvement laws (OR = 1.10, 95% CI = 1.02–1.19) had an increased risk for infant mortality compared to those who lived in states without such laws. No other laws remained significant when tested separately.

When we assessed the crude relationship between the total number of laws, compared to no restrictions, one or two restrictive laws (OR = 1.16, 95%CI = 1.01–1.34), and three or more restrictions (OR = 1.26, 95% CI = 1.11–1.45) were associated with an increased odds for infant mortality (Table 3). Adjusted analysis of the total number of restrictive laws resulted similar trends but resulted in decreased OR estimates. For example, infants born in states with one or two (OR = 1.08, 95%CI = 0.99–1.18) or three or more restrictive laws (OR = 1.10, 95%CI = 1.01–1.20) had greater risk for infant mortality than those born in states with no restrictive laws (Table 3).

Table 2. Characteristics of mothers and US infants born 2008–2010.

Individual Level Characteristics	n	Percentage
Birth year		
2008	4,109,463	34.3
2009	3,993,282	33.4
2010	3,869,884	32.3
Mother's Race		
White	6,410,568	53.5
Black	1,773,995	14.8
Native	118,496	1.0
Asian	692,386	5.8
Latin	2,922,361	24.4
Other	54,822	0.5
Education		
Less than high school	2,396,987	20.2
High School	3,277,870	27.4
Post-secondary	6,297,772	52.5
Marital Status		
Single	4,913,782	41.0
Married	7,058,847	59.0
Birth order		
First	3,986,136	33.3
Second	3,355,612	28.0
Third	2,155,132	18.0
Fourth or more	2,475,749	20.7
Mother's nativity		
Foreign-born	2,829,786	23.6
US Born	9,142,843	76.4
US State-Level Characteristics 2010	**Mean (SD)**	**Max, Min**
Median Income, USD	49973.65 (8130.60)	(36,851, 68,854)
Population, 2010	6,054,080 (6,824,211)	(563,767, 37,300,000)
African American (%)	12 (11.8)	(0.40, 0.55)
Proportion in poverty (%)	14.8 (3.1)	(8.3, 22.4)

In analyses stratified by age, in comparison to those infants born in states with no parental involvement laws, those who were born in states with this restrictive law were at greater risk for infant mortality among mothers aged ≤19 years of age (OR = 1.09, 95% CI = 1.00–1.19), and 20–25 years (OR = 1.10, 95% CI = 1.03–1.17), but not among mothers in older age categories (Table 4). No other findings were significant. When stratified by race, White infants born in states with one or two restrictions were more likely to die than those born in states with no restrictions (OR = 1.15, 95% CI = 1.04–1.27). Infants born to White mothers who were living in states with mandatory wait periods were significantly less likely to die (OR = 0.93, 95% CI = 0.88–0.99). Among Blacks, infants born in states with Medicaid Restrictions (OR = 1.08, 95% CI = 1.00–1.16) and Parental Involvement (OR = 1.08, 95% CI = 0.99–1.17) were more likely to die. Blacks were more likely to experience infant mortality in states with one or two restrictions (OR = 1.07, 95% CI = 0.97–1.19) and in states with 3 or more restrictions (OR = 1.09, 95% CI = 0.99–1.19) in comparison to those infants born in states with no restrictions. When analyses were stratified by education, infants whose mothers had less than high school education born in states with Medicaid Restrictions (OR = 1.08, 95% CI = 1.01–1.17) and one or two restrictions (OR = 1.13, 95% CI = 1.04–1.12) were more likely to die compared to those born in states with no such restrictions. Infants born to mothers with a high school education in states with Parental Involvement laws (OR = 1.11, 95% CI = 1.03–1.21) or in states with 3 or more restrictions were more likely to die (OR = 1.11, 95 %CI = 1.01–1.21) than were those born in states with no restrictions.

Table 3. The relationship between abortion laws and infant mortality controlling for individual and state-level characteristics, 2008–2010.

	Odds for Infant Mortality						
	OR 95%CI	OR 95%CI	OR 95%CI	OR 95%CI	OR 95%CI	OR 95%CI	OR 95%CI
State Characteristics							
Abortion Laws							
one or two restrictions (reference group: no restrictions)	1.08 (0.99,1.18)						
3 or more restrictions (reference group: no restrictions)	1.10 (1.01,1.20)						
Medicare restrictions (reference group: no)							
Yes		1.03 (0.96,1.11)					1.00 (0.93,1.08)
Parental Involvement (reference group: no)							
Yes			1.10 (1.02,1.19)				1.11 (1.01,1.21)
Mandatory counselling (reference group: no)							
Yes				1.04 (0.98,1.09)			1.04 (0.96,1.13)
Mandatory wait period (reference group: no)							
Yes					1.03 (0.98,1.09)		0.97 (0.89,1.06)
Two visit-law (reference group: no)							
Yes						1.01 (0.93,1.09)	0.98 (0.90,1.06)
State Median Income Z-score	0.97 (0.93,1.02)	0.97 (0.92,1.02)	0.97 (0.93,1.02)	0.97 (0.92,1.01)	0.96 (0.92,1.01)	0.96 (0.92,1.01)	0.98 (0.93,1.03)
Proportion Black Z-score	0.98 (0.94,1.01)	0.97 (0.94,1.01)	0.98 (0.95,1.02)	0.98 (0.94,1.02)	0.98 (0.94,1.02)	0.98 (0.94,1.02)	0.98 (0.95,1.03)
Proportion Poor Z-score	0.98 (0.93,1.04)	0.98 (0.92,1.04)	0.98 (0.93,1.04)	0.98 (0.92,1.03)	0.97 (0.92,1.03)	0.97 (0.92,1.03)	0.99 (0.93,1.05)
State Population Z-score	0.99 (0.97,1.01)	0.99 (0.97,1.01)	0.99 (0.97,1.01)	0.99 (0.97,1.01)	0.99 (0.97,1.01)	0.99 (0.97,1.01)	0.99 (0.97,1.01)
Census Division (reference group: New England)							
Middle Atlantic	1.13 (0.98,1.29)	1.08 (0.95,1.23)	1.08 (0.95,1.21)	1.08 (0.95,1.23)	1.06 (0.93,1.21)	1.08 (0.95,1.23)	1.10 (0.96,1.25)
East North Central	1.25 (1.11,1.41)	1.24 (1.11,1.39)	1.24 (1.08,1.42)	1.24 (1.11,1.40)	1.22 (1.08,1.38)	1.25 (1.10,1.41)	1.24 (01.09,1.40)
West North Central	1.10 (0.98,1.23)	1.11 (0.99,1.24)	1.06 (0.95,1.19)	1.21 (1.06,1.37)	1.09 (0.97,1.23)	1.12 (1.00,1.25)	1.08 (0.95,1.21)
South Atlantic	1.21 (1.08,1.36)	1.22 (1.09,1.37)	1.16 (1.03,1.31)	1.22 (1.09,1.37)	1.21 (1.08,1.36)	1.24 (1.10,1.39)	1.16 (1.02,1.31)
East South Central	1.29 (1.13,1.48)	1.30 (1.14,1.49)	1.24 (1.08,1.42)	1.30 (1.14,1.49)	1.30 (1.14,1.49)	1.32 (1.16,1.51)	1.24 (1.08,1.42)
West South Central	1.20 (1.05,1.37)	1.21 (1.06,1.38)	1.15 (1.01,1.31)	1.21 (1.06,1.37)	1.19 (1.04,1.37)	1.23 (1.08,1.39)	1.16 (1.01,1.33)
Mountain	1.08 (0.97,1.20)	1.07 (0.96,1.19)	1.05 (0.94,1.17)	1.08 (0.96,1.20)	1.06 (0.95,1.19)	1.07 (0.96,1.20)	1.06 (0.95,1.19)
Pacific	1.07 (0.94,1.21)	1.03 (0.91,1.16)	1.06 (0.94,1.19)	1.03 (0.91,1.16)	1.02 (0.91,1.15)	1.02 (0.91,1.15)	1.07 (0.95,1.20)

Table 3. *Cont.*

	Odds for Infant Mortality						
	OR	OR	OR	OR	OR	OR	OR
Individual Characteristics	95%CI	95%CI	95%CI	95%CI	95%CI	95%CI	95%CI
Birth cohort							
2009 (reference group: 2008)	0.97	0.97	0.97	0.97	0.97	0.97	0.97
	(0.95,0.99)	(0.95,0.99)	(0.95,0.99)	(0.95,0.99)	(0.95,0.99)	(0.95,0.99)	(0.95,0.99)
2010 (reference group: 2008)	0.94	0.94	0.94	0.94	0.94	0.94	0.94
	(0.93,0.96)	(0.93,0.96)	(0.93,0.96)	(0.93,0.95)	(0.93,0.96)	(0.93,0.96)	(0.93,0.96)
Mother's Age (years)							
	(0.99,1.01)	(0.99,1.01)	(0.99,1.01)	(0.99,1.01)	(0.99,1.01)	(0.99,1.01)	(0.99,1.01)
Mother's Race ((reference group): white)							
Black	1.88	1.88	1.88	1.88	1.88	1.88	1.88
	(1.84,1.92)	(1.84,1.92)	(1.84,1.92)	(1.84,1.92)	(1.84,1.92)	(1.84,1.92)	(1.84,1.92)
Native	1.31	1.31	1.31	1.31	1.31	1.31	1.31
	(1.23,1.40)	(1.23,1.40)	(1.23,1.40)	(1.23,1.40)	(1.23,1.40)	(1.23,1.40)	(1.23,1.40)
Asian	1.15	1.15	1.15	1.15	1.15	1.15	1.15
	(1.10,1.20)	(1.10,1.20)	(1.10,1.20)	(1.10,1.20)	(1.10,1.20)	(1.10,1.20)	(1.10,1.20)
Hispanic	0.98	0.98	0.98	0.98	0.98	0.98	0.98
	(0.96,1.01)	(0.96,1.01)	(0.96,1.01)	(0.96,1.01)	(0.96,1.01)	(0.96,1.01)	(0.96,1.01)
Other	1.25	1.25	1.25	1.25	1.25	1.25	1.25
	(1.12,1.39)	(1.12,1.39)	(1.12,1.39)	(1.12,1.39)	(1.12,1.39)	(1.12,1.39)	(1.12,1.39)
Education (ref: less than high school)							
High School	0.89	0.89	0.89	0.89	0.89	0.89	0.89
	(0.87,0.91)	(0.87,0.91)	(0.87,0.91)	(0.87,0.91)	(0.87,0.91)	(0.87,0.91)	(0.87,0.91)
Post-Secondary	0.68	0.68	0.68	0.68	0.68	0.68	0.68
	(0.66,0.69)	(0.66,0.69)	(0.66,0.69)	(0.66,0.69)	(0.66,0.69)	(0.66,0.69)	(0.66,0.69)
With Partner (reference group: coupled)							
Single	1.33	1.33	1.33	1.33	1.33	1.33	1.33
	(1.30,1.35)	(1.32,1.35)	(1.30,1.35)	(1.30,1.35)	(1.30,1.35)	(1.30,1.35)	(1.30,1.35)
Nativity (reference group: born outside USA)							
US born	1.30	1.30	1.30	1.31	1.30	1.30	1.30
	(1.27,1.33)	(1.27,1.33)	(1.27,1.33)	(1.23,1.40)	(1.27,1.33)	(1.27,1.33)	(1.27,1.33)
Birth Order (reference group: first born)							
Second	1.04	1.04	1.04	1.04	1.04	1.04	1.04
	(1.02,1.06)	(1.02,1.06)	(1.02,1.06)	(1.02,1.06)	(1.02,1.06)	(1.02,1.06)	(1.02,1.06)
Third	1.11	1.11	1.11	1.11	1.11	1.11	1.11
	(1.08,1.14)	(1.08,1.14)	(1.08,1.14)	(1.08,1.14)	(1.08,1.14)	(1.08,1.14)	(1.08,1.14)
	1.39	1.39	1.39	1.39	1.39	1.39	1.39
Fourth or more	(1.36,1.42)	(1.36,1.42)	(1.36,1.42)	(1.36,1.42)	(1.36,1.42)	(1.36,1.42)	(1.36,1.42)

Table 4. The relationship between abortion laws and infant mortality controlling for individual and state-level characteristics, 2008–2010.

| | Odds for Infant Mortality | | | | | | | | | |
| | ≤19 Years Age | | 20 to 25 Years | | 25 to 30 Years | | 30 to 34 Years | | 35 and Older | |
	OR	95%CI	OR	95%CI	OR	95%CI	OR	95%CI	OR	95%CI
State Characteristics										
Parental Involvement (ref: no)										
Yes	1.09	(1.00,1.19)	1.10	(1.03,1.17)	1.03	(0.94,1.12)	1.07	(0.97,1.18)	1.07	(0.98,1.16)
State Median Income Z-score	0.91	(0.84,0.99)	0.98	(0.92,1.04)	0.95	(0.89,1.02)	1.04	(0.95,1.14)	0.89	(0.82,0.96)
Proportion Black Z-score	1.05	(0.99,1.11)	0.97	(0.93,1.01)	1.00	(0.94,1.05)	0.95	(0.89,1.03)	1.05	(0.98,1.12)
Proportion Poor Z-score	0.88	(0.81,0.96)	0.99	(0.93,1.06)	0.99	(0.91,1.07)	0.99	(0.89,1.11)	0.87	(0.79,0.96)
State Population Z-score	0.99	(0.96,1.02)	0.99	(0.97,1.01)	1.00	(0.98,1.03)	0.98	(0.95,1.02)	1.02	(0.99,1.05)
Census Division (ref: New England)	1.00	1.00	1.00	1.00	1.00	1.00	1.00	1.00	1.00	1.00
Middle Atlantic	1.14	(0.96,1.35)	1.12	(0.99,1.27)	1.00	(0.87,1.15)	1.11	(0.95,1.30)	0.93	(0.82,1.07)
East North Central	1.17	(0.99,1.38)	1.21	(1.07,1.37)	1.18	(1.03,1.35)	1.40	(1.20,1.63)	1.15	(1.01,1.32)
West North Central	1.06	(0.89,1.26)	1.07	(0.93,1.22)	1.03	(0.89,1.19)	1.16	(0.98,1.37)	0.98	(0.85,1.14)
South Atlantic	1.06	(0.89,1.26)	1.20	(1.05,1.37)	1.11	(0.96,1.29)	1.26	(1.05,1.50)	0.97	(0.83,1.13)
East South Central	1.18	(0.98,1.41)	1.24	(1.06,1.44)	1.26	(1.07,1.49)	1.42	(1.18,1.70)	1.22	(1.03,1.44)
West South Central	1.16	(0.97,1.38)	1.14	(0.99,1.30)	1.11	(0.96,1.29)	1.38	(1.16,1.65)	1.17	(1.01,1.36)
Mountain	1.15	(0.97,1.36)	0.98	(0.86,1.13)	1.05	(0.91,1.21)	1.24	(1.06,1.45)	1.12	(0.97,1.29)
Pacific	1.14	(0.95,1.37)	1.08	(0.94,1.24)	0.99	(0.85,1.15)	1.11	(0.93,1.33)	1.02	(0.88,1.19)
Individual Characteristics										
Birth cohort										
2008 (ref)	1.00	1.00	1.00	1.00	1.00	1.00	1.00	1.00	1.00	1.00
2009'	0.95	(0.91,1.00)	0.98	(0.95,1.01)	0.98	(0.94,1.01)	0.99	(0.94,1.04)	0.94	(0.89,0.98)
2010'	0.91	(0.87,0.96)	0.95	(0.92,0.98)	0.96	(0.92,0.99)	1.00	(0.95,1.05)	0.89	(0.85,0.93)
Mother's Race (ref: white)	1.00	1.00	1.00	1.00	1.00	1.00	1.00	1.00	1.00	1.00
Black	1.44	(1.37,1.51)	1.72	(1.66,1.78)	2.01	(1.93,2.10)	2.29	(2.16,2.42)	2.35	(2.22,2.49)
Native	0.99	(0.85,1.16)	1.34	(1.21,1.49)	1.34	(1.17,1.55)	1.39	(1.13,1.71)	1.45	(1.16,1.81)
Asian	1.26	(1.06,1.49)	1.07	(0.98,1.18)	1.18	(1.09,1.29)	1.26	(1.15,1.38)	1.14	(1.04,1.25)
Hispanic	0.83	(0.78,0.89)	0.93	(0.89,0.97)	1.04	(0.99,1.10)	1.10	(1.03,1.18)	1.13	(1.06,1.22)
Other	1.02	(0.76,1.36)	1.16	(0.96,1.41)	1.20	(0.96,1.51)	1.44	(1.10,1.88)	1.44	(1.13,1.84)
Education (ref: less than high school)	1.00	1.00	1.00	1.00	1.00	1.00	1.00	1.00	1.00	1.00
High School	0.86	(0.82,0.90)	0.88	(0.85,0.91)	0.93	(0.89,0.97)	1.01	(0.94,1.08)	0.98	(0.92,1.05)
Post-Secondary	0.82	(0.76,0.88)	0.71	(0.69,0.74)	0.73	(0.69,0.76)	0.74	(0.69,0.79)	0.68	(0.63,0.73)
Martial Status (ref: coupled)	1.00	1.00	1.00	1.00	1.00	1.00	1.00	1.00	1.00	1.00
Single	1.11	(1.04,1.19)	1.24	(1.21,1.28)	1.34	(1.30,1.39)	1.39	(1.32,1.45)	1.27	(1.21,1.33)
Nativity (ref: born outside USA)	1.00	1.00	1.00	1.00	1.00	1.00	1.00	1.00	1.00	1.00
US born	1.28	(1.18,1.38)	1.32	(1.26,1.38)	1.36	(1.30,1.43)	1.32	(1.24,1.40)	1.22	(1.15,1.29)
Birth Order (ref: first)	1.00	1.00	1.00	1.00	1.00	1.00	1.00	1.00	1.00	1.00
Second	1.49	(1.42,1.56)	1.07	(1.03,1.10)	0.96	(0.91,1.00)	0.83	(0.78,0.88)	0.79	(0.74,0.85)
Third	1.83	(1.70,1.98)	1.22	(1.18,1.27)	1.04	(1.00,1.10)	0.90	(0.84,0.96)	0.78	(0.73,0.84)
Fourth or more	1.76	(1.58,1.96)	1.54	(1.48,1.60)	1.37	(1.31,1.43)	1.12	(1.06,1.19)	0.99	(0.93,1.05)

6. Discussion

Since the federal legalization of abortion in 1973, many US states have successfully enacted laws restricting access to abortion, the majority of which started in the past two decades [25–27]. As a result, a woman's access to abortion services varies greatly across US states. We exploited variation in state-level restrictive abortion laws and the individual risk for infant mortality, with special attention to differential effects by maternal characteristics. We observed a significant relationship between the number of restrictive abortion laws and infant mortality risk, indicating a potential additive effect. More specifically, infants born in states with three or more restrictive laws were significantly more likely to die before their first birthday than were those born in states with no restrictions.

We posit that the 10% increase that we observed is meaningful from a population health standpoint. There were 22,000 infant deaths in the United States in 2017, a disproportionate number of which occurred in states with restrictive abortion laws [28]. A 10% reduction in infant mortality in these states could eliminate hundreds of excess infant deaths per year.

These findings are consistent with ecological studies that have identified a significant relationship between the legalization of abortion within the US and state funding for family planning and abortion services and infant mortality rates. For example, Krieger et al. observed US infant death rates declined most quickly between 1970 and 1973 in states that legalized abortion in 1970 [29]. They also found that, since 2000, the rate ratio for infant death comparing states in the top funding quartile to states with no funding for abortion services revealed an average 15% reduction in risk among the top funders [7].

The Sexual and Reproductive Framework (SRJ) defines reproductive rights as human rights and recognizes the multiple forms of oppression that impact individuals' decisions about their sexual and reproductive health [30]. By interpreting our results using a SRJ framework, we further emphasize the need for legislation to acknowledge how women's sexual and reproductive health decisions are shaped by social and contextual factors as well as by individual-level resources [30]. Our results support SRJ assertions by suggesting that unintended, and differentially strong adverse effects associated with abortion policies that restrict women's reproductive decision-making. Governments should promote sexual health policies that respect women's reproductive decisions as part of their efforts to enhance population health [31,32]. Furthermore, it could be argued that preventing unwanted pregnancies itself and the consequent need for abortions is crucial for reproductive, maternal, and infant health. Targeting upstream factors—e.g., improved and more expansive reproductive services, especially access to contraception—could decrease the number of unwanted pregnancies, thus eliminating the need for abortions.

One reason for the observed findings could be that restrictions on abortion, through their limiting of the right to make a medical choice, could themselves have detrimental effects on the health of the mother, and consequently on the infant's health as well. Abortion restrictions impede a woman's ability to make health decisions and to exercise autonomy over her reproductive life. This autonomy has been identified as a fundamental human right and an important determinant of women's health [33]. Thus, when women are compelled to carry a pregnancy to term, giving birth may have detrimental effects on the woman's own health and, therefore, on their infant's well-being. Additionally, social stigma attached to abortion may also have an effect on mental health of women seeking to terminate their pregnancy. For example, researchers observed that women who were denied an abortion and, therefore, carried their pregnancies to term experienced psychological distress a year later [34], which can detrimentally affect the health of their infants. Researchers may want to explore casual methods to understand whether abortion laws cause infant mortality. However, given how such laws are implemented the common assumptions of causal methods may be violated. Furthermore, future research is needed to determine whether the mechanisms linking abortion restriction laws leads to adverse infant health independently or through its effects on the mother.

In addition to directly impacting health and wellness, the state laws analyzed here may be markers for an array of factors comprising a state's socio-political environment that fosters inequality and harms population health. In other words, states who enact restrictive abortion laws are more likely to support and fund reproductive and women's health. This can have detrimental consequences on women's health and therefore on, maternal and infant health [35].

We also observed significant heterogeneous associations across socio-demographic groups. When stratified by mother's age (21 and younger and 22 and older), mandatory parental involvement, was significantly associated with an increased risk for mortality among infants of mothers from both age groups. One explanation may be that states with parental involvement laws are associated with an increase in the price of an abortion by 14% [36]. This increase in price may further act as a barrier for young mothers in obtaining an abortion. Among infants born to Black mothers, those born in states with Medicaid restrictions and parental involvement laws were more likely to die than were those born in states without these laws. Black mothers are disproportionately more likely to be from lower socioeconomic status backgrounds and are therefore less apt to have access to health services. Additionally, Black girls are more than twice as likely as White girls to become pregnant [37]. Low socioeconomic status and teenage motherhood are each risk factors for infant mortality.

One unexpected finding was that among infants born to White mothers, those in states with mandatory wait periods were significantly less likely to die in comparison to those states without this law. Mandatory wait periods may be effective at delaying an abortion, which in theory could encourage women to use unsafe practices [38]. The protective finding of mandatory wait periods against infant mortality among White mothers requires further study, including, potentially a qualitative analysis.

Infant mortality rates have been on the decline in recent years. The drop from 6.1/1000 live births in 2009 to 5.8/1000 in 2015 may be explained by decreasing birth rates among young teenage girls [39]. Infants of adolescent mothers are at greater risk for mortality than are those born to adult mothers [39]. Younger mothers may not have sufficient access to resources and health care that are essential for optimal health for their newborns. As births to teen mothers decrease, infant mortality rates could drop as a result.

Strengths and Limitations

This study's results should be interpreted in light of several limitations. Data analyzed for this study was collected from infants born 2008–2010, which was the most recent data available at the time of this investigation. However, one reason for using not using the most recent data is that the CDC does not release individual-level data immediately since it takes years for the data to be prepared, cleaned, and de-identified. Once this has been completed, the data is made available for public use. A delay in our investigation is further caused since obtaining state residence information requires additional paperwork and approval. Thus, linked state-individual data from 2008 to 2010, is the most temporally appropriate data to be utilized for this investigation. Since the study was not a randomized controlled trial (i.e., we could not randomize mothers into states with and without restrictions on abortion) we cannot assess whether observed associations are causal. Furthermore, because we only had data from 2008–2010 and were limited by the number of states changing or adding restrictive abortion laws, we did not utilize a quasi-experimental approach. Furthermore, endogeneity could be an issue due to residual confounding: potential confounders such as individual household income and other socioeconomic conditions were not available. Therefore, the inability to use a quasi-experimental study design and endogeneity due to residual confounding limits our ability to draw causal inferences. Lastly, although we were able to examine the moderating effects of education, we could not test whether household income acts as an effect modifier of the relationship between abortion restrictions and infant mortality.

7. Conclusions

Although we cannot conclude a causal relationship, results from this investigation indicate that the number of state-level restrictions on abortion may be a significant risk factor for infant mortality. In particular, mandatory parental laws that require either parental permission or notification for a minor to have an abortion are significantly associated with infant mortality. Future studies should identify the extent to which this relationship is causal. For example, researchers can take advantage of the passage of new laws against abortion, such as banning abortions once a fetal heartbeat is detected, or restricting providers with unnecessary requirements. These changes present a unique opportunity to better understand how and why these restrictions may cause adverse health outcomes for mothers and affect infant mortality risk.

Author Contributions: R.P. conceptualized and designed the study, conducted the analyses, drafted the initial manuscript, and reviewed and revised the manuscript. S.Y.L. helped to conceptualize and design the study, interpret the findings, wrote subsequent drafts of the manuscript, and critically reviewed the manuscript for important intellectual content. A.E., D.M.C., M.R., and P.M., helped to interpret the results, reviewed and revised the manuscript, and critically reviewed the manuscript for important intellectual content. All authors approved the final manuscript as submitted and agree to be accountable for all aspects of the work. All authors have read and agreed to the published version of the manuscript.

Funding: National Institutes of Health Research, National Institute on Minority Health and Health Disparities 1R15MD010223-01. Roman Pabayo is a Tier II Canada Research Chair in social and health inequities throughout the lifespan. The authors have no financial relationships relevant to this article to disclose.

Acknowledgments: We thank Elizabeth Nash, Senior State Issues Manager at the Guttmacher Institute for the data on State-level Abortion Restrictive Laws.

Conflicts of Interest: The authors declare no conflict of interest.

Abbreviations

OR—Odds ratio. AOR—Adjusted odds ratio. OECD—Organization for Economic Co-operation and Development. SD—Standard deviation. LBID—Linked birth and infant death records. NCHS—National Center for Health Statistics. Using the 2008–2010 U.S. Cohort linked birth/infant death data files, this study identifies the relationship between state-level restrictive laws and infant mortality risk.

References

1. Anderson, R.N. A method for constructing complete annual U.S. life tables. *Vital Health Stat. 2* **2000**, *129*, 1–28.
2. OECD. *Health at a Glance 2013: OECD Indicators*; OECD Publishing: Paris, France, 2013. [CrossRef]
3. Kim, D.; Saada, A. The social determinants of infant mortality and birth outcomes in Western developed nations: A cross-country systematic review. *Int. J. Environ. Res. Public Health* **2013**, *10*, 2296–2335. [CrossRef] [PubMed]
4. Mathews, T.J.; MacDorman, M.F. Infant mortality statistics from the 2009 period linked birth/infant death data set. National vital statistics reports: From the Centers for Disease Control and Prevention, National Center for Health Statistics. *Natl. Vital Stat. Rep.* **2013**, *61*, 1–27. [PubMed]
5. Murphy, S.L.; Xu, J.; Kochanek, K.D.; Arias, E. Mortality in the United States, 2017. *NCHS Data Brief.* **2018**, *328*, 1–8.
6. Espey, E.; Dennis, A.; Landy, U. The importance of access to comprehensive reproductive health care, including abortion: A statement from women's health professional organizations. *Am. J. Obstet. Gynecol.* **2019**, *220*, 67–70. [CrossRef] [PubMed]
7. Krieger, N.; Gruskin, S.; Singh, N.; Kiang, M.V.; Chen, J.T.; Waterman, P.D.; Beckfield, J.; Coull, B.A. Reproductive justice & preventable deaths: State funding, family planning, abortion, and infant mortality, US 1980–2010. *SSM Popul. Health* **2016**, *2*, 277–293. [PubMed]
8. Glasier, A.; Gulmezoglu, A.M. Putting sexual and reproductive health on the agenda. *Lancet* **2006**, *368*, 1550–1551. [CrossRef]
9. Haddad, L.B.; Nour, N.M. Unsafe abortion: Unnecessary maternal mortality. *Rev. Obstet. Gynecol.* **2009**, *2*, 122.
10. Finer, L.; Fine, J.B. Abortion law around the world: Progress and pushback. *Am. J. Public Health* **2013**, *103*, 585–589. [CrossRef]
11. Lobel, M.; DeVincent, C.J.; Kaminer, A.; Meyer, B.A. The impact of prenatal maternal stress and optimistic disposition on birth outcomes in medically high-risk women. *Health Psychol.* **2000**, *19*, 544–553. [CrossRef]

12. Rini, C.K.; Dunkel-Schetter, C.; Wadhwa, P.D.; Sandman, C.A. Psychological adaptation and birth outcomes: The role of personal resources, stress, and sociocultural context in pregnancy. *Health Psychol.* **1999**, *18*, 333–345. [CrossRef] [PubMed]

13. Baibazarova, E.; van de Beek, C.; Cohen-Kettenis, P.T.; Buitelaar, J.; Shelton, K.H.; van Goozen, S.H. Influence of prenatal maternal stress, maternal plasma cortisol and cortisol in the amniotic fluid on birth outcomes and child temperament at 3 months. *Psychoneuroendocrinology* **2013**, *38*, 907–915. [CrossRef] [PubMed]

14. Larsen, A.D. The effect of maternal exposure to psychosocial job strain on pregnancy outcomes and child development. *Dan. Med. J.* **2015**, *62*, 2.

15. Pfost, K.S.; Stevens, M.J.; Lum, C.U. The relationship of demographic variables, antepartum depression, and stress to postpartum depression. *J. Clin. Psychol.* **1990**, *46*, 588–592. [CrossRef]

16. Gelaye, B.; Rondon, M.B.; Araya, R.; Williams, M.A. Epidemiology of maternal depression, risk factors, and child outcomes in low-income and middle-income countries. *Lancet Psychiatry* **2016**, *3*, 973–982. [CrossRef]

17. Medoff, M.H. The relationship between restrictive state abortion laws and postpartum depression. *Soc. Work Public Health* **2014**, *29*, 481–490. [CrossRef]

18. White, S.E.; Gladden, R.W. Maternal mental health and infant mortality for healthy-weight infants. *Am. J. Manag. Care* **2016**, *22*, e389–e392.

19. Sanderson, C.A.; Cowden, B.; Hall, D.M.; Taylor, E.M.; Carpenter, R.G.; Cox, J.L. Is postnatal depression a risk factor for sudden infant death? *Br. J. Gen. Pract.* **2002**, *52*, 636–640.

20. Yazdkhasti, M.; Pourreza, A.; Pirak, A.; Abdi, F. Unintended Pregnancy and Its Adverse Social and Economic Consequences on Health System: A Narrative Review Article. *Iran. J. Public Health* **2015**, *44*, 12–21.

21. Muennig, P.; Reynolds, M.M.; Jiao, B.; Pabayo, R. Why Is Infant Mortality in the United States So Comparatively High? Some Possible Answers. *J. Health Polit. Policy Law* **2018**, *43*, 877–895. [CrossRef]

22. Foster, D.G.; Biggs, M.A.; Raifman, S.; Gipson, J.; Kimport, K.; Rocca, C.H. Comparison of Health, Development, Maternal Bonding, and Poverty among Children Born after Denial of Abortion vs. after Pregnancies Subsequent to an Abortion. *JAMA Pediatr.* **2018**, *172*, 1053–1060. [CrossRef] [PubMed]

23. Nour, N.M. An introduction to maternal mortality. *Rev. Obstet. Gynecol.* **2008**, *1*, 77–81. [PubMed]

24. Latt, S.M.; Milner, A.; Kavanagh, A. Abortion laws reform may reduce maternal mortality: An ecological study in 162 countries. *BMC Women's Health* **2019**, *19*, 1. [CrossRef] [PubMed]

25. Guttmacher Institute. *State Policies in Brief*; Guttmacher Institute: New York, NY, USA, 2008.

26. Guttmacher Institute. *State Policies in Brief*; Guttmacher Institute: New York, NY, USA, 2009.

27. Guttmacher Institute. *State Policies in Brief*; Guttmacher Institute: New York, NY, USA, 2010.

28. Centers for Disease Control and Prevention: Reproductive Health, Maternal and Infant Health. Available online: https://www.cdc.gov/reproductivehealth/maternalinfanthealth/infantmortality.htm (accessed on 17 May 2020).

29. Krieger, N.; Gruskin, S.; Singh, N.; Kiang, M.V.; Chen, J.T.; Waterman, P.D.; Gottlieb, J.; Beckfield, J.; Coull, B.A. Reproductive justice and the pace of change: Socioeconomic trends in US infant death rates by legal status of abortion, 1960–1980. *Am. J. Public Health* **2015**, *105*, 680–682. [CrossRef]

30. Luna, Z.; Luker, K. Reproductive Justice. *Annu. Rev. Law Soc. Sci.* **2013**, *9*, 327–352. [CrossRef]

31. Verbiest, S.; Malin, C.K.; Drummonds, M.; Kotelchuck, M. Catalyzing a Reproductive Health and Social Justice Movement. *Matern Child. Health J.* **2016**, *20*, 741–748. [CrossRef]

32. Siegel, R.; Greenhouse, L. *The Unfinished Story of Roe V. Wade*; Public Law Research Paper No. 643; Reproductive Rights and Justice Stories; Murray, M., Shaw, K., Siegel, R., Eds.; Yale Law School: New Haven, CT, USA, 2018; (2019 Forthcoming).

33. Childress, J.F.; Beauchamp, T.L. *Principles of Biomedical Ethics*, 5th ed.; Oxford University Press: New York, NY, USA, 2001.

34. Biggs, M.A.; Brown, K.; Foster, D.G. Perceived abortion stigma and psychological well-being over five years after receiving or being denied an abortion. *PLoS ONE* **2020**, *15*, e0226417. [CrossRef]

35. Gerdts, C.; Dobkin, L.; Foster, D.G.; Schwarz, E.B. Side Effects, Physical Health Consequences, and Mortality Associated with Abortion and Birth after an Unwanted Pregnancy. *Womens Health Issues* **2016**, *26*, 55–59. [CrossRef]

36. Medoff, M.H. The Spillover Effects of Restrictive Abortion Laws. *Gender Issues* **2008**, *25*, 1–10. [CrossRef]

37. Romero, L.; Pazol, K.; Warner, L.; Cox, S.; Kroelinger, C.; Besera, G.; Brittain, A.; Fuller, T.R.; Koumans, E.; Barfield, W. Reduced Disparities in Birth Rates among Teens Aged 15–19 Years—United States, 2006–2007 and 2013–2014. *MMWR Morb. Mortal. Wkly. Rep.* **2016**, *65*, 409–414. [CrossRef]
38. World Health Organization. *Safe Abortion: Technical and Policy Guidance for Health Systems*, 2nd ed.; World Health Organization: Geneva, Switzerland, 2012.
39. He, X.; Akil, L.; Aker, W.G.; Hwang, H.M.; Ahmad, H.A. Trends in infant mortality in United States: A brief study of the Southeastern states from 2005–2009. *Int. J. Environ. Res. Public Health* **2015**, *12*, 4908–4920. [CrossRef] [PubMed]

International Journal of
*Environmental Research
and Public Health*

Article

Psychometric Properties of the Condom Use Self-Efficacy Scale among Young Colombians

Vanessa Sanchez-Mendoza [1,2,*], Encarnacion Soriano-Ayala [3] and Pablo Vallejo-Medina [4]

[1] International PhD School, Universidad de Almeria, 04120 La Cañada, Spain
[2] School of Psychology, Fundación Universitaria Konrad Lorenz, Bogotá 110231, Colombia
[3] Department of Humanities and Education, Universidad de Almeria, 04120 La Cañada, Spain;
     esoriano@ual.es
[4] SexLab KL—Human Sexuality Laboratory, School of Psychology, Fundación Universitaria Konrad Lorenz,
     Bogotá 110231, Colombia; pablo.vallejom@konradlorenz.edu.co
*   Correspondence: vsm646@inlumine.ual.es

Received: 24 March 2020; Accepted: 21 May 2020; Published: 26 May 2020

Abstract: (1) Background: This study evaluated the psychometric properties of the Condom Use Self-Efficacy Scale among Colombian youth. (2) Method: A total of 2873 men and women between 18 and 26 years old ($M = 21.45$, $SD = 2.26$) took part in this study. All participants answered a socio-demographic survey, the Condom Use Self-Efficacy Scale, the UCLA Multidimensional Condom Attitudes Scale, The Condom Use Errors and Problems Scale, and the Sexual Assertiveness Scale. Sampling was web-based, and the survey was distributed via Facebook. (3) Results: The Condom Use Self-Efficacy Scale demonstrated adequate reliability (ordinal α ranged = 0.76 to 0.92). Exploratory and confirmatory factor analysis suggested a four-factor structure with an explained variance of 69%. This dimensionality was also invariant across gender. Moreover, positive attitudes toward condom use were significantly associated with appropriation and assertiveness. Two dimensions (appropriation and partner disapproval) showed significant gender differences. (4) Conclusions: The Spanish–Colombian version of the Condom Use Self-Efficacy Scale is a psychometrically adequate instrument to measure perceived condom use self-efficacy. This scale can be used in both research and professional settings to measure self-efficacy at using condoms in young people.

Keywords: HIV prevention; unintended pregnancies; condom use; sexual risk; Latins; psychometric; validity; sexual behavior; STI prevention

---

1. Introduction

Sexual health is a significant interest area for public health in Colombia. The Colombian government's 2030 Agenda for Sustainable Development has established goals and indicators concerning health and well-being. Some of those goals are related to sexual and reproductive health: (1) to reduce the HIV/AIDS deaths tolls from 4.9 per 1000 people to 2.4 per 1000 people, and (2) to increase the percentage of women between the ages of 15 and 49 years old who use contraceptive methods from 68.1% to 81.4%. Those goals will be achieve over the next ten years [1]. Additionally, the government created the Ten-Year Public Health Plan 2012–2022 and The National Sexual and Reproductive Rights Policy. Both documents are written understanding health as a human right, gender differences in access to healthcare, taking a life course approach, and attending to social and behavioral factors [2].

In 2018, 36.2 million people lived with HIV/AIDS around the world, 1.7 million individuals became infected in 2018. Although the Caribbean, Central Africa, Europe, and North America reported significant decreases in new infections between 2010 and 2018, Latin America saw a 7% increase in cases [3]. In Colombia, HIV increased by even more than Latin America as a whole. Between 2015 and 2019, new HIV infections increased by 17.3%. The distribution of new infections by population group (among people aged 19–45 years) in Latin America shows that 40% of new infections occur amongst men who have sex with other men, 25% occur amongst sex workers, illegal drug users, and transgender people, and 35% occur in other segments of the population. In 2019, Colombia had 17,502 new cases. Males comprised 80.8%, and sexual intercourse was the cause of infection in most cases (17,219) [4].

Other sexually transmitted infections (STIs) are also a major problem. Global data show that more than a million people are infected daily, and an estimated 376 million new infections occur each year [5]. More than 500 million people are estimated to suffer from the herpes simplex virus, and more than 290 million women have one kind of human papillomavirus infection [6]. In Colombia, 94,000 people sought medical attention due to an STI between 2009 and 2011; according to individual service provision records, the age of highest prevalence was between 20 and 29 years of age, and the female population was the most affected [7]. The impacts of these infections vary across populations, depending on the biological, social, behavioral, and economic factors affecting also the reactions to and physiological results of STIs. Some of these are infertility [8], social inequality [5], depression and associated mental health disorders [9], and increases in healthcare expenditures [10].

Finally, another problem related to sexual health is unintended pregnancies. Latin America is one of the regions with the highest rate of unintended pregnancies; 122 out of 1000 women aged 15–44 are pregnant, but 55% were unintended. Latin America is the only region in the world where the percentage of unplanned pregnancies exceeds planned ones [11]. In Colombia, poverty and social factors are the major reasons to explain the high rates of teen pregnancy, one in every five women between 15 and 19 years of age are or have been pregnant and just 33.6% are planned [12]. Unintended pregnancies in women cause problems such as maternal death (830 women die during childbirth every day), it perpetuates the cycle of poverty in developing countries where 99% of mortality occurs and 7% of women drop out of school due to pregnancy, and finally it increases the risk of stillbirth due to preeclampsia, placental abruption, or maternal malnutrition [6].

Condoms are inexpensive and highly effective means at preventing HIV, STIs, and unintended pregnancies. In Colombia, consistent condom use is uncommon (22%) [13], its use oftentimes is doubtful and incorrect [14]. The greatest predictor of regular, consistent condom use is a person's attitudes towards condoms (i.e., personal positive evaluation that condoms actually work) [15,16], subjective norms or beliefs (i.e., perceived normative support for condom use) [17–19] and self-efficacy (i.e., their belief that they can actually use condoms) [16,20]. Self-efficacy is supported by the social cognitive theory, according to which people's behavior depends on how they evaluate their own abilities [21,22]. Thus, one of the reasons why young people take high risks when they engage in sexual activity may be associated with low perceived self-efficacy about their competence to negotiate condom use. Unlike the attitudes and subjective norms, there are no validated or translated standardized scales in Colombia to evaluate this key construct of sexual health.

Previous studies described above underscore the importance of socio-cognitive variables, such as self-efficacy, to increase condom use frequency and contribute to the prevention of STIs [23,24]. One of the scales used to assess self-efficacy at using condoms is the Condom Use Self-Efficacy Scale (CUSES) [25], which assesses a person's feelings of confidence about buying condoms, putting them on and taking them off, and also the act of negotiating the use of it with a sexual partner. This scale has been used with college students from different cultural backgrounds [26] and other young people's samples [23,24,27]. Validation studies have reported on the adequate psychometric properties of the CUSES as a valid and reliable instrument to measure condom use self-efficacy in different populations. Although developing a scale using the intended respondents' native language remains the best option to measure self-efficacy adequately in a given context, validating and using an existing scale is not only economical, it also provides backgrounds for comparison in other settings [28].

The present study evaluated the psychometric properties of the CUSES in Colombia and established relationships between this scale and results from other validated instruments administered to Colombian participants.

2. Method

2.1. Sample

Participants were recruited through Facebook. Initially, 3649 young people clicked on the scale; 281 of these acknowledged informed consent but failed to respond, seven failed to indicate their sex, 15 were over 26 years of age, 24 were not Colombian, and 449 did not answer the CUSES fully. A total of 2873 young people completed the sociodemographic questionnaire and the CUSES. Inclusion criteria were being between 18 and 26 years of age and being Colombian. Exclusion criteria were failure to provide informed consent or not completing the questionnaires. The mean age was 21.45 years (SD = 2.26). A total of 33.8% (n = 971) were male and 66.2% (n = 1902) were female. The characteristics of the sample are described in Table 1.

Table 1. Sample description.

	Males (n = 971)	Females (n = 1902)	Total (n = 2873)
	N %	*N %*	*N %*
Age			
18	81 (8.3)	230 (12.1)	311 (10.8)
19	109 (11.2)	267 (14)	376 (13.1)
20	134 (13.8)	282 (14.8)	416 (14.5)
21	131 (13.5)	287 (15.1)	418 (15.5)
22	123 (12.7)	236 (12.4)	359 (12.5)
23	127 (13.1)	239 (12.6)	366 (12.7)
24	128 (13.2)	154 (8.1)	282 (9.8)
25	107 (11)	163 (8.6)	270 (9.6)
26	31 (3.2)	44 (2.3)	75 (2.6)
Schooling level N (%)			
Basic Elementary	1 (0.1)		1 (0.0)
Secondary	52 (5.4)	101 (5.3)	153 (5.3)
First Technical level	59 (6.1)	142 (7.5)	201 (7)
Second Technical level	51 (5.3)	85 (4.5)	136 (4.7)
College undergraduate	573 (59)	1159 (60.9)	1732 (60.3)
College graduate	183 (18.8)	323 (17)	506 (17.6)
Postgraduate candidate	41 (4.2)	48 (2.5)	89 (3.1)
Postgraduate	11 (1.1)	44 (2.3)	55 (1.9)
Marital Status N (%)			
Married	14 (1.4)	26 (1.4)	40 (1.4)
Single	867 (89.5)	1659 (87.5)	2526 (88.1)
Widowed	1 (0.1)		1 (0.0)
Co-habiting	81 (8.4)	206 (10.9)	287 (10)
Separated/Divorced	6 (0.6)	6 (0.3)	12 (0.4)
Sexual orientation N (%)			
Exclusively heterosexual	738 (76.2)	1431 (75.4)	2169 (75.6)
Predominantly heterosexual, only incidentally homosexual	66 (6.8)	297 (15.6)	363 (12.7)
Predominantly heterosexual, but more than incidentally homosexual	10 (1)	60 (3.2)	70 (2.4)
Equally heterosexual and homosexual	15 (1.5)	66 (3.5)	81 (2.8)
Predominantly homosexual, but more than incidentally heterosexual	10 (1)	9 (0.5)	19 (0.7)
Predominantly homosexual, only incidentally heterosexual	28 (2.9)	12 (0.6)	40 (1.4)
Exclusively homosexual	98 (10.1)	13 (0.7)	111 (3.9)
Asexual	4 (0.4)	11 (0.6)	15 (0.5)

Table 1. *Cont.*

	Males (*n* = 971)	Females (*n* = 1902)	Total (*n* = 2873)
	N %	*N* %	*N* %
Do you have a partner that you have been seeing for more than 6 months?			
Yes	491 (50.7)	1179 (62.1)	1670 (58.1)
No	477 (49.3)	719 (37.9)	1196 (41.6)
How often do you use condoms during sexual intercourse?			
Every time	271 (28)	334 (17.6)	605 (21.1)
Usually	248 (22.6)	345 (18.2)	593 (20.7)
Frequently	105 (10.8)	184 (9.7)	289 (10.1)
Sometimes	55 (5.7)	125 (6.6)	180 (6.3)
Occasionally	92 (9.5)	253 (13.3)	345 (12)
Rarely	105 (10.8)	321 (16.9)	426 (14.9)
Never	93 (9.6)	336 (17.7)	429 (15)

Three subsamples were randomly obtained from the total sample; data from 873 participants were subjected to exploratory factor analysis (EFA), data from 1000 participants were subjected to confirmatory factor analysis (CFA), and data from 1000 participants were subjected to invariance analysis. Remaining analyses used the full sample.

2.2. Instruments

Sociodemographic variables. An ad hoc questionnaire was created in order to register sex orientation, age, nationality, and city of residence, schooling, gender, sexual orientation, and marital status of the participants. This questionnaire has previously been used in the Colombian context [13].

Condom Use Self-Efficacy Scale (CUSES) [25]. The scale evaluates a person's feelings of confidence about buying condoms, putting them on and taking them off, and negotiating their use with a sexual partner. It is composed of 15 statements with five Likert-type response options from Strongly disagree = 0 to Strongly agree = 4; seven items are reverse-coded. Its original version includes four factors: (a) appropriation, with four items related to confidence in performing the technique necessary to use a condom in a sexual intercourse; (b) partner's disapproval, including five items related to confidence in dealing with possible rejection from a sexual partner due to request to use a condom; (c) assertiveness, including three items related to an individual's ability to ask a partner to use a condom; and (d) intoxication, with three items referred to the respondent's confidence in their ability to use a condom while under the influence of alcohol or other drugs or when they are over-excited. The present study applied for first time the Colombian validated scale for Colombian youth [28] 29. Once reversed items (items 5 to 9) are re-coded, higher scores mean higher self-efficacy.

The UCLA Multidimensional Condom Attitudes Scale (MCAS) [29]. The MCAS scale evaluates people's attitudes toward condom use. It is composed of five dimensions consisting of five items each (reliability, pleasure, stigma, negotiation, and shame). It is answered using Likert-type response options from 1 = totally disagree to 7 = totally agree. The present study applied the Colombian validated scale [30]. Higher scores indicate better attitudes toward the use of condoms. Obtained data in this study showed alphas ranging from 0.71 to 0.87.

Condom Use Error/problems Survey (CUES) [31]. The scale measures errors and problems associated with the use of condoms, considering the last three times a condom was used during the past three months as the recall period. Male (putting on a condom themselves) and female (putting the condom on their partner) versions are available. The scale is composed of 16 items with four Likert-type response options ranging from failure to perform an action to always performing it. High scores indicate a high frequency of errors/problems during any of the last three occasions or during the last three months. The present study applied the Colombian validated scale. [14].

Sexual Assertiveness Scale (SAS) [32]. The scale measures assertiveness with respect to initiation, refusal, and sexually transmitted disease/pregnancy prevention (STD-P P) with a regular partner.

It is composed of 18 Likert-type items (Never = 0 to Always = 4). The brief (9 items) version was used, which has been validated to Colombia [33]. A high score represents high sexual assertiveness. The present study used only the initiation and STD-P P items. Initiation (assertiveness) is related to a person's feelings of confidence about initiating sexual intercourse and making suggestions about physical intimacy. Alphas in the present study for those scales were 0.74 (initiation) and 0.85 (STD-P P).

2.3. Procedure

The questionnaire was published and distributed using Facebook. Payment of USD 150 was made to the virtual platform to promote the scale from 2 October to 17 October 2019. Responses were collected using a secure third-party survey provider (Survey Monkey, https://www.surveymonkey.com). The targeted population was young people aged 18–26. Online participants confirmed their willingness to participate through an online consent procedure, and questionnaire completion was also online. All internet protocol (IP) addresses were logged to discourage multiple responses from a single individual. By limited responses to one per IP address, the quality of the online data were higher.

Data analysis. Results were processed using R [34] (Version 3.6.0) and the R Studio interface [35] (Version 1.1.463). A polychoric matrix was for reliability and factorial analysis calculation; thus, the α presented is not Cronbach's but ordinal. The number of dimensions to be extracted was calculated with the following methods: optimal coordinates, acceleration factor, parallel analysis, Eigenvalues (Kaiser criterion), Velicer Minimum Average Partial MAP, Bayesian Information Criterion BIC, sample size adjusted BIC, Very Simple Structure VSS complexity, and VSS complexity 2. The mode and the quality of the indicators indicated the n of factors. EFA was computed through an maximum likelihood robust ML-R) method using varimax rotation on the polychoric matrix of sub-sample 1. CFA was performed using a Weighted Least Square Mean and Variance Adjusted-Robust (WLSMV-R) estimator on a polychoric matrix based on sub-sample 2. Five different models were tested. The fit indexes used were root mean square error approximation (RMSEA) [36] and its 90% confidence interval (90% CI), the comparative fit index (CFI) [37], and Tucker Lewis Index (TLI) [38]. Values up to 0.08 for RMSEA are usually considered as acceptable, but it is desirable not to exceed a 0.06 threshold; while a value above 0.90 is acceptable, one higher than 0.95 is desirable for CFI and TLI [39]. For the invariance across gender, as before, data were derived from a polychoric matrix, in this case of sub-sample 3. WLSMV-R was also used. Invariance indicators were: a −0.01 change in CFI, paired with changes in RMSEA of +0.015 concerning the least restrictive model [40] and a non-significant increase of the χ^2 using the Scaled Chi-Squared Difference Test [41]. Progressive invariance was tested for four models (configural, metric, scalar, and strict).

The following packages were also used: ggplot2 for data visualization [42] (Version 3.1.1), psych (Version 1.8.12) psychometric (Version 2.2) and psycho (Version 0.4.9.1) [43–45] were used for estimating some psychometric properties. While lavaan [46] (Version 0.6–5), semPlot [47] (Version 1.1.2), and semTools [48] were used for calculating and plotting the Structural Equation Model.

3. Results

The number of dimensions to isolate was evaluated using sub-sample 1. The methods (optimal coordinates, parallel analysis, Kaiser, and VSS complexity 2) suggested a three-dimensional model. Some of these methods are popular (Kaiser) and often recommended (optimal coordinates and parallel analysis). Nevertheless, we decided to compare the original four-dimensional structure of the scale with the three-dimensional structure (Table 2). The three-dimension (3-D) proposal explains 62% of the variance and has RSMEA = 0.127 and TLI = 0.85 as exploratory estimates. The four-dimension (4-D) model explains 69% of the variance, and the exploratory indicators RMSEA and TLI were 0.086 and 0.93, respectively. Item complexity was similar in both models; therefore, CFA results were needed to decide on the model.

Table 2. Exploratory factor analysis based on the polychoric matrix using maximum likelihood and varimax rotation.

	\multicolumn 3 Dimensions							\multicolumn 4 Dimensions						
	D3	**D2**	**D1**	**h^2**	**u^2**	**com**		**D2**	**D3**	**D1**	**D4**	**h^2**	**u^2**	**com**
CUSES8	**0.92**			0.89	0.10	10.1	CUSES8	**0.93**				0.89	0.11	1.1
CUSES9	**0.90**			0.88	0.12	10.2	CUSES9	**0.91**				0.88	0.11	1.1
CUSES7	**0.84**			0.78	0.21	10.2	CUSES7	**0.85**				0.78	0.21	1.1
CUSES5	**0.42**			0.26	0.73	10.8	CUSES5	**0.44**				0.26	0.74	1.7
CUSES6	**0.41**		0.32	0.27	0.72	10.9	CUSES6	**0.43**				0.27	0.73	1.9
CUSES1		**0.93**		0.91	0.09	10.1	CUSES1		**0.94**			0.91	0.09	1.1
CUSES2		**0.89**		0.85	0.15	10.2	CUSES2		**0.89**			0.85	0.15	1.1
CUSES4		**0.85**		0.78	0.22	10.1	CUSES4		**0.86**			0.78	0.22	1.1
CUSES3		**0.76**		0.65	0.35	10.2	CUSES3		**0.77**			0.65	0.35	1.2
CUSES11	0.40		**0.80**	0.85	0.15	10.7	CUSES13			**0.96**		0.98	0.02	1.1
CUSES10	0.35		**0.77**	0.77	0.23	10.6	CUSES14			**0.66**		0.47	0.53	1.1
CUSES12	0.39		**0.71**	0.70	0.30	10.8	CUSES15			**0.43**		0.31	0.68	2.5
CUSES13			**0.48**	0.28	0.72	10.4	CUSES11	0.44			**0.75**	0.88	0.11	2.2
CUSES15			**0.41**	0.23	0.76	10.8	CUSES10	0.40			**0.71**	0.77	0.22	2.1
CUSES14			**0.35**	0.15	0.85	10.4	CUSES12	0.43			**0.60**	0.69	0.31	2.7
Variance	0.22	0.21	0.19			*M* = 10.4	Variance	0.23	0.22	0.12	0.12			*M* = 1.5 *

* *D* = Number of dimensions' model; *h^2* = communality of the item; *u^2* = uniqueness of the item; *com*= Hoffmann's item complexity. Weights lower than 0.30 are hidden; boldface represent correct item-factor weight.

CFA was performed using the second sub-sample. Five different models were tested: (1) a one-dimension model (1-D); (2) the three-dimension model previously explored by independent factors (3D I); (3) the three-dimension model previously explored with related factors (3D R); (4) four independent factors explored previously and proposed by previous theory (4D I); and finally, (5) four related factors explored previously and proposed by previous theory [25] (4D R). The main results are shown in Table 3. Model 4D R seems to have the best fit; it is the only one in which all fit indexes are acceptable. Standardized weights for the chosen model are shown in Figure 1.

Table 3. Fit indexes for the model tested.

Models	χ^2	df	*p*	CFI	TLI	RMSEA	90% CI RMSEA
D-1	4284.53	90	<0.01	0.81	0.78	0.216	0.211–0.222
D-3 I	4146.70	90	<0.01	0.82	0.79	0.212	0.207–0.218
D-3 R	920.65	87	<0.01	0.96	0.95	0.098	0.092–0.104
D-4 I	4242.12	90	<0.01	0.81	0.78	0.215	0.209–0.220
D-4 R	432.33	84	<0.01	0.98	0.98	0.064	0.058–0.071 *

* χ^2 = Chi-Square statistic; *Df* = degrees of freedom; *CFI* = Comparative Fit Index; *TLI* = Tucker Lewis Index; *RMSEA* = root mean square error of approximation.

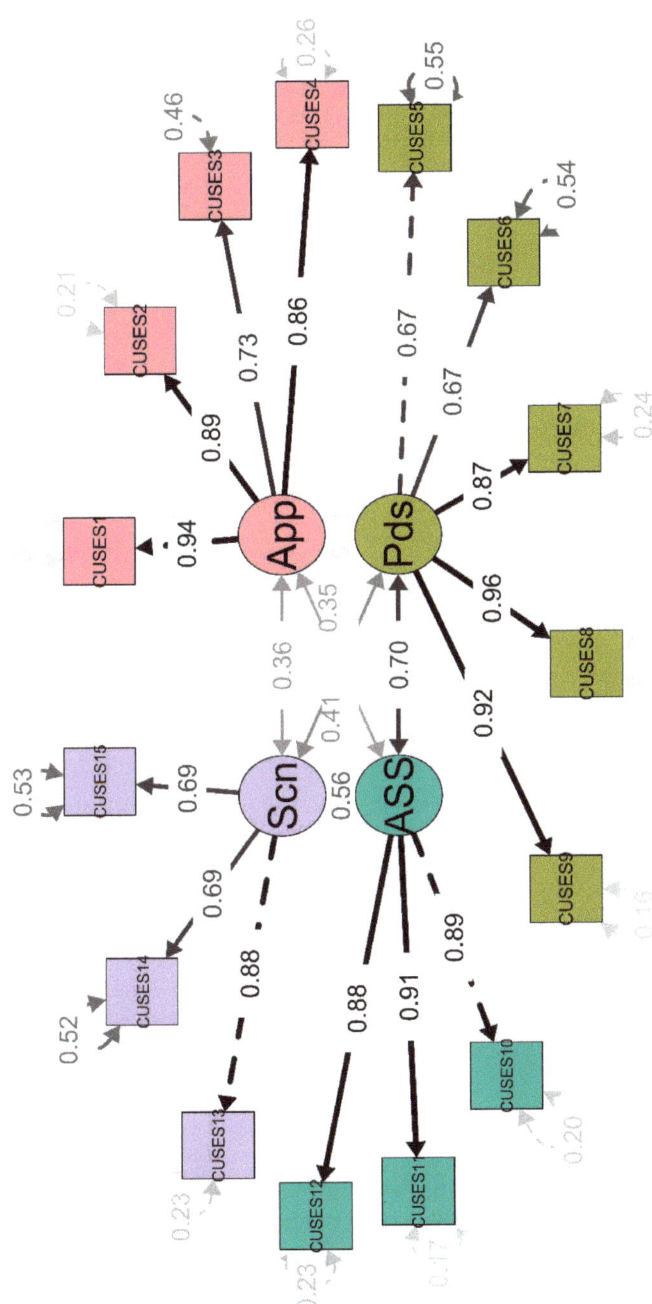

Figure 1. Confirmatory factor analysis (CFA) path diagram on four proposed dimensions. Standardized weights are presented. *App* = appropriation; *Pds* = partner's disapproval; *Ass* = assertiveness; *Scn* = self-control.

Once the four-dimension model was confirmed to be the best, an analysis of gender invariance was performed. For that purpose, we used sub-sample 3. As shown in Table 4, a strict level of invariance was achieved. Based on the three indicators taken into account for the four levels, just a significant increase of the χ^2 was observed for the weak invariance. However, the other indicators (ΔCFI and ΔRMSEA) seem to behave appropriately to suggest gender invariance.

Table 4. Fit indexes for the invariance 4-D R model.

Models	χ^2	$p > \chi^2$	df	p	CFI	ΔCFI	RMSEA	ΔRMSEA
Conf	295.57	-	168	<0.01	0.939	-	0.039	-
Metric	322.49	0.012 *	179	<0.01	0.932	−0.007	0.040	0.001
Scalar	332.95	0.561	190	<0.01	0.932	0	0.039	−0.001
Strict	354.49	0.054	205	<0.01	0.929	−0.003	0.038	−0.001 *

* χ^2 = Chi-square statistic; *Df* = degrees of freedom; *Conf* = configural invariance; *Metric* = metric invariance; *Scalar* = strong invariance; *Strict* = strict invariance.

The psychometric properties of the items began to be obtained. Once strict invariance across gender was tested and no differences were found, descriptive psychometric properties were assessed. In Table 5, ordinal alphas above 0.75 for all dimensions are shown. A couple of items (3 and 15) improve α if eliminated, but no further actions were considered necessary. Overall, corrected correlations between total and items are higher than 0.50. Item distributions cannot be considered normal, especially when considering kurtosis.

Table 5. Some psychometric item properties.

Dim	Item	*M*	*SD*	Skew	Kurtosis	Ci-tc	α-Item	α
App	1	3.18	0.99	−1.23	1.08	0.88	0.88	
	2	3.31	0.88	−1.48	2.30	0.86	0.89	0.92
	3	3.25	0.92	−1.30	1.54	0.73	0.93	
	4	2.80	1.09	−0.68	−0.30	0.82	0.90	
Pds	5	3.40	0.94	−1.75	2.60	0.61	0.89	
	6	3.25	0.97	−1.30	1.09	0.59	0.89	
	7	3.58	0.84	−2.49	6.24	0.82	0.84	0.89
	8	3.62	0.77	−2.57	7.13	0.83	0.84	
	9	3.57	0.84	−2.30	5.14	0.81	0.84	
Ass	10	3.60	0.69	−2.18	5.95	0.84	0.88	
	11	3.67	0.60	−2.25	6.87	0.87	0.85	0.92
	12	3.58	0.71	−2.11	5.47	0.80	0.91	
Scn	13	2.99	0.96	−0.70	−0.01	0.89	0.51	
	14	2.71	1.13	−0.55	−0.34	0.82	0.69	0.76 *
	15	3.09	0.99	−0.96	0.35	0.76	0.81	

* *M* = mean; *SD* = standard deviation; *CI-TC* = corrected item-total correlation; *App* = appropriation; *Pds* = partner's disapproval; *Ass* = assertiveness; *Scn* = self-control.

Criterion validity was also explored. To this aim, a correlation matrix between the CUSES subscales and other theoretically related instruments was created. Positive and significant correlation were observed and confirmed. Significantly low and moderate correlations were observed (see Table 6). Finally, the CUSES sub-scales were compared for gender. Figure 2 shows significant differences in appropriation and partner's disapproval (see Figure 2A). Men (M = 13.86; SD = 2.68) have more appropriation of condom use than women (M = 11.85; SD = 3.45), and women's score (M = 17.68; SD = 3.07) is higher in partner's disapproval (women are better at handling disapproval) than men's score (M = 16.9; SD = 3.45) (see Figure 2B). Neither assertiveness (men (M = 10.89; SD = 1.70); women (M = 10.82; SD = 1.77)) nor self-control (men (M = 8.77; SD = 2.47); women (M = 8.80; SD = 2.40)) show significant differences between men and women (see Figure 2C,D).

Table 6. Means, standard deviations, and correlations with confidence intervals.

Variable	M	SD	1	2	3	4	5	6	7	8	9	10	11
1. App	12.53	3.35											
2. Pds	17.42	3.23	0.08 ** [0.04, 0.12]										
3. Ass	10.85	1.75	0.27 ** [0.24, 0.31]	0.48 ** [0.45, 0.51]									
4. Scn	8.79	2.43	0.27 ** [0.24, 0.31]	0.24 ** [0.20, 0.27]	0.36 ** [0.33, 0.40]								
5. Neg	9.18	4.56	0.25 ** [0.28, 0.21]	0.54 ** [0.57, 0.52]	0.54 ** [0.57, 0.52]	0.30 ** [0.33, 0.26]							
6. Reli	11.87	5.11	0.20 ** [0.23, 0.16]	0.11 ** [0.14, 0.07]	0.15 ** [0.19, 0.11]	0.14 ** [0.18, 0.10]	0.20 ** [0.16, 0.24]						
7. Plea	16.85	6.08	0.14 ** [0.17, 0.10]	0.23 ** [0.26, 0.19]	0.20 ** [0.24, 0.16]	0.29 ** [0.32, 0.25]	0.25 ** [0.22, 0.29]	0.21 ** [0.17, 0.24]					
8. Sham	13.72	7.32	0.28 ** [0.32, 0.25]	0.19 ** [0.23, 0.15]	0.26 ** [0.29, 0.22]	0.21 ** [0.25, 0.17]	0.32 ** [0.29, 0.36]	0.15 ** [0.11, 0.18]	0.17 ** [0.13, 0.20]				
9. Stig	7.43	3.30	0.07 ** [0.11, 0.03]	0.48 ** [0.51, 0.45]	0.38 ** [0.41, 0.34]	0.20 ** [0.24, 0.17]	0.48 ** [0.45, 0.51]	0.15 ** [0.12, 0.19]	0.26 ** [0.23, 0.30]	0.21 ** [0.17, 0.25]			
10. CUEP	9.67	4.46	−0.15 ** [−0.21, −0.09]	−0.19 ** [−0.25, −0.13]	−0.11 ** [−0.17, −0.04]	−0.16 ** [−0.22, −0.10]	0.24 ** [0.18, 0.30]	0.15 ** [0.09, 0.21]	0.44 ** [0.39, 0.49]	0.12 ** [0.06, 0.19]	0.18 ** [0.12, 0.24]		
11. Init	7.86	3.13	0.15 ** [0.12, 0.19]	0.19 ** [0.15, 0.23]	0.19 ** [0.15, 0.22]	0.11 ** [0.07, 0.14]	−0.23 ** [−0.26, −0.19]	−0.07 ** [−0.11, −0.03]	−0.07 ** [−0.10, −0.03]	−0.15 ** [−0.19, −0.11]	−0.17 ** [−0.21, −0.13]	0.04 [−0.03, 0.10]	
12. STI–P	6.37	4.11	0.06 ** [0.02, 0.10]	0.22 ** [0.18, 0.26]	0.20 ** [0.17, 0.24]	0.24 ** [0.20, 0.28]	−0.24 ** [−0.28, −0.21]	−0.16 ** [−0.20, −0.12]	−0.39 ** [−0.42, −0.35]	−0.04 [−0.08, 0.00]	−0.18 ** [−0.21, −0.14]	−0.32 ** [−0.37, −0.26]	−0.02 ** [−0.06, 0.02]

M and SD are used to represent mean and standard deviation, respectively. Values in square brackets indicate the 95% confidence interval for each correlation. The confidence interval is a plausible range of population correlations that could have caused the sample correlation (Cumming, 2014). ** indicates $p < 0.01$. App = appropriation CUSES; Pds = partner's disapproval CUSES; Ass = assertiveness CUSES; Scn = self-control CUSES; Neg = embarrassment about negotiation and use of condoms UCLA; Reli = reliability and effectiveness of condoms UCLA; Plea = sexual pleasure associated with condom use UCLA; Sham = embarrassment about the purchase of condoms UCLA; Stig = stigma attached to persons who use condoms UCLA; CUEP = condom use errors/problems; Init = initiation Sexual Assertiveness Scale (SAS); STI_P_P = STI and pregnancy prevention SAS.

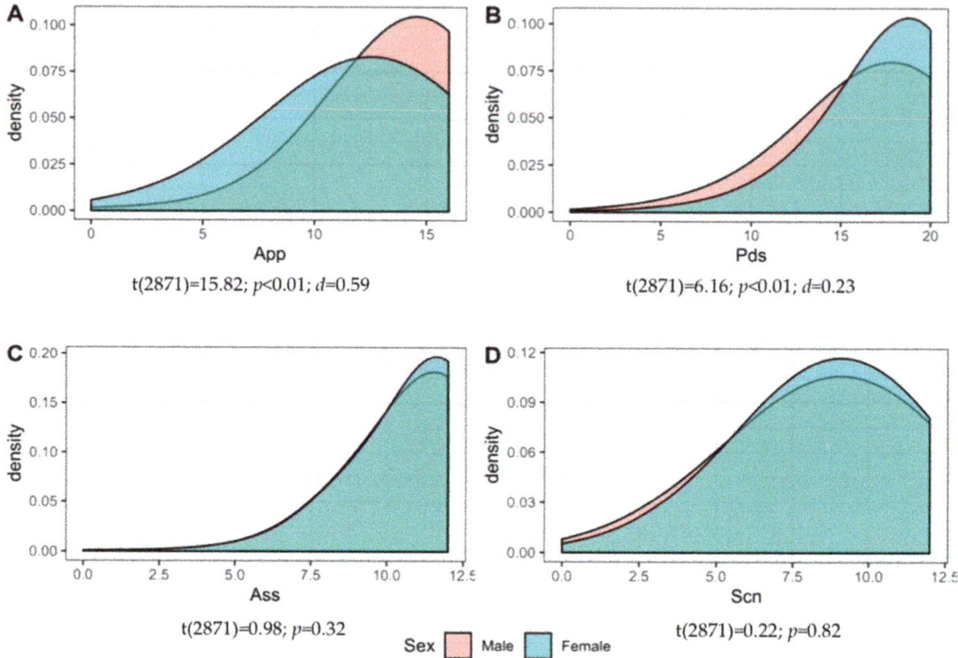

Figure 2. Densiogram distribution across gender. (**A**) gender differences in appropriation (App); (**B**) gender differences in partner's disapproval (Pds); (**C**) gender differences in assertiveness (Ass); (**D**) gender differences in self-control (Scn).

4. Discussion

This study assessed construct and criterion validity of the CUSES [25] adaptation into Colombian Spanish language, as well its reliability and certain psychometric properties. From the socio-cognitive perspective, studies on the reduction of sexual risk behaviors among young people are widespread and cover previously mentioned factors, such as knowledge about condom use, stigma related to its acquisition, and others. The use of CUSES in the present study contributes to the recognition of perceived self-efficacy as a factor mediated by aspects such as self-control, partner's disapproval, and condom use appropriation.

Exploratory and confirmatory factor analyses suggested a four-factor model (appropriation, partner's disapproval, assertiveness, and self-control). This four-factor structure was found to be invariant across gender. Reliability indexes and factors related with other similar variables were adequate. Gender-based differences were also observed. These results show that the current version of the CUSES is valid and reliable for its use in Colombia.

Our first factorial analysis suggested a three-factor scale. Shaweno and Tekletsadik [24] also observed a three-dimensional structure in a brief 9-item version. However, the majority of previous studies [25,26,28] observed four dimensions. Therefore, we also considered a four-dimension structure with even better indicators. This model explained 69% of the variance. Other studies have shown a range of variance ranging from 48.2% [23] to 73.72% [28]. Furthermore, the confirmatory analysis presented better-fit indexes for this four-dimension model that was finally settled down with a strict invariance across gender. To the best of our knowledge, this is the first CUSES study including a strict invariance test. This level of invariance allowed for the comparison of the mean of the factors with a minimal bias across gender due to variance and covariance errors [49]. In other words, items were measured with the same precision in each group and no gender biases were found, this factor is key to better understand sexual behaviors [50]. Our fourth factor was termed self-control, which is consistent with the "intoxicants" [26], "pleasure and intoxicants", [28] or "intoxicant control" [24] factors. The different label used in our study was due to the fact that, in the cultural adaptation, we used an adjustment of cultural words and context [51]. Thus, "heat of passion" is not related to the consumption of alcohol or psychoactive substances in the Colombian cultural context, although, taken together, all these factors are associated with self-control.

All items show adequate psychometric properties and total item correlations above 0.50, as recommended by the literature [52]. In addition, slight improvements can be observed when removing item 3 of the appropriation factor ("I feel confident I could gracefully remove and dispose of a condom when we have intercourse") or item 15 of the self-control factor ("I feel confident I could stop to put a condom on myself or my partner even in the heat of passion"). However, its level of adjustment is good and the improvements if the items were eliminated are considered negligible. Another indicator is item improvements. Item distribution is not normal and a remarkably high kurtosis is found; however, standard deviations are close to 1, indicating adequate score variability [53]. Alphas are observed to be similar to those reported in other studies, with values ranging from 0.76 to 0.92. These values are considered adequate for both research and professional practice.

Consistent with prior research, we find that the CUSES is related with attitudes toward condom use [27]. Concerning relationships between CUSES and other measures (i.e., criterion validity), they are found to be low or moderate in most cases. We find associations with the UCLA Multidimensional Condom Attitudes Scale (MCAS) factors, in which higher scores represent better attitudes toward condom use. Ritchwood, Penn, Peasant, Albritton, and Corbie-Smith [54] observed that greater condom use self-efficacy was predicted by favorable attitudes toward condom use, which is similar to the association between partner´s disapproval, and condom use negotiation observed in the present study. Likewise, the present study finds an association between partner´s disapproval and attitudes toward condom use stigma, which has been related with reduced safer sex practices [55]. These findings highlight the necessity of interventions or training programs to improve young people's abilities to cope with factors that could result in not using a condom. Although the Colombian Ministry of Health has designed advertising campaigns and prevention programs aimed to young people population, it is important to design strategies considering academic curricula. Additionally, a low and inverse association between self-efficacy factors and Condom Use Errors and Problems (CUEP) factors is found: scores on condom use self-efficacy increase as those associated with errors and problems with condom use decrease, which has previously been reported [56]. This observation is interesting because actual problems and errors seem to differ from one's perception of the correct use of condoms, or self-efficacy. Significant relationship with variables such as initiation assertiveness (Init) and sexually transmitted disease/unintended pregnancy prevention assertiveness (STD-P P) were also verified according to assumptions. Thus, condom use self-efficacy certainly has an impact in delaying young people's sex initiation and reducing sexually transmitted infections and unplanned pregnancies [57].

Gender comparisons show that men scored higher in appropriation and lower in partner´s disapproval than women. This may suggest a need for sexual health intervention programs aimed at increasing condom use self-efficacy differentially for men and women. For men, programs should emphasize how to deal with rejection by the sexual partner, and for women, training should be oriented toward appropriation. Differentiated interventions should also be designed to prevent STIs and unintended pregnancies. In this regard, the instrument's subscales provide information on the effectiveness of negotiating condom use that could be associated with cultural factors in the Colombian context [58]. Highly typified gender roles, benevolent sexism, and prejudices associated with female sexuality and women's sexual autonomy [59,60] have also been described by Peasant, Parra, and Okwumabua [61] as factors affecting condom use self-efficacy among women, including their ability to successfully negotiate condom use without compromising their relational goals.

A relevant strength of the present study is the sample size and its heterogeneity across the cities, that means that the CUSES can be used to obtain a reliable and valid measurement of condom use self-efficacy among Colombian young people, and it can be used as a baseline characterization scale for other Latin American Spanish-speaking populations to help in the early identification of youth condom use self-efficacy. Finally, this is the first successful adaptation in the continent.

Limitations of the present study is the non-randomized sampling approach used, the sample was skewed female and heterosexual an, the study excluded people without internet access because of the web-based survey method used. Future studies can further explore the differences in scores obtained by Spanish speaking groups at high risk of acquiring HIV/STI (an example, homosexual males, men who have sex with men and sex workers).

5. Conclusions

The CUSES is the first instrument used in Colombia to measure condom use self-efficacy in Spanish language. The reliability of the CUSES has been established in different studies since 1994 [23–27]. The present study allows CUSES use in Colombia with enough psychometric guarantees for its use for both, social/health and research interest. Some gender differences in condom self-efficacy and uncommon condom use were observed, suggesting the need to implement sexual health programs in Colombia considering these differences.

Author Contributions: V.S.-M. and P.V.-M.: design of the study; acquisition, analysis, and data interpretation; formal analysis of data; drafting—original draft preparation; drafting—final version; have approved the submitted version. E.S.-A.: design of the study; drafting—review and editing; has approved the submitted version. All authors have read and agreed to the published version of the manuscript.

Funding: This research received no external funding.

Conflicts of Interest: The authors declare no conflict of interest.

References

1. CONPES 3918. *Estrategia Para la Implementación de Los Objetivos de Desarrollo Sostenible (ODS) en Colombia*; Departamento Nacional de Planeación: Bogotá, Colombia, 2018. Available online: https://colaboracion.dnp. gov.co/CDT/Conpes/Econ%C3%B3micos/3918.pdf (accessed on 20 February 2020).
2. Ministerio de Salud y Protección Social. *Plan Decenal de Salud Pública*; Ministerio de Salud y Protección Social: Bogotá. Colombia, 2012. Available online: https://www.minsalud.gov.co/plandecenal/Paginas/home2013. aspx (accessed on 20 February 2020).
3. UNAIDS. *Communities at the Centre. Global AIDS Update 2019*; Joint United Nations Programme on HIV/AIDS (UNAIDS): Geneva, Switzerland, 2019.
4. Instituto Nacional de Salud. *Informe de Evento VIH/SIDA. Periodo Epidemiológico XIII Colombia*; Ministerio de Salud: Bogotá, Colombia, 2019. Available online: https://www.ins.gov.co/buscador-eventos/ Informesdeevento/VIH-SIDA%20PE%20XIII%202019.pdf (accessed on 20 February 2020).
5. Rowley, J.; Vander Hoorn, S.; Korenromp, E.; Low, N.; Unemo, M.; AbuRaddad, L.; Chico, R.M.; Smolak, A.; Newman, L.; Gottlieb, S.; et al. Chlamydia, gonorrhea, trichomoniasis and syphilis: Global prevalence and incidence estimates, 2016. *Bull. World Health Organ.* **2019**, *97*, 548–562. [CrossRef] [PubMed]
6. WHO. Report on Global Sexually TRANSMITTED Infection Surveillance. 2018. Available online: https: //www.who.int/reproductivehealth/publications/stis-surveillance-2018/en/ (accessed on 20 February 2020).
7. Ministerio de Salud y Protección Social. *Situación de las Infecciones de Transmisión Sexual Diferentes al VIH. Colombia 2009–2011*; Ministerio de Salud y Protección Social: Bogotá, Colombia, 2011. Available online: https://www.minsalud.gov.co/salud/Documents/observatorio_vih/documentos/monitoreo_evaluacion/1_ vigilancia_salud_publica/a_situacion_epidimiologica/SITUACION%20DE%20LAS%20INFECCIONES% 20DE%20TRANSMISION1.pdf (accessed on 20 February 2020).
8. Tsevat, D.G.; Wiesenfeld, H.C.; Parks, C.; Peipert, J.F. Sexually transmitted diseases and infertility. *Am. J. Obstet. Gynecol.* **2017**, *216*, 1–9. [CrossRef] [PubMed]
9. Jackson, J.M.; Seth, P.; DiClemente, R.J.; Lin, A. Association of depressive symptoms and substance use with risky sexual behavior and sexually transmitted infections among African American female adolescents seeking sexual health care. *Am. J. Public Health* **2015**, *105*, 2137–2142. [CrossRef] [PubMed]

10. Unemo, M.; Bradshaw, C.S.; Hocking, J.S.; de Vries, H.J.; Francis, S.C.; Mabey, D.; Peeling, R.W. Sexually transmitted infections: Challenges ahead. *Lancet Infect. Dis.* **2017**, *17*, e235–e279. [CrossRef]

11. Sedgh, G.; Singh, S.; Hussain, R. Intended and unintended pregnancies worldwide in 2012 and recent trends. *Stud. Fam. Plan.* **2014**, *45*, 301–314. [CrossRef]

12. ENDS. *Componente de Salud Sexual y Reproductiva*; Profamilia: Bogotá, Colombia, 2015; Available online: https://dhsprogram.com/pubs/pdf/FR334/FR334.2.pdf (accessed on 20 February 2020).

13. Morales, A.; Vallejo-Medina, P.; Abello-Luque, D.; Saavedra-Roa, A.; García-Roncallo, P.; Gomez-Lugo, M.; Espada, J.P. Sexual risk among Colombian adolescents: Knowledge, attitudes, normative beliefs, perceived control, intention, and sexual behavior. *BMC Public Health* **2018**, *18*, 1377. [CrossRef]

14. González-Hernández, A.M.; Escobar-Estupiñan, J.; Vallejo-Medina, P. Condom use errors and problems in a simple of Young Colombian. *J. Sex Res.* **2020**. [CrossRef]

15. Albarracin, D.; Johnson, B.; Fishbein, M.; Muellerleile, P. Theories of reasoned action and planned behavior as models of condom use: A meta-analysis. *Psychol. Bull.* **2001**, *127*, 142–161. [CrossRef]

16. Sheeran, P.; Abraham, C.; Orbell, S. Psychosocial correlates of heterosexual condom use: A meta-analysis. *Psychol. Bull.* **1999**, *125*, 90–132. [CrossRef]

17. Armitage, C.J.; Conner, M. Social cognition models and health behaviour: A structured review. *Psychol. Health* **2000**, *15*, 173–189. [CrossRef]

18. Ellis, E.M.; Homish, G.G.; Parks, K.A.; Collins, R.L.; Kiviniemi, M.T. Increasing condom use by changing people's feelings about them: An experimental study. *Health Psychol.* **2015**, *34*, 941–950. [CrossRef] [PubMed]

19. Vincent, W.; Gordon, D.M.; Campbell, C.; Ward, N.L.; Albritton, T.; Kershaw, T. Adherence to traditionally masculine norms and condom-related beliefs: Emphasis on African American and Hispanic men. *Psychol. Men Masc.* **2016**, *17*, 42–53. [CrossRef] [PubMed]

20. Zhang, C.Q.; Zhang, R.; Schwarzer, R.; Hagger, M.S. A meta-analysis of the health action process approach. *Health Psychol.* **2019**, *38*, 623–637. [CrossRef] [PubMed]

21. Bandura, A. Self-efficacy: Toward a unifying theory of behavioural change. *Psychol. Rev.* **1977**, *84*, 191–215. [CrossRef] [PubMed]

22. Bandura, A. Perceived self-efficacy in the exercise of control over AIDS infection. *Eval. Program Plan.* **1990**, *13*, 9–17. [CrossRef]

23. Barkley, T.W., Jr.; Burns, J.L. Factor analysis of the Condom Use Self-Efficacy Scale among multicultural college students. *Health Educ. Res.* **2000**, *15*, 485–489. [CrossRef]

24. Shaweno, D.; Tekletsadik, E. Validation of the condom use self-efficacy scale in Ethiopia. *BMC Int. Health Hum. Rights* **2013**, *13*, 22. [CrossRef]

25. Brafford, L.; Beck, K. Development and Validation of a condom self-efficacy scale for college students. *J. Am. Coll. Health* **1991**, *39*, 219–225. [CrossRef]

26. Brien, T.M.; Thombs, D.L.; Mahoney, C.A.; Wallnau, L. Dimensions of self-efficacy among three distinct groups of condom users. *J. Am. Coll. Health* **1994**, *42*, 167–174. [CrossRef]

27. Forsyth, A.D.; Carey, M.P.; Fuqua, R.W. Evaluation of the validity of the Condom Use Self-Efficacy Scale (CUSES) in young men using two behavioral simulations. *Health Psychol.* **1997**, *16*, 175–178. [CrossRef]

28. Asante, K.; Doku, P. Cultural adaptation of the Condom Use Self-Efficacy Scale (CUSES) in Ghana. *BMC Public Health* **2010**, *10*, 227. [CrossRef] [PubMed]

29. Sanchez-Mendoza, V.; Soriano-Ayala, E.; Vallejo-Medina, P. A visual way to represent content validity and item properties. 2020; Manuscript under review.

30. Vallejo-Medina, P.; Gómez-Lugo, M.; Marchal-Bertrand, L.; Saavedra-Roa, A.; Soler, F.; Morales, A. Desarrollo de Guías para Adaptar Cuestionarios Dentro de una Misma Lengua en Otra Cultura. *Ter. Psicológica* **2017**, *35*, 159–172. [CrossRef]

31. Crosby, R.A.; Graham, C.A.; Milhausen, R.R.; Sanders, S.A.; Yarber, W.L. Condom use errors/problems survey. In *Handbook of Sexuality-Related Measures*; Fisher, T.D., Davis, C.M., Yarber, W.L., Eds.; Routledge: London, UK, 2010; pp. 153–159.

32. Morokoff, P.J.; Quina, K.; Harlow, L.L.; Whitmire, L.; Grimley, D.M.; Gibson, P.R.; Burkholder, G.J. Sexual Assertiveness Scale (SAS) for women: Development and validation. *J. Personal. Soc. Psychol.* **1997**, *73*, 790–804. [CrossRef]

33. Plaza-Vidal, R.; Ibagón-Parra, M.; Vallejo-Medina, P. Translation, adaptation and validation of the Spanish version of the multidimensional condom attitudes scale for Colombian young. *Manuscr. Rev.* **2020**, under review.

34. R Core Team. R: A Language and Environment for Statistical Computing Vienna: R Foundation for Statistical Computing [Computer Software]. 2017. Available online: https://www.R-project.org/ (accessed on 24 February 2020).

35. RStudio Team. *RStudio: Integrated Development for R*; RStudio, Inc. [Computer Software]: Boston, MA, USA, 2016; Available online: http://www.rstudio.com/ (accessed on 24 February 2020).

36. Browne, M.W.; Cudeck, R. Alternative ways of assessing model fit. In *Testing Structural Equation Models*; Bollen, K.A., Long, J.S., Eds.; Sage: Newbury Park, CA, USA, 1993; pp. 136–162.

37. Bentler, P.M. Comparative fit indexes in structural models. *Psychol. Bull.* **1990**, *107*, 238–246. [CrossRef]

38. Tucker, L.R.; Lewis, C. A reliability coefficient for maximum likelihood factor analysis. *Psychometrika* **1973**, *38*, 1–10. [CrossRef]

39. Hu, L.T.; Bentler, P.M. Cut-off criteria for fit indexes in covariance structure analysis: Conventional criteria versus new alternatives. *Struct. Equ. Modeling* **1999**, *6*, 1–55. [CrossRef]

40. Chen, F.F. Sensitivity of goodness of fit indexes to lack of measurement invariance. *Struct. Equ. Modeling* **2007**, *14*, 464–504. [CrossRef]

41. Satorra, A. Scaled and adjusted restricted tests in multi-sample analysis of moment structures. In *Innovations in Multivariate Statistical Analysis: A Festschrift for Heinz Neudecker*; Heijmans, D.D.H., Pollock, D.S.G., Satorra, A., Eds.; Kluwer Academic: Dordrecht, The Netherlands, 2000; pp. 233–247.

42. Wickham, H. *Ggplot2: Elegant Graphics for Data Analysis [Computer Software]*; Springer: New York, NY, USA, 2009; Available online: https://cran.r-project.org/web/packages/ggplot2/index.html (accessed on 20 February 2020).

43. Revelle, W. *Psych: Procedures for Personality and Psychological Research, [Computer Software]*; Northwestern University: Evanston, IL, USA, 2018; Available online: https://CRAN.R-project.org/package=psych (accessed on 24 February 2020).

44. Fletcher, T.D. Psychometric: Applied Psychometric Theory. R package version 2.2. 2010. Available online: https://CRAN.R-project.org/package=psychometric (accessed on 24 February 2020).

45. Rosseel, Y. lavaan: An R Package for Structural Equation Modeling. *J. Stat. Softw.* **2012**, *48*, 1–36. Available online: http://www.jstatsoft.org/v48/i02/ (accessed on 24 February 2020). [CrossRef]

46. Makowski, D. The Psycho Package: An Efficient and Publishing-Oriented Workflow for Psychological Science. *J. Open Source Softw.* **2018**, *3*, 470. Available online: https://github.com/neuropsychology/psycho.R (accessed on 24 February 2020). [CrossRef]

47. Epskamp, S. SemPlot: Path Diagrams and Visual Analysis of Various SEM Packages' Output. R package Version 1.1.2. 2019. Available online: https://CRAN.R-project.org/package=semPlot (accessed on 24 February 2020).

48. Jorgensen, T.D.; Pornprasertmanit, S.; Schoemann, A.M.; Rosseel, Y. SemTools: Useful tools for Structural Equation Modeling. R Package Version 0.5–2. 2019. Available online: https://CRAN.R-project.org/package=semTools (accessed on 24 February 2020).

49. Dimitrov, D.M. Testing for factorial invariance in the context of construct validation. *Meas. Eval. Couns. Dev.* **2010**, *43*, 121–149. [CrossRef]

50. Sakaluk, J.K. Expanding statistical frontiers in sexual science: Taxometric, invariance, and equivalence testing. *J. Sex Res.* **2019**, *56*, 475–510. [CrossRef] [PubMed]

51. Guillemin, F. Cross-cultural adaptation and validation of health status measures. *Scand. J. Rheumatol.* **1995**, *24*, 61–63. [CrossRef] [PubMed]

52. Stevens, J.P. *Applied Multivariate Statistics for the Social Sciences*, 5th ed.; Taylor & Francis Group: New York, NY, USA, 2009.

53. Carretero-Dios, H.; Pérez, C. Standards for the development and review of instrumental studies: Considerations about test selection in psychological research. *Int. J. Clin. Health Psychol.* **2007**, *7*, 863–882.

54. Ritchwood, T.D.; Penn, D.; Peasant, C.; Albritton, T.; Corbie-Smith, G. Condom Use Self-Efficacy among younger rural adolescents: The influence of parent-teen communication, and knowledge of and attitudes toward condoms. *J. Early Adolesc.* **2017**, *37*, 267–283. [CrossRef]

55. Logie, C.; Lys, C.; Okumu, M.; Fujioka, J. Exploring factors associated with condom use self-efficacy and condom use among Northern and Indigenous adolescent peer leaders in northern Canada. *Vulnerable Child. Youth Stud.* **2019**, *14*, 50–62. [CrossRef]

56. Sanders, S.; Hill, B.; Crosby, R.; Janssen, E. Correlates of condom-associated erection problems in young, heterosexual men: Condom fit, self-efficacy, perceptions, and motivations. *AIDS Behav.* **2014**, *18*, 128–134. [CrossRef]

57. Li, J.; Li, S.; Yan, H.; Xu, D.; Xiao, H.; Cao, Y.; Mao, Z. Early sex initiation and subsequent unsafe sexual behaviors and sex-related risks among female undergraduates in Wuhan, China. *Asia Pac. J. Public Health* **2015**, *27*, 21S–29S. [CrossRef]

58. UNFPA. *Análisis de Situación de Condones en Colombia. Ministerio de Salud y Protección Social*; Fondo de población de las Naciones Unidas: Bogotá, Colombia, 2015; Available online: https://colombia.unfpa.org/sites/default/files/pub-pdf/analisiscondonesColombia_web.pdf (accessed on 20 February 2020).

59. Albarracin, J.; Plambeck, C. Demographic factors and sexist beliefs as predictors of condom use among Latinos in the USA. *AIDS Care* **2010**, *22*, 1021–1028. [CrossRef]

60. Unger, J.; Molina, G. Contraceptive use among Latina women: Social, cultural, and demographic correlates. *Women's Health Issues* **1998**, *8*, 359–369. [CrossRef]

61. Peasant, C.; Parra, G.; Okwumabua, T. Condom negotiation: Findings and future directions. *J. Sex Res.* **2015**, *52*, 470–483. [CrossRef] [PubMed]

International Journal of
***Environmental Research
and Public Health***

Article

Son Preference and the Reproductive Behavior of Rural-Urban Migrant Women of Childbearing Age in China: Empirical Evidence from a Cross-Sectional Data

Xiaojie Wang [1], Wenjie Nie [1] and Pengcheng Liu [2,*]

[1] School of Management, Ocean University of China, Qingdao 266100, China; xiaojie_nk@126.com (X.W.);
     nie_wen_jie@163.com (W.N.)
[2] School of Economics, Qingdao University, Qingdao 266100, China
*   Correspondence: liupc_un@163.com

Received: 2 April 2020; Accepted: 1 May 2020; Published: 6 May 2020

Abstract: Son preference has been shown to influence the childbearing behavior of women, especially in China. Existing research has largely focused on this issue using cross-sectional data of urban or rural populations in China, while evidence from the rural-urban migrant women is relatively limited. Based on the data of China Migrants Dynamic Survey in 2015, we used logistic regression models to explore the relationship of son preference and reproductive behavior of rural-urban migrant women in China. The results show that the son preference of migrant women is still strong, which leads women with only daughters to have significantly higher possibility of having another child and results in a higher imbalance in the sex ratio with higher parity. Migrant women giving birth to a son is a protective factor against having a second child compared to women whose first child was a girl. Similarly, the effects of the gender of the previous child on women's progression from having two to three children showed the same result that is consistent with a preference for sons. These findings have implications for future public strategies to mitigate the son preference among migrant women and the imbalance in the sex ratio at birth.

Keywords: son preference; reproductive behavior; migrant women; subsequent parity

1. Introduction

A strong preference for sons over daughters is common in East and Southeast Asia [1,2], notably in China. Son preference is often thought to be an important cause of imbalance in the sex ratio at birth [3–5]. According to the Global Gender Gap Report 2018, China ranked dead last among 149 countries in terms of "sex ratio at birth". This is the result of many factors, such as the Confucian cultural tradition, the socioeconomic system, and gender ideology. In the context of traditional patriarchal, patrilineal and patriarchal systems, sons are considered to have unique value, as they inherit the family name and property and represent an economic value premium to the family and parents [6–10].

Significantly different from that in other countries, the family planning policy in China is one of the limiting factors affecting fertility practices in the country [11–14]. Although the policy has been loosened by the shift from allowing one child to allowing two children for all couples, a clear limitation on the number of children still exists [15]. When there is a difference between the number of children allowed by the family planning policy and the number of children expected by the family, for individuals with a strong preference for boys, there is an incentive to conduct gender selection under the policy conditions or ignore the normal fertility limitation to have more children until a son is born.

It is worth noting that the occurrence of the above two cases is strongly related to the individual's household registration status [16].

Another special policy in China, the household registration policy, also has a significant impact on women's reproductive behavior. The dual structure in urban and rural areas in China leads to significant differences in economic development, children-bearing concepts, and policy supervision [17–19]. The fertility behavior of urban residents is heavily regulated, and the opportunity cost of violating policies is extremely high, so the number of births remains low [20–22], and it is rare to maintain son preference through having more children. However, in addition to urban residents, there are large groups of rural-to-urban migrants in cities. Rural-to-urban migrants are people who have resided in an urban destination for at least 6 months and do not have local household registration [23]. The migrant population is too large to be ignored, accounting for approximately one-sixth of the total population of China [24]. According to the National Bureau of Statistics, the number of migrants reached 241 million in 2018, and half of them were migrant women. A considerable proportion of these women were of child-bearing age [25]. Although they make a living in the city, traditional concepts such as having a son to carry on the family name, raising children to provide for parents in old age and promoting the family status of mothers by giving birth to a boy are deeply entrenched in their minds, which motivates them to have more children and give birth to at least one boy [26]. Furthermore, the opportunity cost of violating policies is lower for migrant women than for urban women, so the fertility behavior of continuing to have children until a son is born is more likely to be observed in this group. However, scholars have pointed out that the well-bear and well-rear concept of urban residents may have a demonstratable effect on migrants and gradually change their fertility conception [19,27,28]. Studies have shown that migrants generally have lower fertility rates than rural residents and higher fertility rates than urban residents [27,29].

To a large extent, the birth conception and behavior of migrant women have a profound impact on the population changes in the places of domicile and migration, which will further affect the implementation of China's family planning policy, the overall fertility level and the changes in the gender structure [30]. Moreover, due to the high mobility of this type of population, the government has difficulties conducting cross-regional fertility regulations. Therefore, the constraints of the family planning policy on migrants are greatly decreased, so fertility intention is more likely to be implemented; that is, there may be a high degree of consistency between the migrating population's reproductive preference for sons and future fertility behaviors [31]. In summary, under the special policy background of China, choosing migrant women as the research object has certain practical significance for studying the relationship between the son preference and reproductive behavior of migrant women of childbearing age, predicting future population development trends and reversing the gender imbalance.

Studies on women's reproductive behavior in the existing literature focus on family reproductive intention, family planning, number of sons and contraceptive usage [8,12,14,32,33]. For example, prospective analysis is performed by directly asking questions such as "How many children do you expect to have?", "Would you like to have another child?" and other questions at the level of fertility intention to conduct a prospective analysis of fertility behavior. However, whether fertility intention can be transformed into actual fertility behavior is still limited by a variety of factors, such as national policies, family economic conditions and physical conditions of the couple, and the internal deviation between fertility intention and behavior cannot be effectively estimated. In particular, there is also a large deviation between son preference in terms of fertility intention and gender selection in actual reproductive behavior. Therefore, the direct investigation of fertility behavior can better reflect the issue of son preference.

The influence of son preference on the number of children and fertility behavior remains a controversial topic. Some studies suggest that a strong preference for sons will lead to the increases in fertility level desired by the state [34,35]. Others have shown that using sex-selective technologies will lead to a reduction in the number of children and thus a lower fertility level. In terms of gender composition, although evidence from India, South Korea, and Vietnam suggests that the preference for sons is closely related to family fertility behavior, couples with only girls are likely to have more children than families with only boys or gender-balanced families [2,8,36–39]. However, in the context of China's special policy, relatively few studies have focused on rural-urban migrant women to investigate the individual characteristics and social factors that may affect fertility behavior. The complex relationship between son preference and actual fertility behavior has not been effectively tested, and it remains to be explored whether the migrant experience can influence the son preference and fertility behavior of immigrant women. In addition, this article can help overcome the adverse bias between fertility intention and fertility behavior in previous studies, and the retrospective research method adopted largely makes the research conclusions more reliable and convincing.

Regarding the research on the son preferences of migrant women, this paper conducts a retrospective study based on the 2015 national monitoring survey data of the China Migrants Dynamic Survey (CMDS) and focuses on two aspects. The first is the pursuit of the subsequent parity progression. The second is the relationship between the sex composition of existing children and subsequent childbearing behavior. That is, in the context of China's special fertility policy, what are the factors that influence migrant women to have a higher number of children? If the gender composition of existing children fails to satisfy migrant women's son preference, will the women violate national policies and have more children? As the number of children increases, will the likelihood of having boys be increased through prenatal sex identification technology? Having more children can reduce the quality of childcare, while choosing to have boys following prenatal sex identification may cause severe gender imbalances, both of which pose challenges to improving the quality of the population and ensuring the gender balance in China.

2. Materials and Methods

2.1. Data Source

The data in this paper are derived from the China Migrants Dynamic Survey in 2015 (CMDS2015). This is an annual large-scale national migrant population sampling survey initiated by the National Health Commission of the People's Republic of China and coordinated by the China Population and Development Research Center. CMDS2015 covers 31 provinces (autonomous regions and municipalities) in mainland China. The stratified, multistage and proportional PPS method is adopted for sampling. The survey covers the basic information of the migrant individuals and their family members, the mobility range of the migrant women, employment and social security, income and expenditure, residence, basic public health services, management of marriage and family planning services, children's mobility and education, etc. The survey questionnaire was presented through direct interviews by investigators who had been trained uniformly and reviewed by professional instructors. After the questionnaire was completed, it was properly kept in each city (district) and randomly checked by a panel of experts. Finally, the quality of the submitted questionnaire was monitored by logical verification and telephone return visits. The CMDS2015 was characterized by the full coverage and representativeness of the migrant population, and the questionnaire has good authenticity and reliability.

2.2. Study Design and Participants

Cross-sectional data of rural-to-urban migrant women with one child were derived from CMDS2015 to explore the complex relationship of son preference and reproductive behavior. Given the research question of our study, we limited our sample to migrant women of childbearing age. Specifically, the sample selection criteria in this article are being married (first marriage or remarriage), having had at least one child, being aged 15–49 years old, having rural household registration, and being a non-local resident. Following the exclusion of non-qualifying individuals and invalid data samples resulting from refusal to answer, the number of valid samples was 36,182.

The dependent variable in this study is the fertility behavior of migrant women, that is, whether each respondent progressed from having one, two and three children to having two, three and four children, respectively. The independent variable is the gender of existing children. The control variables involved in this study are grouped into two categories. The first category elicited sociodemographic characteristics of the participants, including age, ethnic group, marriage duration, individual education level, spouse's education level, employment, mobility range and residence intention. The second category assessed the participants' fertility information based on the following items: the number of existing children and the birthplace of the first child. The proportion of boys being born should be around 51 percent without son preference, and if the proportion in our data is higher than this ratio, it means that the boy preference exists. Since the outcome variable of this study is a binary variable, we used logistics regression analysis to explore the effects of motivation and influencing factors, especially the sex composition of current children, on the subsequent childbearing behavior of migrant women.

2.3. Statistical Analysis

A descriptive analysis was used to describe the characteristics of the participants and their reproductive behaviors. Variables that were significantly associated with fertility behavior by chi-square analysis were entered as independent variables in the binary logistic regression, including age, ethnic group, marriage duration, individual education level, spouse's education level, gender of the first child, mobility range and residence intention. The factors which were significantly related to reproductive behavior were then included in the logistic model analysis. A single-factor regression analysis was used to explore the effect of the gender of existing children on having another child. A multivariable logistic regression model was built to identify the determinants of reproductive behaviors. Through these analyses, crude ORs and 95% confidence intervals (CIs) were estimated for the gender preference and fertility behavior of migrant women. Statistical significance was defined as p-values < 0.05. Data analysis was conducted using the STATA 15.0. (Stata, College Station, TX, USA).

3. Results

3.1. Sociodemographic Characteristics

Table 1 presents the descriptive statistics of key variables, showing that in the overall sample of 36,182, most of the respondents were between the ages of 25 and 44 (81.88%), and the majority were ethnic Han, while less than 10% were minorities. Generally, the respondents were not highly educated, with only 6% of migrant women having a bachelor's degree or above and nearly 60% having a junior high school education (57.32%), while those who had no education or only primary education accounted for 2.25% and 16.4%, respectively. Most women had stable jobs (72.77%). Migrant women with one, two and three children accounted for 57.25%, 38.06%, and 4.18%, respectively, while less than 1% of the sample had more than three children. Most of the migrant women chose to have their first child in the hospital (89.36%), while 9.4% and 1.24% of migrant women chose to bear their first child at home or in a private clinic, respectively. Regarding the geographical range of migration, nearly half of them engaged in interprovincial migration (47.22%), 31.44% were intercity migrants, and 21.34% were intercounty migrants. A total of 85.03% of migrant women had the intention to live for a long time in their destinations.

Table 1. Sociodemographic characteristics of the study participants (*N* = 36,182).

Variables	*N*	%
Age of women (years)		
≤24	657	1.82
25–34	15,092	41.71
35–44	14,535	40.17
45–49	5898	16.3
Marriage duration (years)		
<5	4633	12.8
≥5	31,549	87.2
Ethnic group		
Minority	2744	7.58
Han	33,438	92.42
Highest education level		
No education	815	2.25
Primary school	5934	16.4
Junior high school	20,738	57.32
High school/Secondary school	6493	17.95
University and above	2202	6.09
Highest education level of husband		
Junior high school and below	18,669	79.92
High school/Secondary school and above	4690	20.08
Have a job		
No	9853	27.23
Yes	26,329	72.77
The number of existing children		
1	20,714	57.25
2	13,771	38.06
3	1514	4.18
≥4	182	0.50
Gender of the first child		
Female	17,021	47.04
Male	19,161	52.96
Where the first child was born		
Hospital	32,331	89.36
Private clinic	449	1.24
At home	3402	9.4
Scope of migration		
Trans-provincial	17,086	47.22
Intercity	11,374	31.44
Across the county	7722	21.34
Settlement intention		
No	5415	14.97
Yes	30,767	85.03

3.2. Factors Affecting Childbearing Behavior

Variables that were significantly associated with fertility behavior by chi-square analysis were entered as independent variables in the binary logistic regression. We took fertility behavior for the second child as an example and conducted an analysis. As shown in Table 2, the frequency of having a second child for women whose first child was a girl (53.53%) was significantly higher than those whose first child was a boy (33.18%). Migrant women who have been married for more than 5 years have a much higher proportion of progressing to second births (48.10%) than women who have been married for less than 5 years (6.32%), which shows the increased marriage duration improved the chance of having a second child. There is a significant positive correlation between age and having a second child. With increasing age, the proportion of second children gradually increases. In contrast, with the improvement of women's education level, the proportion of women choosing to have a second child gradually decreases from 73.13% to 13.17%. The education level of the husband shows a similar trend. Migrant women who have stable jobs and want to live in the destination cities have higher frequency of progressing to the next birth than women who do not work or want to stay.

Table 2. Distribution of second-child fertility by sociodemographic variables ($N = 36,182$).

Variables	Whether a Second Child Was Born					
	No		Yes		N	X^2 (p)
Gender of the first child	N	%	N	%		
Female	7910	46.47	9111	53.53	17,021	1.5×10^3 (0.0000)
Male	12,804	66.82	6357	33.18	19,161	
Age of women (years)						
≤24	608	92.54	49	7.46	657	3.0×10^3 (0.000)
25–34	10,890	72.16	4202	27.84	15,092	
35–44	6804	46.81	7731	53.19	14,535	
45–49	2412	40.90	3486	59.10	5898	
Marriage duration (years)						
<5	4340	93.68	293	6.32	4633	2.9×10^3 (0.000)
≥5	16,374	51.90	15,175	48.10	31,549	
Ethnic group						
Minority	1343	48.94	1401	51.06	2744	83.7028 (0.000)
Han	19,371	57.93	14,067	42.07	33,438	
Highest education level						
No education	219	26.87	596	73.13	815	2.8×10^3 (0.000)
Primary school	2123	35.78	3811	64.22	5934	
Junior high school	11,770	56.76	8968	43.24	20,738	
High school/Secondary school	4690	72.23	1803	27.77	6493	
University and above	1912	86.83	290	13.17	2202	
Highest education level of husband						
Junior high school and below	13,257	51.42	12,525	48.58	25,783	1.9×10^3 (0.000)
High school/Secondary school and above	7457	71.70	2943	28.30	10,400	
Have a job						
No	5831	59.18	4022	40.82	9853	63.2967 (0.000)
Yes	14,883	56.53	11,446	43.47	26,329	
Where the first child was born						
Hospital	19,626	60.70	12,705	39.30	32,331	1.5×10^3 (0.000)
Private clinic	150	33.41	299	66.59	449	
At home	938	27.57	2464	72.43	3402	
Scope of migration						
Trans-provincial	9115	53.35	7971	46.65	17,086	201.5062 (0.000)
Intercity	6896	60.63	4478	39.37	11,374	
Across the county	4703	60.90	3019	39.10	7722	
Settlement intention						
No	3223	59.52	2192	40.48	5415	13.4120 (0.000)
Yes	17,491	56.85	13,276	43.15	30,767	

3.3. Sex Composition and Subsequent Parity Progression

Table 3 shows the distribution of the sex composition of existing children. Generally, the proportion of migrant women who progressed from having one, two, and three children to having two, three, and four children, respectively, is gradually decreasing. That is, among the children with a total sample size of 53,530, first and second children represented 36,182 and 15,468, accounting for 67.59% and 28.90%, respectively, while the number of third and fourth children sharply reduced to 1697 (3.17%) and 183 (0.34%), respectively. In terms of sex composition, the sex ratio significantly increases with the birth of more children. The sex ratio of the second children (1.42) is higher than that of the first children (1.13). Although the number of third and fourth children is relatively small, the sex ratio reaches a staggering 1.81 and 2.27, respectively. All the above are significantly higher than the natural sex ratio (1.06).

Table 3. Distribution of the sex composition of existing children.

	Female	%	Male	%	N	Sex Ratio
Gender of the first children	17,021	47.04%	19,161	52.96%	36,182	1.13
Gender of the second children	6404	41.40%	9064	58.60%	15,468	1.42
Gender of the third children	603	35.53%	1094	64.47%	1697	1.81
Gender of the fourth children	56	30.60%	127	69.40%	183	2.27
Total	24,084	44.99%	29,446	55.01%	53,530	1.22

Table 4 shows the effects of the sex composition of children at baseline on women's parity progression from having one, two, and three children to having two, three, and four children, respectively. Considering that younger women may not have completed their childbearing progression [40], we further limited the sample to migrant women of childbearing age over 35, and the sample size narrowed to 20,433. It can be seen that in the transition from the first to the second children, 68.60% of those whose first child was a girl chose to have a second child, while only 43.02% of those whose first child was a boy chose to have a second child. This suggests that the gender of the first child is significantly associated with the probability of progressing to the next parity. In the transition to having a third child, among women whose children were both daughters, the proportion of those who had a third child was 32.16%, significantly higher than the proportion among women whose children were both sons (less than 10%). Among women with three children, relative to women with three sons, those with three daughters had nearly four times the frequency of progressing to the next parity. In addition, women who had a son and a daughter were the most likely to stop having children. Similarly, among families with three children, women with two sons and one daughter were most likely to stop having children. Our findings show patterns of association between the sex composition of existing children and subsequent childbearing behaviors that are consistent with a preference for sons. (All the above model test results are statistically significant).

Table 4. Baseline sex composition on subsequent childbearing (N = 20,433).

Sex Composition of Previous Children at Baseline	Stopped Childbearing	%	Continued Childbearing	%	N	X^2 (*p*)
Gender of the first children						
0 male and 1 female	2979	31.40%	6509	68.60%	9488	1.3×10^3 (0.000)
1 male and 0 female	6237	56.98%	4708	43.02%	10,945	
Gender of the first two children						
0 male and 2 female	1500	67.84%	711	32.16%	2211	977.7714 (0.000)
1 male and 1 female	6187	92.62%	493	7.38%	6680	
2 male and 0 female	2130	91.57%	196	8.43%	2326	
Gender of the first three children						
0 male and 3 female	94	59.12%	65	40.88%	159	145.3611 (0.000)
1 male and 2 female	704	91.91%	62	8.09%	766	
2 male and 1 female	364	92.39%	30	7.61%	394	
3 male and 0 female	72	88.89%	9	11.11%	81	

Table 5 shows the results of the single-factor regression analysis of the gender of existing children on having another child. In the transition from the first to the second child, migrant women with one son is a protective factor against having a second child compared to women whose first child was a girl (OR: 0.43; 95% CI: 0.413–0.450). Among the migrant women with two children, those with one (OR: 0.19; 95% CI: 0.168–0.211) or two sons (OR: 0.21; 95% CI: 0.180–0.246) had a significantly reduced chance of progressing to the next parity compared to women with two daughters. Similarly, the effects of the gender of the previous child on women's progression from having three to four children show the same result that are consistent with a preference for sons. In addition, the son preference effect is significantly enhanced with the birth of more girls.

Table 5. Single-factor regression analysis of the gender of the previous child on having another child.

Sex Composition of Previous Children at Baseline	OR	Coef	Std. Err.	Z	*p*	95% CI
gender of the first children						
0 male and 1 female	1					
1 male and 0 female	0.43	−0.84	0.01	−38.75	0.000	(0.413, 0.450)
gender of the first two children						
0 male and 2 female	1					
1 male and 1 female	0.19	−1.67	0.01	−28.75	0.000	(0.168, 0.211)
2 male and 0 female	0.21	−1.56	0.02	−19.64	0.000	(0.180, 0.246)
gender of the first three children						
0 male and 3 female	1					
1 male and 2 female	0.15	−1.88	0.03	−9.78	0.000	(0.104, 0.222)
2 male and 1 female	0.14	−1.94	0.03	−8.32	0.000	(0.091, 0.227)
3 male and 0 female	0.19	−1.67	0.07	−4.42	0.000	(0.089, 0.393)

3.4. The Effect of Relevant Factors on Subsequent Childbearing

Furthermore, we took fertility behavior for the second child as an example and conducted a multifactor analysis. The results are reported in Table 6. Binary logistic regression analysis was used to explore various factors influencing whether migrant women chose to have a second child. The results showed that ethnic group, education level, residence intention, mobility range and having a job had a significant influence on the fertility behavior of migrant women ($p < 0.05$). Women with one son is a protective factor against having a second child compared to women whose first child was a girl (OR = 0.35, $p < 0.05$). Compared with minority women, a smaller proportion of Han women chose to have a second child due to the restriction of family planning policy. Women over 25 were more likely to have a second child than women aged 24 or younger. Migrant women with age of more than 35 years old increased the chance of progressing to the next parity by more than three times compared to women less than 25 years old (OR = 3.75, $p < 0.05$) (OR = 3.53, $p < 0.05$). Migrant women with education is a protective factor; those with higher education levels reported lower possibility of having another child than those who had never gone to school. The marriage duration more than 5 years increased the chance of having a second child by more than seven times compared to a marriage duration less than 5 years (OR = 7.42, $p < 0.05$). Having a job and the education level of the spouse were negatively correlated with having another child. Women with residence intention were more likely than women without residence intention to have a second child (OR = 1.19, $p < 0.05$). In addition, those who chose to have their first child at home (OR = 1.89, $p < 0.05$) or in a private clinic (OR = 2.08, $p < 0.05$) increased the chance of having a second child than those who had their first child in the hospital.

Table 6. Binary logistic regression analysis of factors influencing reproductive behavior.

Variables	Coef.	OR	Std. Err.	z	p	95% CI
Gender of the first children (Female as ref)	−1.05	0.35	0.01	−43.06	0.000	(0.334, 0.368)
Ethnic group (Minority as ref)	−0.24	0.79	0.04	−5.10	0.000	(0.719, 0.864)
Marriage duration (years) (<5 as ref)	2.00	7.42	0.49	30.11	0.000	(6.512, 8.453)
Age of women (years) (≤24 as ref)						
25–34	0.68	1.97	0.32	4.13	0.000	(1.429, 2.724)
35–44	1.32	3.75	0.62	7.98	0.000	(2.708, 5.128)
45–49	1.26	3.53	0.59	7.53	0.000	(2.541, 4.900)
Highest education level (No education as ref)						
Primary school	−0.23	0.80	0.07	−2.55	0.011	(0.668, 0.949)
Junior high school	−0.69	0.50	0.04	−7.76	0.000	(0.424, 0.599)
High school/Secondary school	−1.08	0.34	0.03	−11.19	0.000	(0.282, 0.411)
University and above	−1.61	0.20	0.02	−13.47	0.000	(0.158, 0.253)
Highest education level of husband (Junior high school and below as ref)	−0.16	0.85	0.02	−7.26	0.000	(0.820, 0.892)
Have a job (No as ref)	−0.20	0.82	0.02	−7.16	0.000	(0.775, 0.865)
Settlement intention (No as ref)	0.17	1.19	0.04	5.11	0.000	(1.113, 1.271)
Where the first child was born (Hospital as ref)						
Private clinic	0.64	1.89	0.20	5.93	0.000	(1.531, 2.330)
At home	0.73	2.08	0.09	16.44	0.000	(1.910, 2.275)
Scope of migration (Trans-provincial as ref)						
Intercity	−0.31	0.74	0.02	−10.98	0.000	(0.697, 0.778)
Across the county	−0.35	0.70	0.02	−11.26	0.000	(0.659, 0.746)
_cons	−1.28	0.28	0.05	−6.81	0.000	(0.193, 0.402)

4. Discussion

China's migrant population is large, and there is a trend of further growth in the future [23,24,41]. The fertility concepts of migrant women, to some extent, aggravate the complexity of the fertility level and gender structure of cities and even the whole of China, which will have a profound impact on future marriage, family structure, family planning policy and social development [19,27]. Based on the data of CMDS2015, this paper puts forward new views on the son preference and reproductive behavior of rural-urban migrant women of childbearing age in China.

The fertility concepts of migrant women tend to be markedly different from those of their rural or urban counterparts, which are influential in women's preference regarding childbearing. Although the emerging literature indicated that the modern outlook of fertility in urban areas might reduce the parental preference for sons [27,42], some scholars argued that the son preference remained strong among migrants [7]. Our analysis confirmed the latter view: migrants still have a strong son preference in the context of China's special policy, which makes having only daughters less desirable and results in a serious imbalance in the sex ratio.

Our findings showed the patterns of association between the sex composition of existing children and subsequent childbearing behaviors that are consistent with the preference for sons, which leads women with only daughters to have significantly higher intentions of having another child and results in a higher imbalance in the sex ratio with higher parity. To elaborate, our study found that migrant women whose first child was a girl were more likely to have another child compare with women whose first child was a boy. The same is true for the second and third children. It can be said that although the migrant women live in the city, the traditional concept of son preference had not changed significantly, which motivated them to have more children and give birth to at least one boy [7]. In addition, this study found that migrant women with both sons and daughters had the lowest odds of having another child. This also indicates that migrant women have a significant preference for sons. However, if the son preference is satisfied, the sex composition of children including both boys and girls is ideal for migrant women.

Socioeconomic characteristics can explain why some migrant women were more inclined to have more than one child. The results showed that with the increase in age, the likelihood of having another child gradually increased. A possible reason is that with the increase in age, individuals obtain a certain amount of savings and a stable work and residence environment that ease the pressure of raising another child, on the basis of which son preference can be achieved. Education level was found to have a significant effect on fertility. For better-educated migrants, the likelihood of having another child decreased greatly, and the evidence of a son preference declined. On the one hand, the increase in education improves migrant women's awareness of and compliance with national policies. On the other hand, highly educated women pay more attention to the quality of childrearing than to the quantity of children [27]. Therefore, encouraging women to receive higher education is an important measure to alleviate China's sex ratio imbalance.

In addition, the sex ratio of first children is slightly higher than the sex ratio under natural conditions (1.04–1.06) [7], indicating that the motivation for the sex selection of first children is not significant. The sex ratio of the second children (1.42) was significantly higher than that of the first children (1.13). Moreover, under family planning policy restrictions, the gender imbalance among third and fourth children worsened further, and the fourth children were more than twice as likely to be boys than girls. This result indicated an obvious tendency for gender selection and a motivation to ignore family planning restrictions and have more children in the case that the gender preference for sons was not satisfied. Therefore, migrant women who gave birth to the first child at home or in a private clinic had significantly higher odds of having a second child than those who gave birth to the first child in the hospital because the former can avoid inspection by family planning authorities.

Our study makes a possible contribution to the literature regarding the effect that son preference and the sex composition of existing children has on the reproductive behavior of women of childbearing age in the rural-urban migrant population. First, most studies of son preference focus on rural areas or locally representative samples, and there are relatively few studies on migrant women, a special group of women who differ from women in rural and urban areas. This paper expands the research scope of relevant fields. Second, most studies tend to explore fertility behaviors through fertility intentions, which are measured at a particular time point and could be unreliable because intentions are fluid. With data and retrospective birth histories, we can use logistic regression analysis to estimate the parity progression among migrant women conditional on sex composition.

The limitations of the study also should be highlighted. First, our study is based on a cross-sectional study that focuses on the number and gender of existing children, which makes it difficult to analyze the dynamics of female fertility. Women's reproductive behavior could be tracked using more effective methods. If large-scale data covering migrant women can be collected in future studies, through a longitudinal study, a more comprehensive understanding of the gender preference and reproductive behavior will become possible. Second, married women rarely migrate independently. In fact, they often migrant with their husbands [43]. Under the gender norms of a patriarchal society, immigrant women may have little influence on family decisions, such as those involving the optimal number of births. Future research could also examine gender preference of husbands in the rural-urban migrant population. Third, there are various differences in the inflow and outflow areas, and the lack of regional cultural surveys have limited the conduct of our research. This requires more detailed data, which are not typically covered by the China Migrants Dynamic Survey. Thus, future research and evaluation studies with a special focus on these issues will be critical to push research in such a direction.

5. Conclusions

China's long-term gender imbalance deserves much attention. Gender imbalance is closely related to son preference. Therefore, the change in the fertility concept of migrant women plays an important role in reversing the gender imbalance. However, under the existing household registration system, migrant women have not been truly integrated into urban society, which makes it difficult for them to adapt to both the urban and the rural cultures. They have become a marginal group that is separated from rural and urban areas but also closely related to these areas, which complicates the transformation of their fertility concept. Therefore, how to carry out selective intervention through public policies, especially aiming to reverse the son preference of migrant women, is a problem worthy of attention in the future.

Author Contributions: X.W.—conceived and designed the study protocol, literature review, revised the manuscript, and approved the final manuscript. W.N. —data analysis, prepared the initial manuscript draft. P.L.—cleaned and analyzed the data, review and edit the manuscript. All authors have read and agreed to the published version of the manuscript.

Funding: This work was supported by National Social Science Foundation of China (19CRK016).

Conflicts of Interest: The authors declare that they have no competing interests.

References

1. Mussino, E.; Miranda, V.; Ma, L. Changes in sex ratio at birth among immigrant groups in Sweden. *Genus* **2018**, *74*, 13. [CrossRef] [PubMed]

2. Rajan, S.; Nanda, P.; Calhoun, L.M.; Speizer, I. Sex composition and its impact on future childbearing: A longitudinal study from urban Uttar Pradesh. *Reprod. Health* **2018**, *15*, 1–9. [CrossRef] [PubMed]

3. Kumar, S.; Sinha, N. Preventing more "missing girls": A review of policies to tackle son preference. *World Bank Res. Obs.* **2019**, *35*, 87–121. [CrossRef]

4. Howell, E.M.; Zhang, H.; Poston, D.L. Son preference of immigrants to the United States: Data from U.S. birth certificates, 2004–2013. *J. Immigr. Minor. Heal.* **2018**, *20*, 711–716.

5. Robitaille, M.C.; Chatterjee, I. Sex-selective abortions and infant mortality in India: The role of parents' stated son preference. *J. Dev. Stud.* **2018**, *54*, 47–56. [CrossRef]

6. Larsen, U.; Chung, W.; Gupta, M. Fertility and son preference in Korea. *Popul. Stud. (NY)* **1998**, *52*, 317–325. [CrossRef]

7. Das Gupta, M.; Jiang, Z.; Li, B.; Xie, Z.; Woojin, C.; Bae, H. Why is son preference so persistent in East and South Asia? A cross-country study of China, India and the Republic of Korea. *J. Dev. Stud.* **2003**, *40*, 153–187.

8. Chaudhuri, S. The desire for sons and excess fertility: A household-level analysis of parity progression in India. *Int. Perspect. Sex. Reprod. Health* **2012**, *38*, 178–186. [CrossRef]

9. Sandström, G.; Vikström, L. Sex preference for children in German villages during the fertility transition. *Popul. Stud. (NY)* **2015**, *69*, 57–71. [CrossRef]

10. Tilt, B.; Li, X.; Schmitt, E.A. Fertility trends, sex ratios, and son preference among Han and minority households in Rural China. *Asian Anthropol.* **2019**, *18*, 110–128. [CrossRef]

11. Baochang, G.; Feng, W.; Zhigang, G.; Erli, Z. China's local and national fertility policies at the end of the twentieth century. *Popul. Dev. Rev.* **2007**, *33*, 129–148. [CrossRef]

12. Cai, Y. China's below-replacement fertility: Government policy or socioeconomic development. *Popul. Dev. Rev.* **2010**, *36*, 419–440. [CrossRef] [PubMed]

13. Cai, Y. China's demographic reality and future. *Asian Popul. Stud.* **2012**, *8*, 1–3. [CrossRef]

14. Basten, S.; Jiang, Q. China's family planning policies: Recent reforms and future prospects. *Stud. Fam. Plann.* **2014**, *45*, 493–509. [CrossRef]

15. Zhou, Y. The dual demands: Gender equity and fertility intentions after the one-child policy. *J. Contemporary China* **2019**, *28*, 367–384. [CrossRef]

16. Chan, K.W.; Li, Z. The Hukou system and rural-urban migration in China: Processes and changes. *China Q.* **1999**, *160*, 818–855. [CrossRef]

17. Ji, Y.; Zhao, X.; Wang, Z.; Liu, S.; Shen, Y.; Chang, C. Mobility patterns and associated factors among pregnant internal migrant women in China: A cross-sectional study from a National Monitoring Survey. *BMC Pregnancy Childbirth.* **2018**, *18*, 165. [CrossRef]

18. Knight, J.; Song, L. The rural-urban divide: Economic disparities and interactions in China. *J. Asian Stud.* **2001**, *60*, 169–170.

19. Hou, H. The diversities of Migrants' fertility intention among the cities during the urbanization. *Urban. Environ. Stud.* **2017**, *1*, 60–72.

20. Guo, Z.; Wu, Z.; Schimmele, C.M. The effect of urbanization on China's fertility. *Popul. Res. Policy Rev.* **2012**, *31*, 417–434. [CrossRef]

21. Chen, W.; Wu, L. Research on the relationships between migration and fertility in China. *Popul. Res.* **2006**, *30*, 13–20.
22. Liu, N.; Lu, L. The effect of children on wage of Chinese urban female workers in the state system. *Popul. Econ.* **2012**, *5*, 10–19.
23. Liang, Z.; Ma, Z. China's floating population: New evidence from the 2000 census. *Popul. Dev. Rev.* **2004**, *30*, 467–488. [CrossRef]
24. Luo, J.; Zhang, X.; Wu, Y.; Shen, J.; Shen, L.; Xing, X. Urban land expansion and the floating population in China: For production or for living. *Cities* **2018**, *74*, 219–228. [CrossRef]
25. Zhou, M.; Guo, W. Fertility intentions of having a second child among the floating population in China: Effects of socioeconomic factors and home ownership. *Popul. Space Place* **2019**, *26*, e2289. [CrossRef]
26. Lillehagen, M.; Lyngstad, T.H. Immigrant parents' preferences for children's sex: A register-based study of fertility behavior in ten national-origin groups. *SSRN Electron. J.* **2014**. [CrossRef]
27. Liao, Q.; Cao, G.; Tao, R. A study on the fertility and son preferences of China's migrants: Empirical evidence from 12 cities. *Popul. Dev.* **2012**, *18*, 2–12.
28. You, D.; Zheng, Z. An analysis on fertility desire of the rural-urban migrant women: An empirical research in Sichuang and Anhui Provinces. *Sociol. Stud.* **2002**, *6*, 52–62.
29. Yin, W.; Yao, Y.; Li, F. Evaluation of fertility levels and adjustment of fertility policy: Based on the current situation of the Chinese mainland's provincial fertility levels. *Soc. Sci. China* **2014**, *35*, 83–105.
30. Wang, X.; Yuan, X.; Han, Y. Household fertility decision and second-child policy: Based on the perspective of migrant population. *Nankai Econ. Stud.* **2018**, *2*, 93–109.
31. Zhuang, Y. Fertility willingness of migrant workers in different generations and its influencing factors-based on the empirical study of 912 rural floating population in Xiamen City. *Social* **2008**, *1*, 138–163.
32. Wang, T.; Wang, C.; Zhou, Y.; Zhou, W.; Luo, Y. Fertility intentions for a second child among urban working women with one child in Hunan province, China: A cross-sectional study. *Public Health* **2019**, *173*, 21–28. [CrossRef] [PubMed]
33. Machiyama, K.; Mumah, J.N.; Mutua, M.; Cleland, J. Childbearing desires and behavior: A prospective assessment in Nairobi Slums. *BMC Pregnancy Childbirth* **2019**, *19*, 100. [CrossRef] [PubMed]
34. Morgan, S.P.; Zhigang, G.; Hayford, S.R. China's below-replacement fertility: Recent trends and future prospects. *Popul. Dev. Rev.* **2009**, *35*, 605–629. [CrossRef] [PubMed]
35. Guilmoto, C.Z. Son preference, sex selection, and kinship in Vietnam. *Popul. Dev. Rev.* **2012**, *38*, 31–54. [CrossRef] [PubMed]
36. Bash, D.; DeJong, R. Son targeting fertility behavior: Some consequences and determinents. *Demography* **2010**, *47*, 521–536.
37. Park, C.B.; Cho, N.H. Consequences of son preference in a low-fertility society: Imbalance of the Sex ratio at birth in Korea. *Popul Dev. Rev.* **1995**, *21*, 59–84. [CrossRef]
38. Das Gupta, M.; Mari Bhat, P. Fertility decline and increased manifestation of sex bias in India. *Popul. Stud. (NY)* **1997**, *51*, 307–315. [CrossRef]
39. Kumar, A.; Kshatriya, G.K. Sex preference and fertility: A study among the Ansaris of Meerut District, Uttar Pradesh. *Stud. Home Community Sci.* **2013**, *7*, 109–117. [CrossRef]
40. Wang, J.; Ma, Z.; Li, J. Rethinking China's actual and desired fertility: Now and future. *Popul. Res.* **2019**, *43*, 32–44.
41. Goodburn, C. Learning from migrant education: A case study of the schooling of rural migrant children in Beijing. *Int. J. Educ. Dev.* **2009**, *29*, 495–504. [CrossRef]
42. Wu, H.; Li, S.; Yue, Z. Analysis on the conception and behavior of urban migrant population—Findings from Shenzhen investigation. *Popul. Res.* **2006**, *1*, 61–68.
43. Tong, Y.; Shu, B.; Piotrowski, M. Migration, livelihood strategies, and agricultural outcomes: A gender study in rural China. *Rural Sociol.* **2019**, *84*, 591–621. [CrossRef]

International Journal of
Environmental Research and Public Health

Article

Healthcare Providers' Knowledge and Attitude Towards Abortions in Thailand: A Pre-Post Evaluation of Trainings on Safe Abortion

Rugsapon Sanitya [1], **Aniqa Islam Marshall** [1,*], **Nithiwat Saengruang** [1], **Sataporn Julchoo** [1], **Pigunkaew Sinam** [1], **Rapeepong Suphanchaimat** [1,2], **Mathudara Phaiyarom** [1], **Viroj Tangcharoensathien** [1], **Nongluk Boonthai** [3] **and Kamheang Chaturachinda** [3]

[1] International Health Policy Program, Ministry of Public Health, Nonthaburi 11000, Thailand; rugsapon@ihpp.thaigov.net (Ru.S.); nithiwat@ihpp.thaigov.net (N.S.); sataporn@ihpp.thaigov.net (S.J.); pigunkaew@ihpp.thaigov.net (P.S.); rapeepong@ihpp.thaigov.net (Ra.S.); mathudara@ihpp.thaigov.net (M.P.); viroj@ihpp.thaigov.net (V.T.)
[2] Division of Epidemiology, Department of Disease Control, Ministry of Public Health, Nonthaburi 11000, Thailand
[3] Women's Health and Reproductive Right and Foundation of Thailand, Bangkok 10210, Thailand; nonglukb@hotmail.com (N.B.); gumhang@hotmail.com (K.C.)
* Correspondence: aniqa@ihpp.thaigov.net

Received: 7 April 2020; Accepted: 2 May 2020; Published: 4 May 2020

Abstract: Although physicians in Thailand can carry out abortions legally, unsafe abortion rates remain high and have serious consequences for women's health. Training programs for healthcare providers on the 'Care of unplanned and adolescent pregnancies for the prevention of unsafe abortions' have been implemented in Thailand with the aim of providing information and challenging negative attitudes about abortions. This study investigated the participants of the training courses in order to: (i) evaluate their knowledge and attitudes towards safe abortions; and (ii) investigate the factors that determine their knowledge and attitudes. A pre-post study design was applied. Descriptive statistics were calculated to provide an overview of the data. Bivariate analysis, a Wilcoxon signed rank test and a multivariable analysis using multiple linear regression were applied to determine the changes in attitudes and assess the likelihood of behaviour change towards adolescents and women experiencing unplanned pregnancy and abortions, according to demographic and professional characteristics. Having had the training, healthcare providers' change in attitudes towards adolescents and women experiencing unplanned pregnancies and abortions were found to be 0.67 points for the nine responses of attitudes and 0.79 points for the 14 responses on various abortion scenarios. Changes in attitude were significantly different among the varying health professional types, with non-doctors increasing by 0.53 points, non-obstetricians and non-gynaecologists increasing by 0.46 points and obstetricians and gynaecologists (OBGYN) increasing by 0.32 points. Positive attitudes towards unplanned pregnancies and unsafe abortions and attitudes towards abortion scenarios significantly increased. The career type of the health professional was a significant factor in improving attitudes. The training program was more effective among non-doctor healthcare providers. Therefore, non-doctors could be the target population for training in the future.

Keywords: abortion; training; health professionals; unplanned pregnancy; Thailand; pre-post evaluation

---

1. Introduction

Annually worldwide, approximately 42 million women with unintended pregnancies choose to undergo abortion. Of these, approximately 20 million procedures are classified as unsafe abortions [?].

An unsafe abortion is defined as the "termination of an unwanted pregnancy either by persons lacking the necessary skills or in an environment that does not conform to minimal medical standards or both" [?]. This includes less-safe abortions, which are conducted using outdated methods by a trained provider or conducted using safe methods but not by a trained individual, as well as least safe abortions which are conducted using dangerous methods by untrained individuals [?]. Safe abortion encompasses clinical practice on abortion, such as the use of manual vacuum aspiration (MVA) and many wider social determinants including abortion-related laws and regulations. Women often resort to unsafe abortions due to barriers in accessing safe abortions, such as restrictive laws, unavailability of services, high cost, stigma, unnecessary requirements to obtain services and conscientious objections of healthcare providers [?]. Unsafe abortions can result in severe maternal consequences, including chronic health complications and disabilities, and are one of the leading causes of maternal death [?]. The majority of unsafe abortions, 97% globally, occur in developing countries in Africa, Latin America and Asia [?]. According to the World Health Organization, "As a preventable cause of maternal mortality and morbidity, unsafe abortions must be dealt with" [?].

Maternal health in Thailand has continued to suffer from unsafe abortion-related complications. A study conducted by the Thai Department of Health, found that 28.5% of hospital admissions from 787 government hospitals were a result of induced abortions. One third of these cases developed serious complications, for which over half had undergone abortions by unqualified healthcare providers [?]. Another study conducted in a Thai public hospital found that of all the women admitted for the treatment of complications from abortions, 36.8% underwent unsafe abortions [?]. Similarly, a study in the South of Thailand reported that unsafe abortions accounted for 35.7% of all abortions and were significantly associated with maternal, financial or family problems [?]. The fatality rate of abortions in Thailand is 300 per 100,000 abortions; this is a high rate compared with the fatality rate of 1 per 100,000 abortions in developed countries [?]. Adolescents are particularly vulnerable, as they are more likely to seek unsafe abortions and experience severe complications [?]. In Thailand, approximately 25.9% of all pregnancies are among adolescents, of which 14.4% result in abortions, making up 18% of the total abortions in the country [?]. The adolescent pregnancy rate in Thailand is 39 per 1000 women and according to the Division of Strategy and Planning of the Ministry of Public Health (MOPH) approximately half of adolescent pregnancies are at a high risk of undergoing unsafe abortions [?].

The Thai Penal Code 305 and the Thai Medical Council Regulation 2003 (BE 2548) permits abortions to be carried out by qualified physicians in certain conditions, including where there is risk to the health and life of the mother or child and pregnancies resulting from rape or incest. However, many women still face barriers to safe abortion services for various reasons, including lack of awareness of abortion services, societal stigma and taboo and the perception that abortion is absolutely illegal [? ?]. Healthcare provider attitudes towards abortions are also one of the major barriers preventing women from accessing safe abortion services in Thailand and therefore healthcare providers play a decisive role in ensuring safe abortions. [?]. Providers with negative attitudes towards abortion can lead women with unplanned pregnancies to risk their life and seek unsafe abortions conducted unlawfully by untrained health personnel [? ?]. Therefore, to rectify this problem, efforts to change the notion of abortion among healthcare providers is necessary.

The Women's Health and Reproductive Rights Foundation of Thailand (WHRRF), a non-profit organization that advocates the health and reproductive rights of women in Thailand, was established in 1998. The organization has been working in collaboration with the Thai Health Foundation, the Royal Thai College of Obstetricians and Gynaecologists (RTCOG) and the Thai Medical Council (TMC) to advocate improved health and standards of quality health services for women in the country, including access to safe abortion services. The WHRRF has been conducting a training program for healthcare professionals on the 'Care of pregnant adolescents and women for the prevention of unsafe abortions', to provide information and shift negative attitudes. Over the period of August 2017 to April 2018, three cohorts of health professionals participated in the program, held at the Royal Thai College of Obstetricians and Gynaecologists in Bangkok. The trainings were organized by the president and vice-president of the RTCOG as well as the president, vice-president and the secretary of the WHRRF and the advisor to the Thai Medical Council. The health professionals were invited to attend the training following a formal invitation letter sent to each Provincial Health Office by the WHRRF. Medical officers from the Department of Obstetricians and Gynaecologists of Ramathibodi Faculty of Medicines and Prince Songkla Faculty of Medicines were invited to conduct the training. The three-day training consisted of three modules. The first module explained the current global and national situation on abortion, including the debates on abortions, the effect of unsafe abortions on the socio-economic situation and health of women, the situations in which unsafe abortions occur, the risk of unsafe abortions and the benefits of access to safe abortions. The second module explained the methods used to provide safe abortions, with a focus on the use of manual vacuum aspiration (MVA) [?]. The final module demonstrated the method by instructors for the correct application of MVA's and allowed participants to practice using MVAs on simulation models. Despite the active participation of the program, the impact of the program on the attendees was never evaluated. This study therefore aimed to: (i) evaluate the knowledge and attitudes towards safe abortion among the training courses participants and (ii) investigate the factors that determined the knowledge and attitudes among the training course participants.

2. Methods

2.1. Study Design and Participants

A pre-post study design was applied. A self-administered survey was filled out by each participant before and after the training. The time used for responding to the questionnaire was about 10–15 min for pre-test and for post-test. The course attendees were asked to return the filled questionnaire form to the course facilitator. Data from health professionals who attended the training program on the 'Care of pregnant adolescents and women for the prevention of unsafe abortions' between August 2017 and April 2018 were included.

2.2. Ethics Approval

This study is part of the routine monitoring system by IHPP on progress and access to sexual and reproductive health. IHPP is a research institute of the Ministry of Public Health, and therefore it was not necessary to obtain ethics approval. Despite this, the researchers strictly followed ethical standards where all individual information was strictly kept confidential and not reported in the paper.

2.3. Data Collection and Measures

Data from surveys conducted by the Women's Health and Reproductive Rights Foundation of Thailand were obtained. The survey was composed of four main parts: (1) demographic characteristics; (2) work experience with adolescents and women who have had unplanned pregnancies and undergone abortions; (3) perceptions towards adolescents and women with unplanned pregnancies and unsafe abortions (9 sub-questions); and (4) scenarios on abortions (14 sub-questions). Details of all the sub-sections are presented in the results section.

The demographic measures comprised sex (male, female), age (years), type of health profession (doctor, nurse, pharmacist and welfare workers) and the specialization of doctors (general practice, obstetrics and gynaecology, family medicine, preventative medicine, among others). The measures of experience of working with adolescents and women who had experienced unplanned pregnancies and undergone abortions included: (a) counselling experience with adolescents and women who had unplanned pregnancies (ever, never); (b) treating adolescents and women who had undergone abortions already (ever, never); (c) regulations and guidelines on the termination of pregnancies (know, do not know); (d) knowledge of manual vacuum aspiration (MVA) as a tool for safe abortion (know, do not know; ever seen, ever heard, never seen, never heard); and (f) experience in using manual vacuum aspiration (MVA) (yes, no). The measures of the attitudes of health professionals were rated on a five-point Likert scale ranging from 1 representing 'strongly disagree' to 5 'strongly agree'. Nine items measured the general attitudes towards adolescent pregnancy and fourteen items measured the attitudes towards various scenarios for when abortions may be performed.

2.4. Analysis

Pre- and post-surveys were matched using a participant code number that was provided to each participant when they registered to the training and recorded onto both the pre-survey and post-survey. Participants who did not complete both the pre- and post-surveys or whose information about demographic characteristics was incomplete were excluded from the analysis. All analyses were conducted using the Stata/IC version 14 (StataCorp LLC, Texas, TX, USA) and the statistical significance was assessed at alpha of 0.05. and the statistical significance was assessed at alpha of 0.05.

2.5. Descriptive Statistics

Descriptive statistics were calculated for all the demographic characteristics and experiences of working with adolescents and women with unplanned pregnancies and having undergone abortions, including percentages for categorical variables, and mean and standard deviation (SD) for continuous variables. The mean, SD, median and interquartile range (IQR) were calculated for each item for participant attitudes towards adolescents and women with unplanned pregnancies and abortions and attitudes towards various abortion scenarios.

2.6. Bivariate Analysis

A bivariate analysis was conducted to determine the significance of the change in attitudes towards adolescents and women with unplanned pregnancies and having undergone abortions as well as the attitudes towards various abortion scenarios. The Shapiro–Wilk test was used to examine distribution normality. The Wilcoxon signed rank test was applied to avoid assumptions of normal distribution needed for the paired sample t-tests. The Kruskal–Wallis rank test was applied to analyse the significance of the change in attitudes according to demographic characteristics and professional experiences, including sex, age, profession as well as knowledge and experience on abortion including aspects of regulations, treating, counselling and MVA.

2.7. Multivariable Analysis

Multivariable analysis was applied to assess the likelihood of the change in attitudes towards adolescents and women with unplanned pregnancies and having undergone abortions as well as attitudes towards various abortion scenarios, using demographic and experience characteristics as predicator variables. Multiple linear regression was applied. The outcome variables were the change in the average score between the pre-test and post-test results on attitudes towards adolescents and women with unplanned pregnancies and having undergone unsafe abortions (Y1) and the response to abortion scenarios (Y2) (part 3 and part 4 of the questionnaire). The predictor variables were selected from the results of the bivariate analysis. Those with statistical significance at $p < 0.2$ were included in the multivariable analysis. The authors also conducted a kernel-based regularized least squares method, a special case of linear regression which allows the researchers to account for data with non-normal distribution, in order to assess whether the results would differ from conventional regression analysis.

3. Results

A total of 325 participants attended the training programme: 99, 147 and 79 in the first, second and third batch, respectively. Of these, 250 completed both the pre- and post-test surveys and had a matching survey pair. The demographics and professional experiences section were incomplete for three participants, who were excluded (Figure ??). The data from 247 participants were included and analysed: 43.3% from the second training session, 30.8% from the first and the remaining 25.9% from the third.

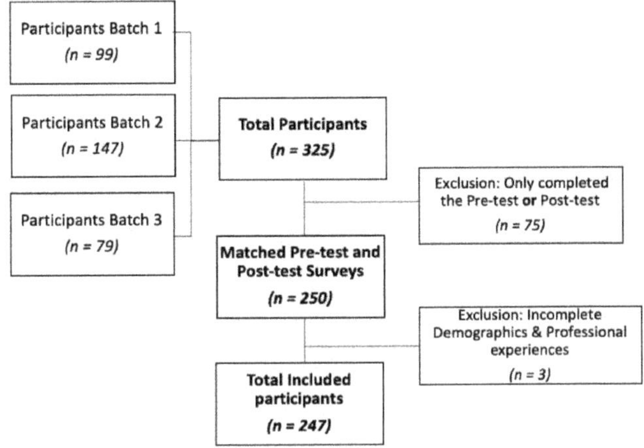

Figure 1. Flow diagram of the included participants.

3.1. Demographics Characteristics and Work Experience

The participant age ranged from 23 to 58 years with an average of 35.9 years. The majority of participants were female (84.6) and were nurses (55.5%). Of the participants that were doctors (43.72%), 52.8% were general practitioners and 36.1% were obstetricians and gynaecologists. The survey found that 63.2% of all participants had prior experience of counselling for unplanned pregnancies and 80.6% had previously treated patients that had undergone abortions. 52.5% of participants had previous knowledge on the regulations of the Medical Council on Practices Regarding Termination of Medical Pregnancy. With regards to manual vacuum aspirations, 65.23% had knowledge of MVA, 64% had knowledge on the requirements of the use of MVA, 67.21% had previously seen an MVA and 50.21% had used an MVA. Details on the demographic characteristics and experiences of all the participants are summarized in Table ??.

Table 1. Participant Characteristics.

Characteristics	Total ($n = 247$)
Sex (male/female) (%)	15.4/84.6
Age (Year) (Mean ± SD) [Min–Max]	35.9 ± 8.8 [23–58]
Profession (%)	
Doctor ($n = 137$) (% of total participants)	43.7
General (% of doctors)	52.8
Obstetrics and gynaecology (% of doctors)	36.1
Family medicine (% of doctors)	4.6
Preventative medicine (% of doctors)	1.9

Table 1. *Cont.*

Characteristics	Total ($n = 247$)
Others (% of doctors)	4.6
Nurse (% of total participants)	55.5
Pharmacist (% of total participants)	0.4
Welfare workers (% of total participants)	0.4
Prior experience counselling for unplanned pregnancies (% with experience)	63.16
Prior experience treating for unplanned pregnancies (% with experience)	80.57
Knowledge of regulations of the Medical Council on the Practices Regarding the Termination of Medical Pregnancies (% with knowledge)	52.46
Manual vacuum aspirations (MVA)	
Know of MVA (%)	65.23
Knowledge of requirements for medical professionals for the use of MVA (%)	64.00
Seen MVA (%)	67.21
Used MVA (%)	50.21

3.2. Comparison of Pre-post Results on Attitudes towards Adolescents and Women Experiencing Unplanned Pregnancies and Unsafe Abortions

The median pre-test and post-test responses were found to be significantly different for each of the nine responses on attitudes towards unplanned pregnancies and unsafe abortions at $p < 0.001$ (Table ??). The median of the combined average of all nine responses was found to have significantly increased by 0.67 points at $p < 0.001$ (Figure ??).

Table 2. The comparison of the pre- and the post-test attitudes towards adolescents and women experiencing unplanned pregnancies and unsafe abortions ($n = 247$).

Questions	Pre-Test	Post-Test
	Mean (SD), Median [IQR]	Mean (SD), Median [IQR]
1. At the present, unplanned pregnancies and unsafe abortions are a major public health problem that should be addressed.	4.62 (0.50), 5 [1]	4.76 (0.43), 5 [0]
2. In your area, unplanned pregnancies and unsafe abortions are a major public health problem, that should be addressed.	4.10 (0.71), 4 [1]	4.39 (0.68), 4 [1]
3. One reason for unsafe abortions is the limited options for pregnant women and the societal pressures pregnant women face.	4.02 (0.78), 4 [1]	4.64 (0.54), 5 [1]
4. Family and society should help unplanned pregnancies.	4.50 (0.54), 5 [1]	4.74 (0.46), 5 [1]
5. Women with unplanned pregnancies should have the right to decide and choose whether to continue or terminate the pregnancy.	4.00 (0.82), 4 [1]	4.61 (0.59), 5 [1]
6. Women that need to terminate their pregnancies, according to the criteria set by the Regulations of Medical Council of Thailand, should receive safe abortion, with the same level of services and benefits as other health problems.	4.38 (0.59), 4 [1]	4.72 (0.46), 5 [1]
7. Doctors and healthcare providers play a major role in addressing unplanned pregnancies and unsafe abortions.	4.49 (0.54), 5 [1]	4.72 (0.48), 5 [1]
8. You are happy to help, advise and provide consultations regarding abortions and places to obtain safe abortion services for those with unplanned pregnancies.	4.05 (0.84), 4 [1]	4.53 (0.58), 5 [1]
9. Thailand should allow the sale of medical abortion drugs as well as emergency contraceptives at pharmacies.	2.38 (1.24), 2 [2]	3.74 (1.20), 4 [2]
Average Score of all questions	4.06 (0.42), 4 [1.6]	4.54(0.41), 4.67 [0.6]

Note: All p-Values were ≤ 0.001 using the Wilcoxon signed rank test.

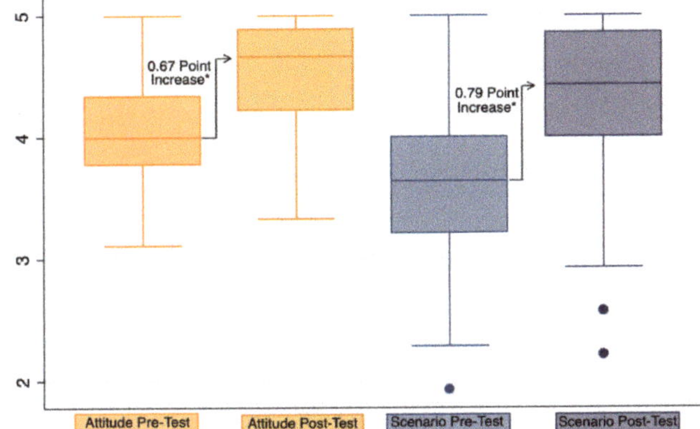

Figure 2. Box plot of the average pre-test and post-test responses. Note: Attitudes towards adolescents and women unplanned for pregnancy, and unsafe abortions. Scenario: Response to example of scenarios on abortions. * p-Value ≤ 0.001

The difference in the average responses for attitudes towards adolescents and women experiencing unplanned pregnancies and having undergone unsafe abortions was found to be significantly different according to career at $p = 0.013$. Although the median pre-test scores were equal for all career types, with an average score of 4, the median post-test scores highly varied (Figure S1). The greatest increase in positive change was found in non-doctor participants with an average increase of 0.53 points, compared with doctors who were not specialized in obstetrics and gynaecology with an average increase of 0.46 points, and obstetricians and gynaecologists with an average increase of 0.32 points. The differences in the average response were also found to be significant among participants who had prior knowledge of the regulations of the Medical Council on Practices Regarding the Termination of Medical Pregnancy ($p = 0.0039$). Those with prior knowledge had an average increase of 0.55 points, while those without prior knowledge increased by an average of 0.42 points (Table ??).

Table 3. Differences in the pre- and the post-test changes in the responses according to demographic and professional characteristics.

Characteristics	Attitudes towards Adolescents and Women Experiencing Unplanned Pregnancies, and Unsafe Abortions	Change in Response to Examples of Scenarios on Abortions
Sex	$p = 0.863$	$p = 0.217$
Female	0.48 (0.43), 0.44 [0.55]	0.75 (0.58), 0.71 [0.71]
Male	0.46 (0.38), 0.44 [0.55]	0.61 (0.40), 0.64 [0.5]
Age	$p = 0.821$	$p = 0.900$
Less than 35	0.46 (0.44), 0.44 [0.55]	0.71 (0.52), 0.71 [0.64]
More than 35	0.49 (0.40), 0.44 [0.55]	0.74 (0.60), 0.71 [0.64]
Career	$p = 0.013$ *	$p < 0.001$ *
OBGYN	0.32 (0.38), 0.33 [0.44]	0.54 (0.46), 0.42 [0.64]
Other doctor	0.46 (0.38), 0.44 [0.44]	0.59 (0.44), 0.57 [0.57]
Non-doctor	0.53 (0.44), 0.55 [0.66]	0.84 (0.61), 0.86 [0.71]
Knowledge of MVA	$p = 0.248$	$p = 0.507$
Knowledge of regulation	$p = 0.004$ *	$p = 0.189$
Prior knowledge	0.55 (0.43), 0.55 [0.61]	
No prior knowledge	0.42 (0.39), 0.33 [0.55]	
Experience in treating	$p = 0.071$	$p = 1.000$
Experience in counselling	$p = 0.087$	$p = 0.157$
Experience in using MVA	$p = 0.846$	$p = 0.698$

Note: Mean (SD), Median [IQR]; * Statistically significant (Defined as $p < 0.05$); OBGYN: Obstetricians and gynaecologists.

The predictor variables incorporated into the regression model included healthcare professions, prior knowledge of the regulations of the Medical Council on Practices Regarding Termination of Medical Pregnancy and experience of treating and counselling teenagers and women with unplanned pregnancies. The analysis determined that non-medical doctor health professionals were most likely to benefit from the training. Although both the multiple linear regression (Table ??) and the kernel-based regularized least squares method (Table S2) showed similar trends, the results were only significant when using multiple linear regression ($p = 0.041$).

Table 4. Least-squares regression analysis on the attitudes towards adolescents and women experiencing unplanned pregnancies and having undergone unsafe abortions.

(Reference: OBGYN)	Coef. (95% CI)	*p*-Value
Non-OBGYN doctor	0.095 (−0.074, 0.264)	0.269
Non-doctor	0.165 (0.006, 0.323)	0.041
Prior knowledge of regulation	−0.067 (−0.182, 0.047)	0.247
Experience in treating	−0.102 (−0.236, 0.032)	0.134
Experience in counselling	0.002 (−0.002, 0.007)	0.317
Constant	0.475 (0.279, 0.671)	0.000

3.3. Comparison of Pre-post Results on Attitudes towards Various Scenarios for Abortions

The median pre-test and post-test responses were found to be significantly different for each of the 14 responses concerning attitudes towards abortion scenarios (Table ??) at $p < 0.001$. The median of the combined average of all 14 responses was found to have significantly increased by 0.79 points at $p < 0.001$ (Figure ??). The difference in the average responses to examples of scenarios on abortions was also found to be significantly different according to career at $p \leq 0.001$. The most significant increase was found in non-doctor participants, with an average increase of 0.84 points, compared with doctors not specialized in obstetrics and gynaecology with an average increase of 0.59 points and then obstetricians and gynaecologists with an average increase of 0.54 points (Table ??). However, it is important to note that the pre-test scores of non-doctors were lower with a median score of 3.57, compared to the median scores of obstetricians and gynaecologists at 3.79 and other doctors with a score of 3.64 (Figure S1).

The predictor variables incorporated into the regression model included healthcare profession, prior knowledge of the regulations of the Medical Council on Practices Regarding Termination of Medical Pregnancy, and experience of counselling teenagers and women with unplanned pregnancies. Similarly to the analysis of attitudes towards adolescents and women experiencing unplanned pregnancies and having undergone unsafe abortions, the regression analysis determined that non-medical doctor health professionals were the most likely to benefit from the training. The results were found to be significant using both multiple linear regression, at $p = 0.003$ (Table ??) and kernel-based regularized least squares method, at $p = 0.006$ (Table S1 and Table S2).

Table 5. The comparison of the pre- and the post-test responses to the examples of abortion scenarios ($n = 247$).

Questions	Pre-Test	Post-Test
	Mean (SD), Median [IQR]	Mean (SD), Median [IQR]
1. If the pregnant woman has underlying diseases and the pregnancy poses serious harm to their health or life.	4.56 (0.59), 5 [1]	4.85 (0.39), 5 [0]
2. If the pregnant woman has physical or intellectual disabilities hindering their ability to care for themselves.	4.45 (0.66), 5 [1]	4.81 (0.49), 5 [0]
3. If the pregnant woman has HIV/AID.	3.40 (1.23), 3 [3]	3.74 (1.25), 4 [2]

Table 5. *Cont.*

Questions	Pre-Test	Post-Test
	Mean (SD), Median [IQR]	Mean (SD), Median [IQR]
4. If the pregnant woman has rubella.	4.03 (0.89), 4 [1]	4.50 (0.71), 5 [1]
5. If the foetus has anomalies that can result in being physically or intellectually disabled.	4.31 (0.81), 4 [1]	4.74 (0.52), 5 [0]
6. If the foetus has genetic disorders or serious diseases.	4.38 (0.81), 4 [1]	4.78 (0.49), 5 [0]
7. If the pregnant woman's mental health is at risk.	3.71 (0.97), 4 [2]	4.53 (0.65), 5 [1]
8. If the pregnant woman is under the age of 15.	3.33 (1.07), 3 [1]	4.32 (0.83), 5 [1]
9. If the pregnant woman is under the age of 20 and still in school.	3.01 (1.04), 3 [2]	3.97 (0.93), 4 [2]
10. If the pregnancy is a result of rape.	4.34 (0.76), 4 [1]	4.78 (0.47), 5 [0]
11. If the pregnancy is a result of incest.	3.39 (1.03), 3 [1]	4.24 (0.91), 5 [1]
12. If the pregnancy is a result of contraceptive failure.	2.97 (1.15), 3 [2]	4.19 (0.87), 4 [1]
13. If the pregnant woman is facing economic problems.	2.76 (1.08), 3 [1]	4.11 (0.93), 4 [1]
14. If the pregnant woman is unmarried.	2.67 (1.11), 3 [1]	3.91 (0.99), 4 [2]
Average Score of all questions	3.67 (0.64), 3.64 [0.79]	4.39 (0.53), 4.43 [0.86]

Note: All *p*-values were ≤ 0.001 using the Wilcoxon signed rank test.

Table 6. Least-squares regression analysis on the abortion scenarios.

(Reference: OBGYN)	Coef. (95% CI)	*p*-Value
Non-OBGYN doctor	0.064 (−0.159, 0.288)	0.573
Non-doctor	0.323 (0.112, 0.534)	0.003 *
Prior knowledge of regulation	0.041 (−0.109, 0.191)	0.54
Experience in counselling	0.003 (−0.004, 0.009)	0.437
Constant	0.489 (0.227, 0.751)	0.000

* Statistically significant (Defined as $p < 0.05$)

4. Discussion

The evidence indicated that following the implementation of the training programme, healthcare providers' positive attitudes towards unplanned pregnancies and unsafe abortions significantly increased. Overall, career types of healthcare providers significantly contributed to changes in attitudes towards unplanned pregnancies and unsafe abortions. Detailed analysis suggested that non-obstetric and non-gynaecology doctors and supporting staff (such as nurses) tended to benefit most from the training. This was evident in the significant increase in their assessment scores relative to the scores for obstetrics and gynaecology doctors (as shown in ????).

Our findings were similar to findings from other studies. For instance, in Zimbabwe, a study found that supporting staff apart from doctors (such as a nurse, midwife, senior nurse or hospital administrator) play an important role in supporting women in accessing safe abortions [?]. A study by Cooper et al. gave a positive view on nurses' and midwives' attitudes towards abortion [?]. However, a systematic literature review on the perceptions and attitudes of healthcare providers in sub-Saharan Africa and Southeast Asia by Ulrilka et al. suggested that nurses and midwives disliked being involved in abortion services, and commonly reported hesitancy in providing these services. The nurses' resistance to providing abortion services was a powerful barrier against access to safe abortion services, with nurses' and midwives' strong opposition to abortion affecting rural women in particular [?]. Such findings however contradicted the findings in this study, as the results did not present strong negative attitudes towards abortion among nurses and supporting staff. In contrast, the findings showed that nurses and supporting staff could be potential target groups for further trainings, as their scores were enhanced the most compared with other health professionals.

This study has some policy implications. Firstly, this kind of training on safe abortion is useful to wider health professionals, institutions and organizations, which in turn can play an important role in creating awareness of unsafe abortion and providing safe abortion services. Secondly, abortion is not only a matter for obstetrics and gynaecologists. The study found that non-obstetric doctors and support staff can play an important role. Finally, this study found that knowledge of regulations on abortions is quite low as less than 50% of the participants had adequate prior knowledge on these (Table **??**). Therefore, the Royal Thai College of Obstetricians and Gynaecologists and the medical council should work together to communicate these regulations to health professionals and the wider public.

Some limitations remain. Firstly, the participation in this training was voluntary and this therefore created a risk of selection bias. Those who opt out from this kind of training may not have similar favourable attitudes towards abortion. Secondly, the exclusions of participants undermined the statistical power as some observations were dropped (Figure **??**). This may be a reason why some factors did not show statistical significance. However, researchers checked the demographic characteristics of the participants that were excluded and found no significant differences compared to the study sample. Thirdly, analyses by different methods can yield different results [?]. The questionnaire in section three and four applied a Likert scale to analyse the attitudes of participants. In this study we opted to use regression analysis; however, if researchers devised and used other analytical methods such as the chi squared or logistic analyses, they may yield different results. However, researchers found the results to be quite valid as the kernel-based regularized least squares method was applied and there was no marked difference in the results. Fourth, participants' locations of work, either urban or rural settings, were not collected, and as the trainings were conducted only in Bangkok, it is expected that most participants were from the Bangkok metropolitan area. Therefore, differences in attitudes between urban and rural areas could not be clearly identified. Further studies should be conducted to analyse the urban and rural area differences in the attitudes by the location of their workplaces. Finally, this assessment is subjective. The results showing the positive attitudes of the participants who attended the training programme do not guarantee good practice in a real-life situation and the study sample does not necessarily reflect the viewpoints of all the health professionals in the country. Further research should assess real-life practices and attitudes, including both qualitative and quantitative components.

5. Conclusions

Following the training, the score for positive attitudes towards unplanned pregnancies and unsafe abortions as well as attitudes towards abortion scenarios significantly increased. The main determinant, which significantly contributed to positive attitudes towards unplanned pregnancies and unsafe abortions, was the career type of the healthcare provider. In particular, non-doctor health professionals were likely to benefit the most from this kind of training and could be the target population for training in the future. Further research using both qualitative and quantitative methods should be conducted to assess the attitudes and real-life practices of healthcare professionals concerning abortions in Thailand.

Supplementary Materials: The following are available online at http://www.mdpi.com/1660-4601/17/9/3198/s1, Figure S1: Box Plot of the Average Pre-test and Post-test Response by career type, Table S1 Regularized least squares analysis on towards adolescents and women unplanned for pregnancy, and unsafe abortions; Table S2: Kernel-based regularized least squares analysis

Author Contributions: All authors designed the study. R.S.(Rugsapon Sanitya), A.I.M., R.S.(Rapeepong Suphanchaimat) and V.T. were responsible for data analysis. R.S.(Rugsapon Sanitya), A.I.M., N.S., S.J., P.S., R.S.(Rapeepong Suphanchaimat) and M.P. crafted the first draft of the manuscript. V.T., K.C., N.B., and R.S.(Rapeepong Suphanchaimat) revised and finalized the manuscript. All authors contributed toward data analysis, drafting and critical revision.

Funding: The authors gratefully acknowledge Walaiporn Patcharanarumol for the funding support through IHPP from the Thailand Science Research and Innovation (TSRI) under the Senior Research Scholar on Health Policy and System Research [Contract No. RTA6280007].

Acknowledgments: The investigators are immensely grateful for the support from IHPP's and WHRRF's staff.

Conflicts of Interest: The authors declare that they have no competing interests.

References

1. Haddad, L.B.; Nour, N.M. Unsafe abortion: Unnecessary maternal mortality. *Rev. Obstet. Gynecol.* **2009**, *2*, 122–126. [PubMed]
2. World Health Organization. *The Prevention and Management of Unsafe Abortion: Report of a Technical Working Group, Geneva, 12–15 April 1992*; World Health Organization: Geneva, Switzerland, 1993.
3. Ganatra, B.; Gerdts, C.; Rossier, C.; Johnson, B.R., Jr.; Tunçalp, Ö.; Assifi, A.; Sedgh, G.; Singh, S.; Bankole, A.; Popinchalk, A.; et al. Global, regional, and subregional classification of abortions by safety, 2010–2014: Estimates from a Bayesian hierarchical model. *Lancet* **2017**, *390*, 10110. [CrossRef]
4. World Health Organization. Preventing Unsafe Abortion: WHO. 2019. Available online: https://www.who.int/news-room/fact-sheets/detail/preventing-unsafe-abortion (accessed on 30 March 2020).
5. World Health Organization. *Department of Reproductive Health and Research. Reproductive Health Strategy to Accelerate Progress Towards the Attainment of International Development Goals and Targets*; World Health Organization: Geneva, Switzerland, 2004.
6. Warakamin, S.; Boonthai, N.; Tangcharoensathien, V. Induced abortion in Thailand: Current situation in public hospitals and legal perspectives. *Reprod. Health Matters* **2004**, *12* (Suppl. 24), 147–156. [CrossRef]
7. Srinil, S. Factors associated with severe complications in unsafe abortion. *J. Med Assoc. Thail.* **2011**, *94*, 408.
8. Chunuan, S.; Kosunvanna, S.; Sripotchanart, W.; Lawantrakul, J.; Lawantrakul, J.; Pattrapakdikul, U.; Somporn, J. Characteristics of Abortions in Southern Thailand. *Pacific Rim Int. J. Nurs. Res.* **2012**, *16*, 97–112.
9. Chaturachinda, K. Unsafe abortion in Thailand: Roles of RTCOG. *Thai J. Obstet. Gynaecol.* **2014**, *2014*, 2–7.
10. Olukoya, A.; Kaya, A.; Ferguson, B.; AbouZahr, C. Unsafe abortion in adolescents. *Int. J. Gynecol. Obstet.* **2001**, *75*, 137–147. [CrossRef]
11. World Health Organization. *World Health Statistics 2019: Monitoring Health for the SDGs, Sustainable Development Goals*; World Health Organization: Geneva, Switzerland, 2019.
12. Ministry of Public Health. Public Health Statistics A.D.2017. In *Nonthaburi: Strategy and Planning Division*; Ministry of Public Health: Bangkok, Thailand, 2017.
13. Whittaker, A. *Abortion, Sin and the State in Thailand*; Routledge Curzon: New York, NY, USA, 2004.
14. Chaturachinda, K.; Boonthai, N. Unsafe Abortion: An Inequity in Health Care. *Thail. Perspect. J. Popul. Soc. Stud.* **2017**, *25*, 287–297. [CrossRef]
15. Praditpan, P.; Chaturachinda, K. *Doctors must Heed Abortion Needs*; Bangkok Post: Bangkok, Thailand, 2016.
16. World Health Organization. *Clinical Practice Handbook for Safe Abortion*; WHO: Geneva, Switzerland, 2014.
17. Madziyire, M.G.; Ann, M.; Taylor, R.; Elizabeth, S.; Tsungai, C. Knowledge and attitudes towards abortion from health care providers and abortion experts in Zimbabwe: A cross sectional study. *Pan. Afr. Med. J.* **2019**, *34*, 94. [CrossRef] [PubMed]
18. Cooper, D.; Dickson, K.; Blanchard, K.; Cullingworth, L.; Mavimbela, N.; von Mollendorf, C.; van Bogaert, L.; Winikoff, B. Medical Abortion: The Possibilities for Introduction in the Public Sector in South Africa. *Reprod. Health Matters* **2005**, *13*, 35–43. [CrossRef]
19. Loi, U.R.; Kristina, G.-D.; Elisabeth, F.; Marie, K.-A. Health care providers' perceptions of and attitudes towards induced abortions in sub-Saharan Africa and Southeast Asia: A systematic literature review of qualitative and quantitative data. *BMC Public Health* **2015**, *15*, 139.
20. Sullivan, G.M.; Artino, A.R., Jr. Analyzing and interpreting data from likert-type scales. *J. Grad. Med. Educ.* **2013**, *5*, 541–542. [CrossRef] [PubMed]

International Journal of
*Environmental Research
and Public Health*

Article

Government Expenditure on Maternal Health and Family Planning Services for Adolescents in Mexico, 2003–2015

Leticia Avila-Burgos [1], Julio César Montañez-Hernández [1,*], Lucero Cahuana-Hurtado [2], Aremis Villalobos [3], Patricia Hernández-Peña [4] and Ileana Heredia-Pi [1]

[1]   Center for Health Systems Research, The National Institute of Public Health, Cuernavaca 62100, Morelos, Mexico; leticia.avila@insp.mx (L.A.-B.); ileana.heredia@insp.mx (I.H.-P.)
[2]   School of Public Health and Administration, Universidad Peruana Cayetano Heredia, Lima 15102, Peru; cahuana.luce@gmail.com
[3]   Center for Population Health Research, The National Institute of Public Health, Cuernavaca 62100, Morelos, Mexico; alvillalobos@insp.mx
[4]   Netherlands Interdisciplinary Demographic Institute-KNAW, University of Groningen, 2511 CV The Hague, The Netherlands; hernandez@nidi.nl
*   Correspondence: julio.montanez@insp.mx

Received: 18 March 2020; Accepted: 27 April 2020; Published: 29 April 2020

Abstract: The purpose of this study was to assess whether government policies to expand the coverage of maternal health and family planning (MHFP) services were benefiting the adolescents in need. To this end, we estimated government MHFP expenditure for 10- to 19-year-old adolescents without social security (SS) coverage between 2003 and 2015. We evaluated its evolution and distribution nationally and sub-nationally by level of marginalization, as well as its relationship with demand indicators. Using Jointpoint regressions, we estimated the average annual percent change (AAPC) nationally and among states. Expenditure for adolescents without SS coverage registered 15% for AAPC for the period 2003–2011 and was stable for the remaining years, with 88% of spending allocated to maternal health. Growth in MHFP expenditure reduced the ratio of spending by 13% among groups of states with greater/lesser marginalization; nonetheless, the poorest states continued to show the lowest levels of expenditure. Although adolescents without SS coverage benefited from greater MHFP expenditure as a consequence of health policies directed at achieving universal health coverage, gaps persisted in its distribution among states, since those with similar demand indicators exhibited different levels of expenditure. Further actions are required to improve resource allocation to disadvantaged states and to reinforce the use of FP services by adolescents.

Keywords: government health expenditure; adolescents; maternal health; family planning

---

1. Introduction

Mexico and other Latin American countries need to invest in the sexual and reproductive health of adolescents [1–4]. Ensuring the availability of healthcare and promoting healthy behaviors in this population group generate economic benefits that improve future labor productivity. Efforts on behalf of adolescents also contribute to reducing risks, such as complications during pregnancy in very young women, as well as preventing premature births and low-birthweight babies [3,5]. From a human rights perspective, adolescents are entitled to receive the information and health services they need to survive, grow, and realize their full potential as individuals [1,3,5].

As a result of Mexican population dynamics, adolescents now exert an unprecedented impact on national demographics. With nearly one-fourth of the population being aged 10–19 years in 2014 [6], the health needs of adolescents have acquired particular relevance. Their behavior has evolved towards early initiation of sexual activity [7,8] with limited use of contraceptive methods (CMs), especially among the poorer groups [9,10]. With a fertility rate that grew from 64 births per 1000 adolescents (15–19 years old) in 2009 to 77 births in 2014 [10,11], Mexico ranks first in adolescent fertility among Organization for Economic Cooperation and Development (OECD) member states [12]. The healthcare system needs to strengthen its response such that the health expectations and requirements of this age group are addressed and their access to health services is expanded.

Over two decades ago, Mexico launched its System of Social Protection in Health (*Sistema de Protección Social en Salud*) with the aim of achieving universal health coverage. Its principal component, the *Seguro Popular* health insurance scheme, was conceived as a mechanism for enhancing service coverage and providing financial protection to the 45% non-salaried population in Mexico without access to social security (SS) [13,14]. Within this scheme, improving maternal health (MH) through greater coverage has been considered a core objective. An example of this commitment was the introduction of the Healthy Pregnancy Strategy (*Estrategia Embarazo Saludable*) in 2008. This program incorporated all pregnant women without SS coverage into the *Seguro Popular*, providing them with a package of MH services free of charge [15]. Although adolescents without SS coverage were not explicitly targeted by these initiatives, they benefited from the increased supply of maternal health (MH) and family planning (FP) services [16,17].

In 2009, Mexico stepped up its efforts to improve availability of sexual and reproductive health services for adolescents. One upshot was the Model of Comprehensive Care for the Sexual and Reproductive Health of Adolescents implemented in 2013. Two years later, having recognized adolescent pregnancy as a national problem, Mexico established the National Strategy for the Prevention of Adolescent Pregnancy. This initiative was designed to reduce the fertility rate among adolescents aged 15–19 years and to eradicate pregnancy among those aged 10–14 years [8]. One of its principal components was greater coverage for sexual and reproductive health services and modern contraception methods (CMs). To this end, the National Strategy engaged numerous public health institutions—those pertaining to the Social Security Institute, which covered 43% of the population, and those serving the population without SS coverage through the *Seguro Popular*, the State Health Services, and the *IMSS PROSPERA* Program, renamed *IMSS Bienestar* at the close of 2018 [18,19].

These initiatives expanded the coverage of antenatal care services [16] and improved the availability of CMs for the adolescent population lacking SS coverage [7,9], thereby mobilizing greater resources from the government. To facilitate the transfer of funds to State Health Services, the *System of Social Protection in Health* introduced a per capita payment scheme based on the number of affiliates. This was intended to overcome the historical inertia in the distribution of funds [20].

The Mexican *Reproductive Health Account* system offers information on public maternal health and family planning (MHFP) spending from 2003 to 2015 [21], which serves to monitor and analyze the resources and performance of these services during this period [17,19,22,23]. However, data are not disaggregated by age group, and the consequences of changes in MHFP spending for adolescents without SS coverage are therefore unknown. To assess whether policies for expanding the coverage of maternal health and family planning (MHFP) are benefiting the adolescents in need, we (1) estimated MHFP expenditure for the entire adolescent population without SS coverage and determined its growth and distribution among MH and FP services and providers for the period 2003–2015; (2) analyzed the expenditure gaps among states by level of marginalization; and (3) examined the relationship between MHFP spending and demand indicators. Given that changes in the health system have centered on the population without SS coverage, we confined our analyses to this population.

2. Material and Methods

We conducted an ecological study in order to estimate government MHFP expenditure for adolescents without SS coverage. To achieve this, we drew data for 2003–2015 from the *Reproductive Health Accounts* [21], constructed according to the *OECD System of Health Accounts* [24] and the *World Health Organization's Guide to Producing Reproductive Health Subaccounts* [25].

2.1. MHFP Expenditure for Adolescents

Analysis included government strategies aimed at the population without SS coverage, namely the *Seguro Popular* insurance and care providers, the State Health Services, and the *IMSS PROSPERA* Program [19,21]. Expenditure for adolescents was grouped according to care providers, specifically hospitals and ambulatory-care centers [21,24]. The selection of beneficiaries focused on young people between 10 and 19 years old, the age range for adolescence established by the Specific Action Program for the Sexual and Reproductive Health of Adolescents [26].

To estimate expenditure for the adolescent population, we carried out the following procedures. First, we identified the number of MH and FP consultations offered by ambulatory-care centers to 10- to 19-year-olds without SS coverage. Calculations were undertaken by type of service (MH or FP), year, and state. Subsequently, we estimated the proportions of these consultations with respect to the total number of MH and FP consultations offered to the total population without SS coverage. Likewise, we weighted MH and FP expenditures incurred by hospitals using the proportions of in-hospital days and consultations offered to adolescents without SS coverage. Based on OECD and WHO methodology [24], we used these proportions to weight the MH and FP expenditures sustained by ambulatory-care centers and hospitals for the total population lacking SS coverage. The underlying assumption was that spending was similar for comparable types of in-hospital days and consultations regardless of the ages of patients. Analyses covered 131.9 million in-hospital days as well as 1571 million general and specialized in-hospital consultations. We obtained the data from the General Directorate of Health Information under the Federal Ministry of Health (*DGIS*) [27], and grouped them according to the International Classification of Diseases, 10th Revision (ICD-10) [28].

MH services included care during pregnancy, childbirth (vaginal or cesarean), and the postpartum period, as well as abortion. FP services included counseling, consultations, the provision of CMs (hormonal or otherwise), and the performance of definitive surgical procedures (tubal ligation and vasectomy) for the entire adolescent population, 10 to 19 years old. Consultations were grouped by provider (hospital or ambulatory-care center).

We calculated MHFP expenditure in constant US dollars and converted the figures to 2015 international dollars (Purchasing Power Parity, PPP, 2015 US$1 = MXN 8.541) [29]. To establish a comparison among states, we adjusted MHFP expenditure by adolescent women aged 10–19 according to *DGIS* data [27].

2.2. Demand Indicators

Demand has been defined as the proportion of a population experiencing health needs and requiring health services to satisfy them. For the purposes of this study, we defined the demand indicator for MH services as the number of pregnant adolescents without SS coverage, and for FP services as the number of sexually active adolescent women without SS coverage. To estimate these indicators, we used data from the 2009 and 2014 National Surveys of Demographic Dynamics (*ENADIDs*) [30]. Although the *ENADID* design identifies fertility and pregnancy at the population level, it captures information only on pregnancies among 15- to 54-year-old women. Thus, our demand indicator for MH services included only 15- to 19-year-old adolescents who were pregnant at the time of the surveys or the year before. Under the demand indicator for FP services, we included adolescent women between the ages of 15 and 19 who were sexually active during the month prior to the surveys and those aged 10 to 19 who became sexually active during the survey period.

To ensure comparability, we constructed the MHFP indicators for adolescents in accordance with international practices. For instance, we considered women as the basis of our CM demand indicators given the distinct impact that the use/non-use of CMs exerts on their reproductive health and risk of pregnancy. Moreover, female contraceptives offer a wider range of methods and costs compared to male contraceptives, limited mostly to condoms available at lower costs [31,32]. For these reasons, most resource-tracking methodologies, including the Health Accounts, base their indicators of FP spending on women as the common denominator [25]. Finally, international recommendations on investment in adolescent FP activities also consider women as the basis, particularly in light of the high-priority problem of adolescent motherhood [33]. We obtained data on the size of the adolescent population lacking SS coverage for the period 2010–2018 from the *DGIS* databases [27]. As information was unavailable for the period prior to 2010, we calculated data from 2003 to 2009 based on the average annual growth of the adolescent population without SS coverage during 2010–2018. States were grouped according to the State Marginalization Index of the National Population Council [34].

2.3. Analytical Strategy

To assess growth in MHFP spending during the period 2003–2015, we estimated the average annual percent change (AAPC) in MHFP expenditure through Jointpoint regression models [35]. Given the inertial allocation in the health budget [36], we adjusted the models using the logarithm of expenditure with autocorrelated errors. For 2009 and 2011, we inserted nodes on the introduction of the Healthy Pregnancy Strategy and its efforts to enroll the entire non-SS population in the *SP*, respectively. For each program, we estimated the ambulatory-care center/hospital ratio as an indicator of the relative growth in expenditure at ambulatory- and primary-care centers.

To analyze the alignment of expenditure with the population requiring MH and FP services among states, we calculated and assessed the concentration indices (CIs) for 2009 and 2014 based on their concentration curves. This involved the following procedures: (a) we calculated the proportions of public MH and FP expenditures spent in state j with respect to total MH and FP expenditures at the national level, respectively; (b) we calculated the proportion of the population requiring these services in state j with respect to the total population requiring these services at the national level; and (c) we generated concentration curves by arranging the states on the x-axis according to the proportions of populations needing these services, from the lowest to the highest, and connecting them with their respective proportions of expenditures on the y-axis. The CI is the area between the concentration curve and the diagonal, and its values range from −1 to 1, with zero denoting equality [37]. Finally, we analyzed the expenditure and population data for 2014 by creating scatter plots and estimating Spearman correlations to ascertain their relationships and statistical significance. We used STATA version 13.0 for the analyses [38].

2.4. Ethical Considerations

To obtain data on service production (in-hospital days as well as consultations offered at hospitals and ambulatory-care centers), we used secondary public sources and the *ENADID* database, neither of which contained personally identifiable information. For expenditures, we used the *Reproductive Health Accounts*. This study was approved by the Research Ethics Committee of the National Institute of Public Health (No. 577-2016).

3. Results

In 2015, MHFP expenditure for the adolescent population without SS coverage totaled US$428 million, 88% of which was spent on MH. Meanwhile, the AAPC for the period 2003–2011 amounted to 15.4% (CI95%: 14.3–16.5) (Figure 1b and Table 1). The MH/FP ratios stood at 21.3 in 2003 and 7.5 in 2015. At the national level, expenditure per adolescent woman in the 10–19 age group rose from US$17 in 2003 to $64 in 2015 (Figure 1a,b and Table 1).

MH and FP services demonstrated different rates of growth for expenditures, which modified the distribution of funds among healthcare providers (Figure 2). From 2003 to 2011, spending on antenatal care in ambulatory-care centers showed an AAPC of 29.3% (CI95%: 23.5–35.4). Accordingly, while 61% of expenditure in ambulatory-care centers was used to finance antenatal care in 2003, this figure had grown to 85% by 2015. Spending on postpartum care grew at an annual rate of 9.3% from 2003 to 2009 (CI95%: 5.4–7.8) and 5.5% during the rest of the period analyzed (CI95%: 4.2–6.8) (Table 1 and Figure 2a).

Hospital spending on MH complications and childbirth registered an AAPC of 14.3% (CI95%: 13.2–15.5 and 11.9–16.8, respectively) from 2003 to 2009; the AAPC for the rest of the period analyzed was lower. Meanwhile, hospital spending on antenatal consultations, at a lower level, showed an annual growth rate of 8.1% (CI95%: 4.0–12.5) from 2003 to 2009, and 3.0% (CI95%: 1.5–4.5) from that year until 2015 (Table 1 and Figure 2b). The ambulatory-care center/hospital ratio of expenditure on MH services reflected a higher growth rate in ambulatory-care centers—in 2003, for each dollar spent by hospitals on MH, ambulatory-care centers spent $0.3. By 2015, the ratio was 0.6.

In contrast with MH, FP expenditure was predominantly spent at ambulatory-care centers (Figure 3). Until 2008, the trends and levels of FP expenditure at hospitals and ambulatory-care centers were similar. However, from 2009 to 2011, spending by ambulatory-care centers experienced its period of greatest growth, with an AAPC of 130.6% (CI95%: 25.5–323.7). Thus, although hospital spending grew at an annual rate of 6.6% (CI95%: 3.7–9.5), by the end of the period analyzed, the ambulatory-care center/hospital spending ratio for FP services was 11 (Figure 3 and Table 1).

Table 2 shows that national expenditure on MHFP per adolescent woman had an AAPC of 11.8% (CI95%: 9.2–14.5) during the period 2003–2015. The increase in spending per adolescent woman reduced the gap between states with higher and lower levels of spending by 48% (from 5.96 in 2003 to 3.15 in 2015). Furthermore, the expenditure gap between states with higher and lower levels of social marginalization diminished by 13% (1.92 in 2003 and 1.68 in 2015). Although regions classified as having moderate, high, and very high levels of marginalization increased their spending per adolescent by 12% per year, no significant differences emerged in the annual rates among the five regions.

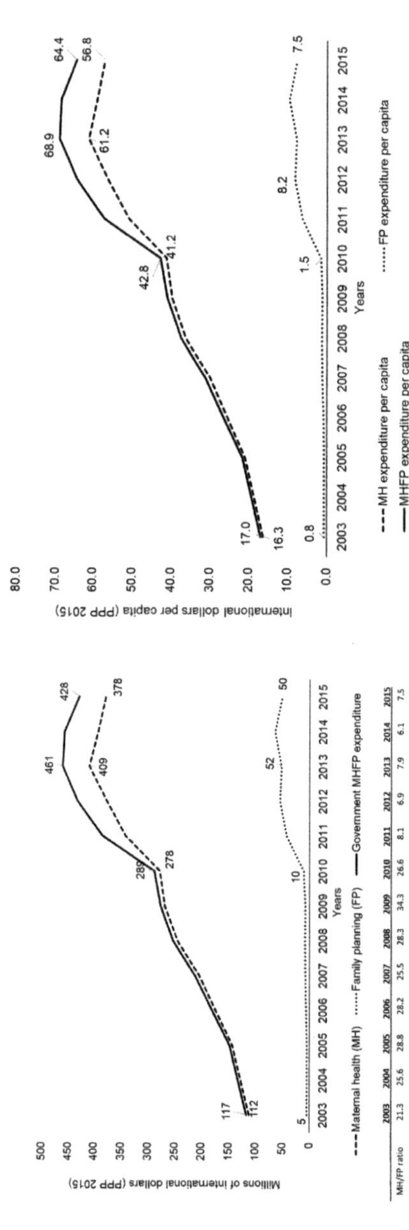

(**a**) MHPF expenditure in millions of US dollar

(**b**) MHPF expenditure per adolescent woman aged 10–19, US dollar

Figure 1. Government Maternal Health and Family Planning (MHFP) expenditure for adolescents aged 10–19 without Social Security coverage, Mexico, 2003–2015, US$ (PPP 2015).

Table 1. Average annual percent change in government Maternal Health and Family Planning (MHFP) expenditure for the entire adolescent population aged 10–19 without Social Security coverage, Mexico, 2003–2015.

	2003–2009 AAPC [a] [CI 95%]	2009–2011 AAPC [a] [CI 95%]	2011–2015 AAPC [a] [CI 95%]
Government MHFP expenditure for adolescents	15.4 [b] [14.3–16.5]	15.4 [b] [14.3–16.5]	3.9 [−3.1–11.4]
Government MH expenditure for adolescent women	16.1 [b] [14.4–17.7]	11.6 [b] [7.7–15.8]	3.2 [−2.2–8.8]
Ambulatory-care centers			
Antenatal care	29.3 [b] [23.5–35.3]	29.3 [b] [23.5–35.3]	6.1 [−17.4–36.4]
Postpartum care	9.7 [b] [3.5–16.3]	5.5 [b] [4.2–6.8]	5.5 [b] [4.2–6.8]
Hospitals			
Complications during pregnancy, childbirth and puerperium	14.3 [b] [13.2–15.5]	8.3 [b] [5.6–11.0]	1.3 –3.0–5.8]
Childbirth care	14.3 [b] [11.9–16.8]	2.7 [b] [0.3–5.2]	2.7 [b] [0.3–5.2]
Antenatal and postpartum consultations	8.1 [b] [4.0–12.5]	3.0 [b] [1.6–4.5]	3.0 [b] [1.6–4.5]
Government FP expenditure for the entire adolescent population	9.3 [b] [2.2–16.9]	92.3 [-8.4-303.9]	16.5 [−7.3–46.5]
Ambulatory-care centers			
FP consultations and counseling	4.6 [-0.4-9.8]	130.6 [b] [25.5–323.7]	19.1 [−0.5–42.5]
Hospitals			
Definitive FP methods	6.6 [b] [3.8–9.5]	6.6 [b] [3.8–9.5]	6.6 [b] [3.8–9.5]

[a] AAPC: Jointpoint models adjusted for autocorrelated errors due to inertial spending behavior (95%). [b] $p < 0.05$ value.

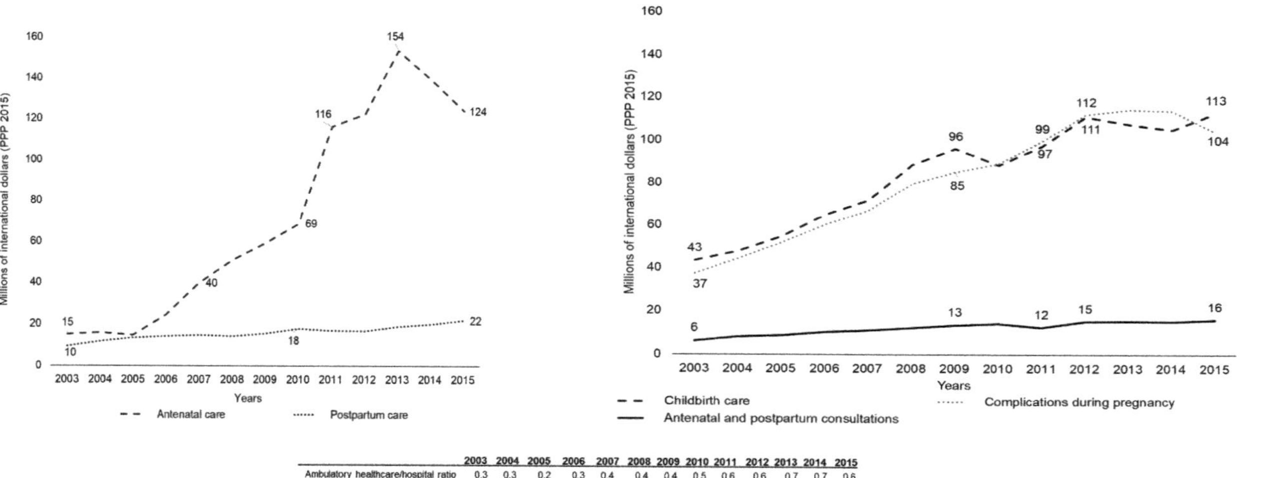

(**a**) MH expenditure at ambulatory-care center   (**b**) MH expenditure at hospital

Figure 2. Government Maternal Health (MH) expenditure for adolescent women aged 10–19, without Social Security coverage, by type of healthcare provider, Mexico, 2003–2015, US$ million (PPP 2015).

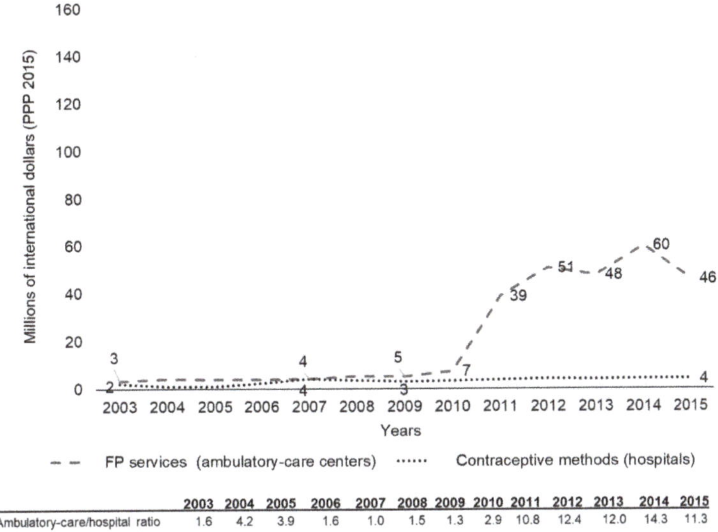

Ambulatory-care/hospital ratio	2003	2004	2005	2006	2007	2008	2009	2010	2011	2012	2013	2014	2015
	1.6	4.2	3.9	1.6	1.0	1.5	1.3	2.9	10.8	12.4	12.0	14.3	11.3

Figure 3. Government Family Planning (FP) expenditure for the entire adolescent population aged 10–19 without Social Security coverage, by type of healthcare provider, Mexico, 2003–2015, $ million (PPP 2015).

Finally, Figure 4 shows the concentration curve for expenditures on MH and FP and the correlations between expenditures and their respective demand indicators. In 2009 and 2014, half of the states (n = 16) concentrated less than 30% of the expenditures (concentration index = 0.32 for both years) (Figure 4a). However, for FP expenditure, the levels of inequality increased between 2009 and 2014 (CI = 0.30 and CI = 0.38, respectively) (Figure 4b). On the other hand, although a positive and significant correlation emerged between expenditure on MH at the state level and its demand indicator (rho > 0.9476, $p < 0.05$), differences were observed in the distribution of spending. For example, states such as Michoacan (MICH) and Guerrero (GRO) exhibited different levels of spending (Figure 4c) despite having comparable proportions of pregnant adolescents. The situation was similar for expenditure on FP (rho > 0.9016, $p < 0.05$), where Guerrero (GRO) and Mexico City (CDMX) presented comparable proportions of sexually active adolescents (Figure 4d) but different expenditure levels. Likewise, Figure 4c indicates that the MH expenditure levels in states with a low marginalization status, such as Jalisco (JAL) were similar to those of highly marginalized states and smaller populations, such as Tabasco (TAB) and San Luis Potosí (SLP).

Table 2. Government MHFP expenditure per adolescent woman aged 10–19. States grouped by level of marginalization, Mexico, 2003–2015 ($ [PPP 2015]).

	2003	2004	2005	2006	2007	2008	2009	2010	2011	2012	2013	2014	2015	AAPC [a] [CI95%]
National spending	17.0	19.3	21.7	26.4	31.1	37.5	41.0	42.8	57.1	64.3	68.9	68.5	64.4	11.8 [9.2–14.5]
Min (state)	7.0	6.5	8.9	11.6	19.8	22.9	26.0	26.3	37.1	38.2	44.8	44.2	38.0	17.2 [10.8–23.9]
Max (state)	42.0	48.9	61.7	84.3	96.6	112.2	106.1	88.9	114.8	158.4	113.8	125.4	119.7	10.1 [4.8–15.7]
Ratio Max/Min	5.96	7.55	6.91	7.25	4.89	4.91	4.08	3.38	3.09	4.15	2.54	2.84	3.15	
Marginalization [b]														
Very low	98	116	123	151	169	229	244	233	274	362	327	306	326	10.0 [6.5–13.6]
Low	198	236	226	289	338	393	460	451	520	569	592	595	572	9.6 [7.2–12.1]
Moderate	193	227	224	274	315	350	395	418	567	657	680	736	693	12.8 [11.3–14.4]
High	160	197	242	308	334	379	386	368	549	582	637	649	616	12.1 [9.8–14.4]
Very high	51	55	62	73	87	104	109	119	159	186	213	212	194	12.6 [9.5–15.8]
Ratio: very high/very low	1.92	2.09	1.98	2.07	1.95	2.19	2.24	1.96	1.72	1.95	1.54	1.44	1.68	

[a] AAPC: (Average Annual Percent Change); Jointpoint models adjusted for autocorrelated errors due to inertial spending behavior (95%) [b] States grouped by level of marginalization [31]: very high marginalization: Chiapas (CHIS), Oaxaca (OAX) and Guerrero (GRO); high marginalization: Puebla (PUE), Hidalgo (HGO), Tabasco (TAB), Veracruz (VER), San Luis Potosi (SLP), Campeche (CAMP), Michoacan (MICH) and Yucatan (YUC); moderate marginalization: Queretaro (QRO), Guanajuato (GTO), Tlaxcala (TLAX), Quintana Roo (QROO), Morelos (MOR), Sinaloa (SIN), Zacatecas (ZAC), Durango (DGO) and Nayarit (NAY); Low marginalization: Jalisco (JAL), Mexico (MEX), Aguascalientes (AGS), Sonora (SON), Colima (COL), Chihuahua (CHIH), Tamaulipas (TAMP) and Baja California Sur (BCS); very low marginalization: Baja California (BC), Coahuila (COAH) Nuevo Leon (NL), and Mexico City (CDMX).

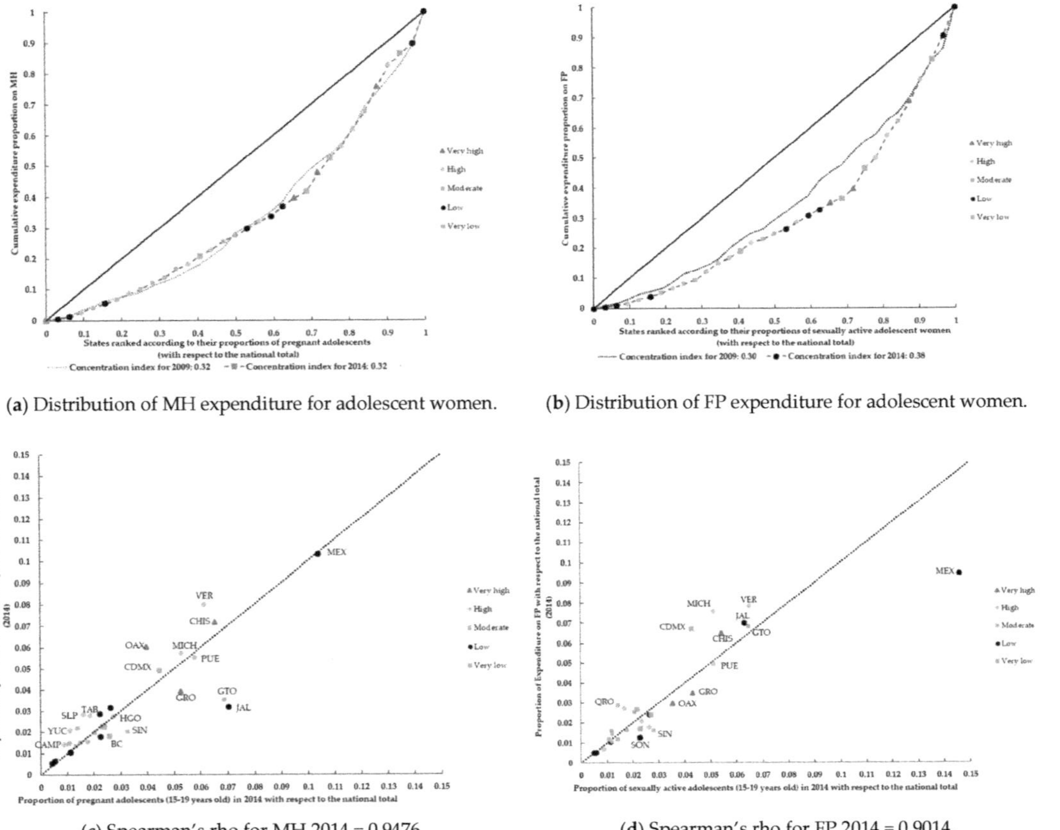

(**a**) Distribution of MH expenditure for adolescent women.

(**b**) Distribution of FP expenditure for adolescent women.

(**c**) Spearman's rho for MH 2014 = 0.9476.

(**d**) Spearman's rho for FP 2014 = 0.9014.

Figure 4. Concentration curves and correlations: government MH and FP expenditure for adolescent women (15–19 years old) without Social Security coverage but with potential demand indicators (Mexico, 2009 and 2014).

4. Discussion

The adolescent population is one of the groups where public policy could have the most dramatic impact given the repercussions for health and wellbeing and the economic benefits these policies would generate [1,4]. In a context of limited resources, it is crucial to identify the areas of opportunity in which investments in the healthy development of this population group would be most efficiently used. In this regard, the results of this study indicate that efforts by the Mexican government to expand healthcare coverage for the population without Social Security (SS) coverage benefited the non-SS adolescent population. Thus, government expenditure on maternal health and family planning (MHFP) showed an average annual percent increase of 15% for the period 2003–2011 but no significant change in the remaining years. During 2003–2015, MHFP expenditure for non-SS adolescents represented 25% of government MHFP expenditure for the total population lacking SS coverage [21]. Nonetheless, despite the growth in spending, inequalities in distribution persisted. The adolescent population living in the most marginalized states and suffering from the greatest level of economic inequality, as well as historical differences in resource allocation [16,17,39] continues to demonstrate below-average levels of per capita spending.

In terms of sexual and reproductive health, it has been documented that the Mexican adolescent population has low rates of contraceptive use and difficulty in planning their sexual lives [7,9,40]. For these reasons, the first contact this population has with health services is generally to receive antenatal care, and their use of CMs frequently begins after the first pregnancy [8]. These patterns result in a situation in which antenatal, childbirth, and postpartum care claim 88% of MHFP expenditure for adolescents. The results show a greater increase in expenditure for MH services in ambulatory-care centers than in hospitals. This difference is explained primarily by a surge in the volume of antenatal consultations offered to adolescents without SS coverage (which rose from 66,436 in 2003 to 1.56 million in 2015) [21]. Despite an improvement in antenatal coverage for pregnant adolescents rising from 61% to 71.8% between 2000 and 2012 [16], the figures continue to be lower, especially in marginalized communities, than those achieved for pregnant adults (20 years and older) [16,41,42]. Because pregnancy during adolescence increases the risk of obstetric complications [1], it is hardly surprising that a third of MH expenditure was spent on hospital care for complications arising during pregnancy, childbirth, and the postpartum period.

With regard to FP expenditure, the observed annual increase of 130% incurred by ambulatory-care centers from 2009 to 2011 was a reflection of various events: (a) an increase in the number of people enrolled in the *Seguro Popular* from 9.1 million in 2009 to 43.5 million in 2010, which enhanced the coverage of health services, FP included, for the population lacking SS coverage [9,13,14,43]; (b) the strengthening of the FP program specifically for adolescents, leading to an increase in post-obstetric-event contraception [42] and which boosted the percentages of sexually active adolescents using a CM in their first sexual intercourse from 43% to 66% in women and from 70% to 85% in men, where the latter was attributed to greater male condom use [44]; and (c) the centralization and increase in the purchase of CMs [8]. As a result of higher FP expenditure, spending per adolescent woman 10–19 years old rose from US$0.8 in 2003 to $7.5 in 2015, reaching the threshold of $2.93 per woman per year since 2011, as recommended by the Guttmacher Institute [33]. In spite of these advances, however, deficiencies in CM coverage persist, and sexually active adolescents continue to be the group with the lowest rate of CM use in Mexico [9,40]. Evidence indicates that investment in FP programs generates significant monetary and social returns [34,45]. It has been estimated that each additional dollar invested in satisfying the demand for modern CMs generates savings of US$2–6 in healthcare spending for pregnant women and newborns [34]. If we also include the long-term effects of reducing maternal and infant mortality and increased economic growth, savings rise to US$60–120 per dollar spent on FP services [45]. This underlines the relevance of continuing to invest in FP for adolescents, as well as the need to make more efficient use of these resources.

Our results document the growing importance of ambulatory-care centers as health-service providers. These providers will no doubt continue to increase in importance, given that the current health reform in Mexico, which has replaced the *Seguro Popular* with the Health Institute for Wellbeing, is oriented towards strengthening the primary-care model for the population lacking SS coverage [46,47]. Nonetheless, the heterogeneity that characterizes ambulatory-care providers in terms of resources and the quality of services delivered [16,48,49] suggests that it will be necessary to improve their capacity to respond, particularly with regard to their supply of modern CMs [49].

Although government MHFP spending has increased across the board nationally, discrepancies persist in its distribution among states, since those with similar adolescent populations in need of MH and FP services exhibit different levels of expenditure. This could be the result of various factors, such as the persistence of inertial allocations: (a) in 2015, 33% of the *Seguro Popular* budget earmarked for payroll continued to be allocated to State Health Services on the basis of long-established, routine procedures [39]; (b) the concentration of infrastructure and personnel have traditionally privileged some states at the expense of others; and (c) managerial capacities have diverged and continue to vary widely among State Health Services [17,48,50].

Our study had the following limitations: (1) our analysis was restricted to government schemes providing coverage for the population without SS coverage, which prevents generalizing results to the entire public health system. However, these schemes provided coverage to 45% of the Mexican population [13,14], and their expenditures represented 46% of total public health spending in 2016 [39]; and (2) using production data to estimate the distribution of expenditure among health conditions and/or diseases, which could lead to estimation errors. Nonetheless, the OECD has evaluated this methodology in various countries, demonstrating its validity and consistency [51]. The WHO [52], as well as the OECD [53] encourage its use. (3) The age range considered for the demand indicators of MH services was also 15–19 years. After reviewing a variety of data sources, such as the Birth Information Subsystem, we decided to use data from the National Survey of Demographic Dynamics (*ENADID*). We arrived at this decision because the Birth Information database omits deceased children and abortions. Information from the *ENADID* thus provided the closest approximation to the adolescent pregnancy phenomenon under study. In addition, it has been documented that pregnant women under 15 register their children with the appropriate authorities later than other women [54]. (4) Finally, it should be noted that expenditure was analyzed at the state level, without considering the wide variability in the distribution of local spending [17,50]. Future studies need to explore in greater detail the relationship among MHFP resources, as well as their distribution and health outcomes at the municipal level.

5. Conclusions

Governments around the world have recognized the need to invest in the sexual and reproductive health of adolescents. Sustainable Development Goal 3.7 [55] calls upon countries to ensure universal access to sexual and reproductive healthcare services, including for FP. This will require additional resources. Financial evidence on the levels of expenditure allocated to these services and on its distribution throughout the population is a key input for planning public investment. It serves as a basis for governments and health authorities to define how much more they must invest, what types of services should be prioritized, and which areas can be improved with regard to equity and efficiency.

The findings of this study fill an information gap on the levels of investment in sexual and reproductive health services in Mexico for a group traditionally lacking visibility—the adolescents. Our results demonstrate that the health policies implemented between 2003 and 2015 increased expenditure on the sexual and reproductive health of adolescents without Social Security coverage; in spite of this, however, problems persist in ensuring an equitable distribution of these resources. Looking ahead, the implementation of specific policies for the prevention of adolescent pregnancy will require special attention as the current health reform evolves. It will be necessary to monitor the financial implications of the ensuing changes and their consequences for adolescent health services. Subsequent analyses will also need to combine the allocation of expenditure with results indicators in this population in order to understand the extent to which investments are provided equitably and efficiently.

Author Contributions: Conceptualization, L.A.-B.; data curation, formal analysis, J.C.M.-H.; validation, methodology, L.A.-B. and J.C.M.-H.; critical review of the manuscript, writing-review and editing, L.A.-B. and J.C.M.-H., I.H.-P., L.C.-H., P.H.-P. and A.V.; funding acquisition, I.H.-P. All authors have read and agreed to the published version of the manuscript.

Funding: This work was supported by the CONACYT Sectoral Fund for Research in Health and Social Security in Mexico (Grant No. 261230, year 2015). The funders had no role in study design, data collection and analysis, decision to publish, or preparation of the manuscript.

Acknowledgments: We thank the National Center for Gender Equity and Reproductive Health of the Ministry of Health for access to and use of data from the Reproductive Health Accounts. We also extend our thanks to Luz Maria Montes Romero for her support in the search for and recuperation of the bibliography related to this article, as well as to Patricia Solis Albarrán for her valuable contribution in the English translation and editing of this work.

Conflicts of Interest: The authors declare having no conflict of interest.

References

1. World Health Organization. *Adolescent Development. Maternal, Newborn, Child and Adolescent Health*; WHO Publishing: Geneva, Switzerland, 2018. Available online: http://www.who.int/maternal_child_adolescent/topics/adolescence/development/en/ (accessed on 3 October 2019).

2. Taghizadeh Maghaddam, H.; Bahreini, A.; Ajilian Abbasi, M.; Fazli, F.; Saeidi, M. Adolescence Health: The Needs, Problems and Attention. *Int. J. Pediatr.* **2016**, *4*, 1423–1438. [CrossRef]

3. Patton, G.C.; Sawyer, S.M.; Santelli, J.S.; Ross, D.A.; Afifi, R.; Allen, N.B.; Kakuma, R. Our future: A Lancet commission on adolescent health and wellbeing. *Lancet* **2016**, *387*, 2423–2478. [CrossRef]

4. Ki-moon, B. Sustainability—engaging future generations now. *Lancet* **2016**, *387*, 2351–2478. [CrossRef]

5. World Health Organization. Strengthening the Health Sector Response to Adolescent Health and Development. 2010. Available online: https://www.who.int/maternal_child_adolescent/documents/cah_adh_flyer_2010_12_en.pdf?ua=1 (accessed on 22 January 2019).

6. Instituto Nacional de Estadística y Geografía (INEGI). *Datos de Población*; INEGI Publishing: Aguascalientes, Mexico, 2015. Available online: https://www.inegi.org.mx/temas/estructura/ (accessed on 11 July 2019).

7. Hernández, M.F.; Muradás, M.C.; y Sánchez, M. Panorama de la Salud Sexual y Reproductiva, 2014. CONAPO. La Situación Demográfica en México 2015. Available online: http://www.conapo.gob.mx/es/CONAPO/Panorama_de_la_salud_sexual_y_reproductiva_2014 (accessed on 8 January 2019).

8. Gobierno de la República. *Estrategia Nacional para la Prevención del Embarazo en Adolescentes*; Gobierno de la República: Mexico City, Mexico, 2018. Available online: https://www.gob.mx/inmujeres/acciones-y-programas/estrategia-nacional-para-la-prevencion-del-embarazo-en-adolescentes-33454 (accessed on 20 December 2018).

9. Allen-Leigh, B.; Villalobos-Hernández, A.; Hernández-Serrato, M.I.; Suárez, L.; De la Vara, E.; De Castro, F.; Schiavon-Ermani, R. Inicio de vida sexual, uso de anticonceptivos y planificación familiar en mujeres adolescentes y adultas en México. *Salud Publica Mex.* **2013**, *55*, S235–S240. [CrossRef] [PubMed]

10. Instituto Nacional de Estadística y Geografía (INEGI). Encuesta Nacional de la Dinámica Demográfica 2014. In *La Anticoncepción: Implicaciones en el Embarazo Adolescente, Fecundidad y Salud Reproductiva en México*; Instituto Nacional de Estadística y Geografía INEGI: Aguascalientes, Mexico, 2017. Available online: http://coespo.qroo.gob.mx/Descargas/doc/8%20FECUNDIDAD/La%20anticoncepci%C3%B3n%20inplicaci%C3%B3n%20en%20el%20embarazo%20en%20adolescentes%20ENADID%202014.pdf (accessed on 12 July 2019).

11. Menkes, C.; Sosa-Sánchez, I.A. Características del Embarazo y de la Fecundidad de las Adolescentes en México. In *Retos del Cambio Demográfico de México*; Avila, J., Bringas, H., López, M., Eds.; UNAM: Mexico City, Mexico, 2016; pp. 179–209. Available online: https://www.researchgate.net/publication/316981804_Caracteristicas_del_embarazo_y_de_la_fecundidad_de_las_adolescentes_en_Mexico (accessed on 1 July 2019).

12. Organización Panamericana de la Salud, Fondo de Población de las Naciones Unidas, Fondo de las Naciones Unidas para la Infancia. *Acelerar El Progreso Hacia La Reducción Del Embarazo En La Adolescencia En América Latina y El Caribe*; Organización Panamericana de la Salud: Washington, DC, USA, 2018. Available online: https://www.unicef.org/lac/informes/acelerar-el-progreso-hacia-la-reducci%C3%B3n-del-embarazo-adolescente-en-am%C3%A9rica-latina-y-el (accessed on 1 July 2019).

13. Knaul, F.M.; Frenk, J. Health insurance in Mexico: Achieving universal coverage through structural reform. *Health Aff.* **2005**, *24*, 1467–1476. [CrossRef] [PubMed]

14. Frenk, J.; Gonzalez-Pier, E.; Gomez-Dantes, O.; Lezana, M.A.; Knaul, F.M. Comprehensive reform to improve health system performance in Mexico. *Salud Publica Mex.* **2007**, *49*, S23–S36. [CrossRef] [PubMed]

15. Gobierno de la República. *Los Objetivos de Desarrollo del Milenio en México*; Informe de Avances 2010; Presidencia de la República: Mexico City, Mexico, 2011. Available online: http://www.objetivosdedesarrollodelmilenio.org.mx/Doctos/Inf2010.pdf (accessed on 19 September 2019).

16. Saavedra-Avendaño, B.; Darney, B.G.; Reyes-Morales, H.; Serván-Mori, E. El Aseguramiento Público en Salud Mejora la Atención en los Servicios? El caso de la Atención Prenatal en Adolescentes en México. *Salud Publica Mex.* **2016**, *58*, 561–568. [CrossRef] [PubMed]

17. Rodríguez-Franco, R.; Serván-Mori, E.; Gómez-Dantés, O.; Contreras-Loya, D.; Pineda-Antúnez, C. Old principles, persisting challenges: Maternal health care market alignment. *PLoS ONE* **2018**, *13*, e0199543. [CrossRef] [PubMed]

18. Instituto Nacional de Salud Pública (INSP). *Encuesta Nacional de Salud y Nutrición de Medio Camino 2016*; Informe de Resultados INSP: Mexico City, Mexico, 2016. Available online: https://www.gob.mx/cms/uploads/attachment/file/209093/ENSANUT.pdf (accessed on 20 February 2019).

19. Avila-Burgos, L.; Cahuana-Hurtado, L.; Montañez-Hernandez, J.; Servan-Mori, E.; Aracena-Genao, B.; del Río-Zolezzi, A. Financing Maternal Health and Family Planning: Are We on the Right Track? Evidence from the Reproductive Health Subaccounts in Mexico, 2003–2012. *PLoS ONE* **2016**, *11*. [CrossRef] [PubMed]

20. Grupo de Trabajo de la Fundación Mexicana Para la Salud. Universalidad en los servicios de salud en México. *Salud Publica Mex.* **2013**, *55*, EE1–EE64. [CrossRef]

21. Ávila-Burgos, L.; Montañez-Hernández, J.C.; Cahuana-Hurtado, L.; Ventura-Alfaro, C.E. *Cuentas en Salud Reproductiva y Equidad de género*; Estimación 2014 y 2015 y evolutivo 2003–2015; Instituto Nacional de Salud Pública (MX)/Secretaría de Salud (MX): Cuernavaca, Mexico, 2017. Available online: https://www.gob.mx/cms/uploads/attachment/file/293674/CuentasSR2014_22enero.pdf (accessed on 12 January 2020).

22. Servan-Mori, E.; Avila-Burgos, L.; Nigenda, G.; Lozano, R. A Performance Analysis of Public Expenditure on Maternal Health in Mexico. *PLoS ONE* **2016**, *11*, e0152635. [CrossRef] [PubMed]

23. Cahuana-Hurtado, L.; Avila-Burgos, L.; Pérez-Núñez, R.; Uribe-Zúñiga, P. Analysis of reproductive health expenditures in Mexico, 2003. *Rev. Panam Salud Publica* **2006**, *20*, 287–298. [CrossRef] [PubMed]

24. Organisation for Economic Co-Operation; Development, Eurostat; World Health Organization. *A System of Health Accounts 2011: Revised Edition*; OECD Publishing: Paris, France, 2017. Available online: https://ec.europa.eu/eurostat/documents/3859598/7985806/KS-05-19-103-EN-N.pdf/60aa44b0-2738-4c4d-be4b-48b6590be1b0 (accessed on 11 April 2019). [CrossRef]

25. World Health Organization. *Guide to Producing Reproductive Health Subaccounts within the National Health Accounts Framework*; World Health Organization: Geneve, Switzerland, 2009. Available online: https://apps.who.int/iris/handle/10665/44181 (accessed on 24 November 2019).

26. Gobierno de la República. Programa de Acción Específico. Salud Sexual y Reproductiva Para Adolescentes 2013–2018. Available online: https://www.gob.mx/salud/acciones-y-programas/programa-de-accion-especifico-salud-sexual-y-reproductiva-para-adolescentes-2013-2018-10072 (accessed on 24 October 2019).

27. Dirección General de Información en Salud, Secretaría de Salud. Cubos Dinámicos. Available online: http://www.dgis.salud.gob.mx/contenidos/basesdedatos/BD_Cubos_gobmx.html (accessed on 24 July 2019).

28. Organización Panamericana de la Salud. *Clasificación Estadística Internacional de Enfermedades y Problemas Relacionados con la Salud, 10th, ed.*; PAHO, Ed.; Organización Panamericana de la Salud: Washington, DC, USA, 1995; Volume 2, Available online: http://ais.paho.org/classifications/Chapters/pdf/Volume2.pdf (accessed on 16 January 2020).

29. Organisation for Economic Co-Operation and Development, OECD Statistics, National Accounts, PPPs and Exchanges Rates. Available online: https://stats.oecd.org/ (accessed on 24 October 2018).

30. Instituto Nacional de Estadística y Geografía, INEGI. Encuesta Nacional de la Dinámica Demográfica 2009 y 2014. Available online: https://www.inegi.org.mx/programas/enadid/2014/default.html (accessed on 30 June 2019).

31. United Nations; Department of Economic and Social Affairs; Population Division. *Contraceptive Use by Method 2019: Data Booklet (ST/ESA/SER.A/435)*; United Nations: New York, NY, USA, 2019; p. 13. Available online: https://www.un.org/development/desa/pd/sites/www.un.org.development.desa.pd/files/files/documents/2020/Jan/un_2019_contraceptiveusebymethod_databooklet.pdf (accessed on 12 April 2020).

32. The DHS Program. The Demographic and Health Surveys (DHS) Program. *DHS Stat Compiler.* Available online: https://www.dhsprogram.com/data/STATcompiler.cfm (accessed on 12 April 2020).

33. Guttmacher Institute. *Adding It Up: Investing in Contraception and Maternal and Newborn Health 2017*; Guttmacher Institute: New York, NY, USA, 2017. Available online: http://progress.familyplanning2020.org/content/finance (accessed on 21 May 2019).

34. Consejo Nacional de Población; CONAPO; Secretaría de Gobernación. Índices de Marginación por Entidad Federativa y Municipio 2010. Available online: http://www.conapo.gob.mx/es/CONAPO/Indices_de_Marginacion_2010_por_entidad_federativa_y_municipio (accessed on 24 July 2019).

35. Division of Cancer Control & Population Sciences. *Joinpoint Regression Program Version 4.2.0.1. Statistical Methodology and Applications Branch, Surveillance Research Program*; National Cancer Institute: Bethesda, MD, USA, 2015.

36. Consejo Nacional de Evaluación de la política de Desarrollo Social (CONEVAL). El Ramo 33 en el Desarrollo social en México: Evaluación de Ocho Fondos de Política Pública. Available online: https://www.coneval.org.mx/Informes/Evaluacion/Estrategicas/Ramo_33_PDF_02032011.pdf (accessed on 8 July 2019).

37. Wagstaff, A.; Paci, P.; Van Doorslaer, E. On the measurement of inequalities in health. *Soc. Sci. Med.* **1991**, *33*, 545–557. [CrossRef]

38. Stata. *Stata Corp. Release 11*; Stata Press: College Station, TX, USA, 2009; Volume 1–4.

39. Secretaría de Salud. Dirección General de Información en Salud. Gasto en Salud en el Sistema Nacional de Salud. 2019. Available online: http://www.dgis.salud.gob.mx/contenidos/sinais/gastoensalud_gobmx.html (accessed on 17 December 2019).

40. Villalobos, A.; Hubert, C.; Hernández-Serrato, M.I.; de la Vara-Salazar, E.; Suárez-López, L.; Romero-Martínez, M.; Ávila-Burgos, L.; Barrientos, T. Maternidad en la adolescencia en localidades menores de 100,000 habitantes en las primeras décadas del milenio. *Salud Publica Mex.* **2019**, *61*, 742–752. [CrossRef] [PubMed]

41. Suárez-López, L.; de Castro, F.; Hubert, C.; de la Vara-Salazar, E.; Villalobos, A.; Hernández-Serrato, M.I.; Shamah-Levy, T.; Ávila-Burgos, L. Atención en salud materno-infantil y maternidad adolescente en localidades menores de 100,000 habitantes. *Salud Publica Mex.* **2019**, *61*, 753–763. [CrossRef] [PubMed]

42. Heredia Pi, I.; Serván Mori, E.; Reyes Morales, H.; Lozano, R. Brechas en la cobertura de atención continua del embarazo y el parto en México. *Salud Publica Mex.* **2013**, *55*, 249–258. [CrossRef]

43. Comisión Nacional de Protección Social en Salud. *Sistema de Protección Social En Salud: Informe de Resultados Enero-Diciembre 2011*; Comisión Nacional de Protección Social en Salud: Mexico City, Mexico, 2012. Available online: http://www.transparencia.seguro-popular.gob.mx/index.php/planes-programas-e-informes/22-planes-rogramas-e-informes/39-informes-de-labores-de-la-cnpss (accessed on 20 October 2019).

44. Gutiérrez, J.P.; Rivera-Dommarco, J.; Shamah-Levy, T.; Villalpando-Hernández, S.; Franco, A.; Cuevas-Nasu, L.; Romero-Martínez, M.; Hernández-Ávila, M. *Encuesta Nacional de Salud y Nutrición 2012*; Resultados Nacionales; Instituto Nacional de Salud Pública (MX): Cuernavaca, Mexico, 2012. Available online: https://ensanut.insp.mx/encuestas/ensanut2012/informes.php (accessed on 13 April 2020).

45. FP2020 Family Planning's Return on Investment. FP2020. 2018. Available online: https://www.familyplanning2020.org/sites/default/files/Data-Hub/ROI/FP2020_ROI_OnePager_FINAL.pdf (accessed on 21 May 2019).

46. *Gobierno de México. Senado de la República. Iniciativa 2019 Que Reforma la Ley General de Salud, de la Ley de Coordinación Fiscal y de la Ley de los Institutos Nacionales de Salud en Materia de Acceso a los Servicios de Salud y Medicamentos Asociados Para las Personas Que no Poseen Seguridad Social*; Gaceta de la Comisión Permanente: Mexico City, Mexico, 2019. Available online: https://infosen.senado.gob.mx/sgsp/gaceta/64/1/2019-07-03-1/assets/documentos/Ini_Delgado.pdf (accessed on 27 January 2020).

47. Secretaría de Salud. *Programa Sectorial de Salud 2019–2024*; Secretaría de Salud: Mexico City, Mexico, 2020. Available online: http://saludsinaloa.gob.mx/wpcontent/uploads/2019/transparencia/PROGRAM_SECTORIAL_DE_SALUD_2019_2024.pdf (accessed on 27 January 2020).

48. Dirección General de Información en Salud (DGIS). *Unidades de Primer Nivel de Atención en los Servicios Estatales de Salud*; Evaluación 2008; Secretaría de Salud: Mexico City, Mexico, 2008. Available online: http://www.dged.salud.gob.mx/contenidos/dess/descargas/upn/upna_sesas_2008.pdf (accessed on 21 July 2019).

49. Torres-Pereda, P.; Heredia-Pi, I.B.; Ibañez-Cuevas, M.; Avila-Burgos, L. Quality of family planning services in Mexico: The perspective. of demand. *PLoS ONE* **2019**, *14*. [CrossRef]

50. Pérez-Pérez, E.; Servan-Mori, E.; Nigenda, G.; Ávila-Burgos, L.; Mayer-Foulkes, D. Government expenditure on health and maternal mortality in México: A spatial-econometric analysis. *Int. J. Health Plann. Manag.* **2019**, *34*, 619–635. [CrossRef] [PubMed]

51. Organisation for Economic Co-operation and Development (OECD). *Extension of Work on Expenditure by Disease, Age and Gender. Eu Contribution Agreement 2011 53 01*; OECD Health Division: Paris, France, 2013. Available online: https://www.oecd.org/els/health-systems/Extension-of-work-on-expenditure-by-disease-age-and-gender_Final-Report.pdf (accessed on 29 January 2020).

52. World Health Organization (WHO). Global Health Expenditure Database. SHA Methodology. Disease Manual Distribution Ratios. Available online: https://apps.who.int/nha/database/DocumentationCentre/Index/en (accessed on 29 January 2020).

53. Organisation for Economic Co-Operation and Development (OECD). Estimating Expenditure by Disease, Age and Gender. 12 April 2016. Available online: https://www.oecd.org/els/health-systems/estimating-expenditure-by-disease-age-and-gender.htm (accessed on 29 January 2020).

54. Consejo Nacional de Población. Fecundidad en niñas y Adolescentes de 10 a 14 Años, Niveles, Tendencias y Caracterización Sociodemográfica de las Menores y de los Padres de Sus Hijos(as), a Partir de las Estadísticas del Registro de Nacimiento, 1990–2016. Available online: https://www.gob.mx/conapo/documentos/fecundidad-en-ninas-y-adolescentes-de-10-a-14-anos (accessed on 28 November 2019).

55. United Nations. Objetivos de Desarrollo Sostenible. Goal 3: Ensure Healthy Lives and Promote Well-Being for all at all Ages. Available online: https://www.un.org/sustainabledevelopment/health/ (accessed on 29 January 2020).

International Journal of
*Environmental Research
and Public Health*

Article

Economic Crisis and Sexually Transmitted Infections: A Comparison Between Native and Immigrant Populations in a Specialised Centre in Granada, Spain

María Ángeles Pérez-Morente [1], Adelina Martín-Salvador [2,*], María Gázquez-López [3], Pedro Femia-Marzo [4], María Dolores Pozo-Cano [4], César Hueso-Montoro [4,*] and Encarnación Martínez-García [4]

1   Faculty of Health Sciences, University of Jaén, 23071 Jaén, Spain; mmorente@ujaen.es
2   Faculty of Health Sciences, University of Granada, 52005 Melilla, Spain
3   Faculty of Health Sciences, University of Granada, 51001 Ceuta, Spain; mgazquez@ugr.es
4   Faculty of Health Sciences, University of Granada, 18016 Granada, Spain; pfemia@ugr.es (P.F.-M.); pozocano@ugr.es (M.D.P.-C.); emartinez@ugr.es (E.M.-G.)
*   Correspondence: ademartin@ugr.es (A.M.-S.); cesarhueso@ugr.es (C.H.-M.)

Received: 13 March 2020; Accepted: 31 March 2020; Published: 5 April 2020

Abstract: This study aimed to analyse the influence of the economic crisis on the prevalence of sexually transmitted infections (STIs) in the immigrant population compared to the native population. A cross-sectional study was conducted by reviewing 441 clinical records (329 Spanish nationals and 112 non-Spanish nationals) of individuals who, between 2000 and 2014, visited an STI clinic in Granada and tested positive for an infection. Descriptive statistical analyses were performed, and infection rates, odds ratios, and 95% confidence intervals (CIs) were calculated. The mean age was 28.06 years (SD = 8.30; range = 16–70). During the period 2000–2014, the risk of being diagnosed with an STI was higher among non-Spanish nationals than among Spanish nationals (odds ratio (OR) = 5.33; 95% CI = 4.78–6.60). Differences between both populations were less marked during the crisis period (2008–2014: OR = 2.73; 95% CI = 2.32–3.73) than during the non-crisis period (2000–2007: OR = 12.02; 95% CI = 10.33–16.17). This may be due to underreporting of diagnoses in the immigrant population. Immigrants visiting the STI clinic in Granada are especially vulnerable to positive STI diagnoses compared to the native population.

Keywords: sexually transmitted infections; economic recession; transients and migrants

---

1. Introduction

Sexually transmitted infections (STIs) are a global public health problem. According to the World Health Organization (WHO), more than one million people contract an STI every day. An estimated 357 million people become infected with chlamydia, gonorrhoea, syphilis, or trichomoniasis each year [1].

According to the latest epidemiological report from the European Centre for Disease Prevention and Control, the number of gonorrhoea infections had increased by 17% in 2017 compared to the previous year. The United Kingdom was the country with the highest proportion of confirmed cases (75 per 100,000 inhabitants), followed by Ireland (47), Denmark (33), Iceland (29), Norway (27), and Sweden (25). The countries with the lowest proportion of reported cases (<1 per 100,000) were Bulgaria, Croatia, Cyprus, Poland, and Romania. There were 33,189 (7.1 cases per 100,000 inhabitants) new cases of syphilis. The highest rate was found in Iceland (15.4 cases per 100,000 inhabitants), followed by Malta (13.5), the United Kingdom (11.8), and Spain (10.3). The lowest rates (<3 cases per 100,000 inhabitants) were observed in Croatia, Cyprus, Estonia, Italy, Portugal, and Slovenia. Furthermore, 25,353 people were diagnosed with Human Immunodeficiency Virus (HIV) in the European Union (6.2 cases per 100,000 inhabitants) [2].

In 2016, in Spain, the STI with the highest incidence rate was chlamydia trachomatis (17.85/100,000), followed by gonorrhoea (13.8/100,000), and syphilis (7.22/100,000). In the case of syphilis, it has been stabilised since 2011. However, gonococcal infection rates have increased steadily from 2001 to 2016, from 2.02/100,000 to 13.89/100,000 [3].

Men who have sex with men (MSM), sex workers, transgender people, intravenous drug users, and immigrants have been identified as the key populations at highest risk of contracting an STI [4].

Increased migration has contributed to the spread of HIV, Hepatitis B Virus (HBV), and other STIs, with the vast majority of cases occurring in migrants from low- and middle-income countries who have moved to high-income countries. Most interventions in the United States of America, Australia, and Europe focus on individual behaviour rather than on broader sociocultural factors [5].

According to the Spanish National Statistics Institute (INE, by its Spanish acronym), the immigrant population in Spain in 2018 increased by 100,764 to 4,663,726, an increase 23% larger than in 2017 [6]. In 2017, 36.1% of new Human Immunodeficiency Virus (HIV) infections were diagnosed in non-Spanish individuals, being more frequent in the Latin American population (19%) [7]. In a study conducted at a hospital in Madrid, half of the 371 new HIV diagnoses were made in immigrants [8]. The immigrant population constitutes a very diverse and particularly vulnerable group due to the socio-cultural context, the language barrier (in some cases), their economic level, and their employment and legal status in Spain. In addition, the WHO points out that insufficient data on STIs at the local level compromises the global response to this problem [4].

Periods of economic crisis represent one of the factors that increase the proportion of socially and economically vulnerable citizens, including immigrants [9]. Different studies have analysed the effects of the economic crisis on the immigrant population in several European countries [10–13]. The economic situation in Spain in recent years has increased the impact of communicable diseases, especially on the most vulnerable populations [14–16]; thus, hindering healthcare delivery to immigrants [17].

The present research is in line with these studies and aims to analyse STI-related differences between the pre-crisis and crisis periods by comparing the native population with the immigrant population, of those who visited the Sexually Transmitted Disease and Sexual Health Clinic in Granada (Spain). Population rates have been taken into account in this comparison, which we consider to be a differentiating feature of this study, with respect to previous studies.

The objectives of this study were as follows: to describe the evolution of STIs in the non-Spanish population in comparison to the native population living in Granada (Spain); to explore, in the former group, the potentially higher risk of contracting some of these infections in comparison to the rest of the population, using the period of economic crisis as a variable of interest.

2. Materials and Methods

An observational study was conducted by analysing the cases of service users diagnosed with STIs who visited the Sexually Transmitted Disease and Sexual Health Clinic in Granada, Spain, between 2000 and 2014, inclusive. This specialised centre, attached to the Andalusian Health Service (SAS, by its Spanish acronym), is the referral service for the entire province of Granada and, according to INE data, during the years analysed, a yearly average total of 550,000 individuals aged between 15 and 64 years old have used their services [18].

These records were taken from a randomly generated database within a larger project from which this study derives. A sample size was calculated to detect differences in the basic variables of STI presence in patients with a new clinical record. This calculation was made in order to detect differences in a binary variable, seeking to detect differences of 20% in two years, with a power of 80% and applying an error of $\alpha = 5\%$ to the test. The sample size required to detect this difference was 97 clinical records per year. The sample was obtained from a database of new records for each year, from which the first and the last record number of each year were taken using systematic random sampling without replacement.

This database contains the clinical records of 1437 adult users without cognitive impairment whose reason for visit was suspicion of, or confirmed presence of, an STI. The sample analysed in the present study, which is a sub-sample of the aforementioned sample, consists of subjects who had been diagnosed with an STI.

Data were collected from the clinical records using four categories: symptoms; control; contact follow-up; and HIV. The country of birth was the independent variable (Spanish nationals vs. non-Spanish nationals), which was determined by means of an official identification document (i.e., national identity card, residence permit, work permit, or passport). Time was the main explanatory covariate (i.e., the 2000–2014 period), in which 2008 was considered to be the onset of the economic crisis in Spain, as numerous studies indicate [9,10,19]. Other variables included were: (a) socio-demographic variables: sex (male/female), age in years (analysed as a continuous variable), occupation (sex worker/former sex worker/other), employment status (employed/unemployed), level of education (no education or primary education/secondary education/higher education), living with a partner (yes/no), and sexual orientation identity (heterosexual/bisexual/homosexual); (b) variables related to clinical care received: reason for visit (according to the reasons provided in the clinical history), previous treatment (yes/no), number of subsequent visits, number of new subsequent episodes; and (c) risk factors for contracting STIs: regular partner having symptoms (yes/no), period of time since last sexual contact without a condom, number of partners in the last month, number of partners in the last year, lifelong sexual history (number of sexual partners throughout lifetime), drug use (yes/no), previous STIs, and age of first sexual intercourse in years (analysed as a continuous variable). The following variables, registered in the clinical records as nominal variables were transformed into quantitative variables for ease of analysis: period of time since last sexual contact without a condom, number of partners in the last month, number of partners in the last year, and lifelong sexual history.

To analyse the effect of the financial crisis on STI diagnoses, the annual rates of STI diagnosis per 100,000 inhabitants were calculated for Spanish nationals and non-Spanish nationals using the direct method, taking as the denominator the number of individuals over 15 years old residing in the province of Granada for each group, according to the data published by the INE in the continuous annual census [18]. These rates were plotted to highlight trends. Odds ratios (ORs) of STI diagnoses between non-Spanish nationals and Spanish nationals and 95% confidence intervals (CIs) were calculated for each year of the pre-crisis (2000–2007) and crisis (2008–2014) periods, as well as for the total study period (2000–2014).

Statistical analyses were conducted using the Statistical Package for the Social Sciences 22.0 (SPSS; International Business Machines Corporation [IBM], 2016, Armonk, NY, USA).

The study protocol was approved by the Biomedical Research Ethics Committee of the province of Granada (research protocol approved on 12 February, 2012 and 1 April, 2015), as well as by the Management Directorate of the Granada-Metropolitano Health District, to which the centre where data were collected is attached.

3. Results

The total number of service users diagnosed with an STI was 441, of whom 329 were Spanish (74.6%) and 112 (25.3%) were immigrants. The mean age was 28.06 years (SD = 8.30; range = 16–70). Table 1 shows the characteristics of the sample separated by nationality (Spanish vs non-Spanish).

Table 1. Sample description of people diagnosed with a sexually transmitted infection (STI) by nationality (Spanish or non-Spanish) ($n = 441$).

	Non-Spanish Nationals ($n = 112$; 25.39%)	Spanish Nationals ($n = 329$; 74.6%)
	n (%)	*n* (%)
Sex ($n = 441$)		
Male	28 (25.0)	201 (61.1)
Female	84 (75.0)	128 (38.9)
Occupation ($n = 419$)		
Sex worker/Former sex worker	64 (59.8)	8 (2.6)
Other	43 (40.2)	304 (97.4)
Employment status ($n = 400$)		
Employed	72 (73.5)	131 (43.4)
Unemployed	26 (26.5)	171 (56.6)
Level of education ($n = 416$)		
No education/Primary education	32 (32.0)	45 (14.2)
Secondary education	39 (39.0)	56 (17.7)
Vocational training/Training module	6 (6.0)	43 (13.6)
Higher education	23 (23.0)	172 (54.4)
Has regular partner ($n = 417$)		
Yes	69 (65.7)	212 (67.9)
No	36 (34.3)	100 (32.1)
Reason for visit ($n = 441$)		
Symptoms	34 (30.4)	243 (73.9)
Other	3 (2.7)	9 (2.7)
HIV	75 (67.0)	77 (23.4)
Previous treatment ($n = 326$)		
Yes	21 (27.6)	85 (34.0)
No	55 (72.4)	165 (66.0)
Sexual orientation identity ($n = 428$)		
Heterosexual	103 (94.5)	248 (77.7)
Bisexual	3 (2.8)	14 (4.4)
Homosexual	3 (2.8)	57 (17.9)
Regular partner has symptoms ($n = 163$)		
Yes	14 (34.1)	53 (43.4)
No	27 (65.9)	69 (56.6)
Drug use ($n = 194$)		
Yes	16 (34.0)	63 (42.9)
No	31 (66.0)	84 (57.1)
Previous STIs ($n = 356$)		
Yes	24 (27.0)	65 (24.3)
No	65 (73.0)	202 (75.7)
	Mean (SD)	Mean (SD)
Age ($n = 441$)	27.36 (6.08)	28.22 (8.76)
Subsequent visits ($n = 439$)	1.45 (1.24)	1.25 (1.46)
New episodes ($n = 440$)	1.06 (1.52)	0.69 (1.12)
Period of time since last sexual contact without a condom ($n = 259$)	2.36 (0.87)	2.40 (0.80)
Number of partners in the last month ($n = 441$)	3.48 (1.88)	1.34 (0.81)
Number of partners in the last year ($n = 419$)	5.03 (2.43)	2.39 (1.42)
Lifelong sexual history ($n = 99$)	2.41 (0.85)	1.84 (0.85)
Age of first sexual intercourse ($n = 233$)	17.13 (2.46)	17.62 (3.33)

SD: Standard Deviation; HIV: Human Immunodeficiency Virus. Period of time since last sexual contact without a condom: 1 = never, 2 = less than one month, 3 = one to six months, 4 = six to 12 months, 5 = more than 12 months; Number of partners in the last month: 1 = 0–1, 2 = 2, 3 = 3–5, 4 = more than 5; Number of partners in the last year: 1 = 0–1, 2 = 2, 3 = 3–5, 4 = 6–10, 5 = 11–20, 6 = more than 20; Lifelong sexual history: 1 = 0–10, 2 = 10–20, 3 = more than 20.

Table 2 shows the main diagnoses identified by population group, the most frequent being Human Papilloma Virus (HPV) infection. Non-Spanish nationals were disproportionately diagnosed with Gardnerella (16.1%) and syphilis (5.4%), whereas Spanish nationals were disproportionately diagnosed with Molluscum contagiosum (10.6%) and gonococcal infection (6.7%). The proportions observed in the case of HIV infection are similar in both groups. However, there is a greater proportion of Herpes simplex virus infection in Spanish nationals and a greater proportion of HBV infection in non-Spanish nationals.

Table 2. Distribution of STIs in non-Spanish and Spanish populations (n = 378). Granada, Spain, 2000–2014.

	Non-Spanish Nationals (n = 112)	Spanish Nationals (n = 329)
	n (%)	*n* (%)
HPV (n = 208)	41 (36.6)	167 (50.8)
Candidiasis (n = 66)	29 (25.9)	37 (11.2)
Molluscum contagiosum (n = 38)	3 (2.7)	35 (10.6)
Gardnerella (n = 31)	18 (16.1)	13 (4.0)
Syphilis (n = 26)	6 (5.4)	20 (6.1)
Gonococcal infection (n = 25)	3 (2.7)	22 (6.7)
Herpes simplex virus (n = 22)	4 (3.6)	18 (5.5)
HIV (n = 15)	4 (3.6)	11 (3.3)
HBV (n = 4)	3 (2.7)	1 (0.3)
Other (n = 6)	1 (0.8)	5 (1.5)

HPV: Human Papilloma Virus. HIV: Human Immunodeficiency Virus. HBV: Hepatitis B Virus.

The rates of STI diagnoses among the non-Spanish population residing in the province of Granada were higher than among the Spanish population in all the years analysed, with the exception of 2014 (2.05 vs. 5.52) (Figure 1).

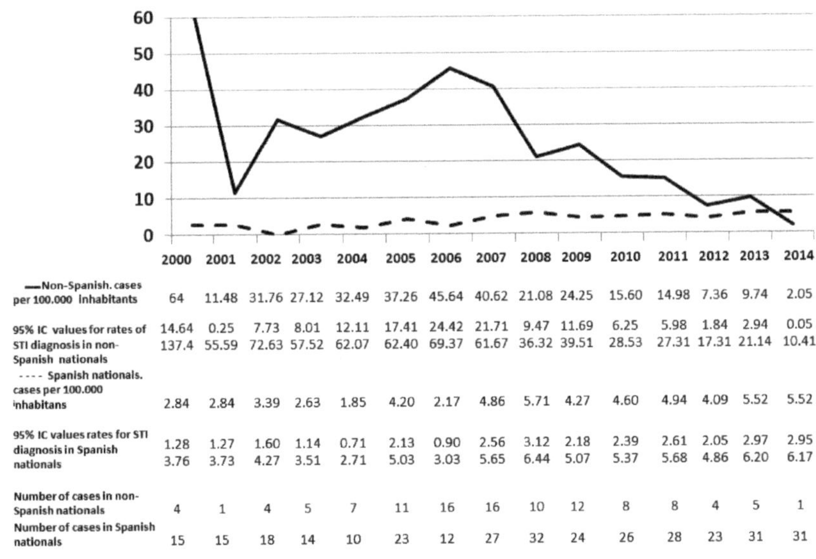

	2000	2001	2002	2003	2004	2005	2006	2007	2008	2009	2010	2011	2012	2013	2014
——Non-Spanish. cases per 100.000 inhabitants	64	11.48	31.76	27.12	32.49	37.26	45.64	40.62	21.08	24.25	15.60	14.98	7.36	9.74	2.05
95% IC values for rates of STI diagnosis in non-Spanish nationals	14.64 137.4	0.25 55.59	7.73 72.63	8.01 57.52	12.11 62.07	17.41 62.40	24.42 69.37	21.71 61.67	9.47 36.32	11.69 39.51	6.25 28.53	5.98 27.31	1.84 17.31	2.94 21.14	0.05 10.41
- - - - Spanish nationals. cases per 100.000 inhabitans	2.84	2.84	3.39	2.63	1.85	4.20	2.17	4.86	5.71	4.27	4.60	4.94	4.09	5.52	5.52
95% IC values rates for STI diagnosis in Spanish nationals	1.28 3.76	1.27 3.73	1.60 4.27	1.14 3.51	0.71 2.71	2.13 5.03	0.90 3.03	2.56 5.65	3.12 6.44	2.18 5.07	2.39 5.37	2.61 5.68	2.05 4.86	2.97 6.20	2.95 6.17
Number of cases in non-Spanish nationals	4	1	4	5	7	11	16	16	10	12	8	8	4	5	1
Number of cases in Spanish nationals	15	15	18	14	10	23	12	27	32	24	26	28	23	31	31

Figure 1. Annual distribution of STI rates for Spanish nationals and non-Spanish nationals aged between 15 and 64 years old. Granada, 2000–2014. STIs: sexually transmitted infections. CI: confidence interval.

In both the pre-crisis (2000–2007) and crisis (2008–2014) periods, as analysed separately, as well as for the entire time period analysed as one, there was a higher risk of being diagnosed with an STI among immigrants than among Spanish nationals. However, the difference observed between the two populations was less pronounced during the crisis period (Table 3).

Table 3. Annual odds ratios for STIs in non-native populations versus native populations aged between 15 and 64 years old. Granada, Spain, 2000–2014.

Year	OR	95% CI
2000	22.51	13.18–64.32
2001	4.05	1.72–21.62
2002	9.37	5.54–26.26
2003	10.31	6.25–27.53
2004	17.54	10.87–44.77
2005	8.87	6.19–17.97
2006	21.01	14.44–43.83
2007	8.36	6.12–15.38
2008	3.69	2.59–7.41
2009	5.68	4.02–11.24
2010	3.39	2.29–7.35
2011	3.03	2.05–6.52
2012	1.80	1.08–4.94
2013	1.76	1.11–4.37
2014	0.37	0.16–1.91
2000–2014	5.33	4.78–6.60
2000–2007	12.02	10.33–16.17
2008–2014	2.73	2.32–3.73

OR: odds ratio. STIs: sexually transmitted infections. CI: confidence interval.

4. Discussion

Regarding the socio-demographic profile of the sample analysed, it is worth noting that, in the immigrant population, a greater proportion of patients were women compared to in the Spanish population. However, the mean age of the sample was very similar in both population groups, being between 27 and 28 years old. There was a higher proportion of individuals with higher education in the native population than in the immigrant population. The high percentage of sex workers or former sex workers in the immigrant population analysed stands out. In reference to the number of partners in the previous month, in the previous year, and throughout their sexually active lives, the mean values of all three of these variables are higher in immigrants than in Spanish nationals.

The most frequently diagnosed STI, in both immigrant and native populations, was HPV infection. This is consistent with a study on STIs conducted in the same region, Andalusia [20], in which HPV infection was the most frequent infection as early as 2009. With respect to other diagnoses, such as hepatitis B and syphilis, also reported in other studies, the results are consistent with another study [21] that reports a higher prevalence of hepatitis B in the immigrant population compared to the native population. In contrast, the pattern is different for syphilis, with a higher prevalence in the native population than in the immigrant population.

Based on the results obtained from the analysis of the crisis and pre-crisis periods, it can be observed that the risk of being infected with an STI is greater in the non-Spanish population throughout the entire period analysed. In addition, the proportion of patients who were not Spanish nationals experienced a linear upward trend until 2012. In this year, the trend seemed to reverse, leading to a progressive reduction of the proportion of patients who were not Spanish nationals, compared to what it had been prior. As a result, the described profile points to a reducing effect on the risk for non-Spanish nationals relative to Spanish nationals during the crisis years. This trend can also be observed in the chart comparing the changes in STI rates. In the Spanish population, the rate of STIs remains relatively stable throughout the period studied. However, the rate of STIs in the non-Spanish population is higher and more erratic in the pre-crisis period than in the crisis period, during which this rate seems to stabilise at a level closer to that of the Spanish population.

There may be several explanations for the apparent drop in risk of STI diagnosis during the crisis period compared to the pre-crisis period, which are consistent with a potential underestimation of STI diagnosis in the immigrant population during the crisis years. For instance, the deteriorating social and working conditions following the onset of the crisis, which fundamentally affected the poorest and most vulnerable areas of society, including immigrants [22,23], may have caused many immigrants in Spain to return to their countries of origin [24] and slowed down the arrival of immigrants in Spain [25]. A previous study concluded that, between 2006 and 2012, the health status of the immigrants who had arrived in Spain prior to 2006 was worse in comparison to that of the native population. A possible explanation for this may be the loss of the healthy immigrant effect during the most severe impact of the economic crisis on immigrants [26]. In addition, the passage in 2012 of the Royal Decree-Law 16/2012, on urgent measures to guarantee the sustainability of the Spanish National Health Service and improve the quality and safety of its services [27], restricted access to health care for illegal immigrants. The fear of being diagnosed with an STI and its potential ramifications (e.g., losing their job and/or residence permit) stands in the way of carrying out diagnostic tests in the immigrant population, especially when access to treatment and healthcare is being restricted.

The effect of austerity policies on the reduction of preventive strategies should also be taken into account. The decrease in use of contraceptives since 2007, the absence of protective measures against STIs in one fifth of occasional or sporadic relationships, as well as the increasing incidence of syphilis, gonorrhoea, and HIV in certain groups of individuals, demonstrate that there is a need to place more emphasis on preventive strategies and to strengthen the commitments made by institutions concerning the most vulnerable areas and groups of individuals. The economic crisis weakens the educational and healthcare systems to the same extent that it weakens prevention and promotion measures relating to sexual health [9].

Finally, the representation of sex workers among the study population must also be considered. A previous study conducted in female sex workers in Spain shows that the prevalence of self-reported STIs experienced a significant increase between 2005 and 2011 (from 14% to 20.6%), pointing to inconsistent condom use as a factor worth considering [28]. In our study, a higher proportion of the immigrant population are sex workers compared to the native population, which may also explain the higher prevalence of STIs. However, further studies are needed to corroborate this association.

Limitations

Caution must be exercised when interpreting the data obtained and generalising them to the immigrant population as a whole due to certain limitations of this study. For example, it should be taken into consideration that this study has been carried out in a specific geographical area. As a result, this study has a low degree of external validity. However, the close relationship that exists between population, culture, healthcare systems, and the use of healthcare services (such as STI-related services) justifies the need to perform field analyses such as this one. Additionally, given that this study has been conducted in a specific healthcare district, it provides varied and accurate local data on the composition of the population attached to this area, data that are not usually available from demographical sources.

The non-Spanish population analysed in this study is over-represented in comparison with the non-Spanish population officially resident in the province of Granada, which may indicate the large number of illegal immigrants that characterises international migration in Southern Europe. In Spain, official sources have estimated that, only during the period analysed, 195,458 undocumented individuals had arrived in Spain [29], with Granada being one of the most affected provinces [30].

Furthermore, despite having analysed a long period of time, the cross-sectional nature of this study does not allow causal associations to be established. As a result, the findings shown in this respect must be regarded as hypotheses to be tested with other more complex designs that facilitate the establishment of stronger causal relationships. There is an inconclusive result suggesting the need to further investigate the results for more conclusive outcomes in future studies.

5. Conclusions

During the period 2000–2014, the risk of being infected with an STI is greater in the non-Spanish population throughout the entire period analysed. There was also a gradual decrease in the rate of STI diagnoses in the immigrant population from 2009 to the lowest level of the time series in 2014, which led to a lower risk of being diagnosed with an STI during the same period. The difference in risk observed between the two populations is less marked during the crisis period (2008–2014) compared to the non-crisis period (2000–2007). This may be attributed to an underreporting of diagnoses in the immigrant population during the crisis period.

Drawing on the epidemiological and social context, the findings of this study show a population profile (of the non-Spanish population), which is more vulnerable to STIs. It should be noted that, for example, the lack of STI diagnosis and treatment in illegal immigrants hinders STI control. This, in turn, may increase STI transmission likelihood and could result in a deterioration in the health of the affected group and, potentially, of the general population [8]. Public health policies must improve the control and treatment of existing cases, allocate more resources for the detection of unreported cases, and put in place more effective preventative measures at a lower cost.

Author Contributions: Conceptualization, M.Á.P.-M., A.M.-S., M.D.P.-C., C.H.-M., and E.M.-G.; Data curation, M.Á.P.-M; Formal analysis, P.F.-M. and E.M.-G.; Methodology, M.Á.P.-M, M.G.-L., C.H.-M., and E.M.-G.; Supervision, C.H.-M. and E.M.G.; Writing—original draft, M.Á.P.M, A.M.S. and E.M.G.; Writing—review & editing, M.Á.P.-M, A.M.S., M.G.-L., P.F.-M., M.D.P.-C., C.H.-M., and E.M.-G. All authors have read and agreed to the published version of the manuscript.

Funding: This research received no external funding.

Acknowledgments: The authors would like to thank all members of the STI clinic team in the province of Granada, as well as Nurses María Teresa Sánchez Ocón, Esperanza Cano Romero and María Visitación Mingorance Ruiz, for their help in collecting the data.

Conflicts of Interest: The authors declare no conflict of interest.

References

1. World Health Organization. Sexually Transmitted Infections (STIs). Available online: https://www.who.int/es/news-room/fact-sheets/detail/sexually-transmitted-infections-(stis) (accessed on 20 May 2019).
2. European Centre for Disease Prevention and Control: An agency of the European Union. Annual Epidemiological Reports. Available online: https://www.ecdc.europa.eu/en/annual-epidemiological-reports (accessed on 26 March 2020).
3. Unidad de vigilancia del VIH y conductas de riesgo. *Vigilancia Epidemiológica de las Infecciones de Transmisión Sexual, 2016*; Centro Nacional de Epidemiología/Subdirección General de Promoción de la Salud y Vigilancia en Salud Pública-Plan Nacional sobre el Sida: Madrid, Spain, 2018.
4. World Health Organization. Global Health Sector Strategy on Sexually Transmitted Infections, 2016–2021. Available online: https://www.who.int/reproductivehealth/publications/rtis/ghss-stis/en/ (accessed on 20 May 2019).
5. Ghimire, S.; Hallett, J.; Gray, C.; Lobo, R.; Crawford, G. What Works? Prevention and Control of Sexually Transmitted Infections and Blood-Borne Viruses in Migrants from Sub-Saharan Africa, Northeast Asia and Southeast Asia Living in High-Income Countries: A Systematic Review. *Int. J. Environ. Res. Public Heal.* **2019**, *16*, 1287. [CrossRef] [PubMed]
6. Instituto Nacional de Estadística. INEbase. Available online: https://www.ine.es/prensa/cp_j2018_p.pdf (accessed on 20 May 2019).

7. Área de Vigilancia de VIH y Comportamientos de Riesgo. *Vigilancia Epidemiológica del VIH y sida en España 2017: Sistema de Información sobre Nuevos Diagnósticos de VIH y Registro Nacional de Casos de Sida*; Plan Nacional sobre el Sida-D.G. de Salud Pública, Calidad e Innovación / Centro Nacional de Epidemiología–ISCIII: Madrid, Spain, 2018.

8. García, F.; Rubio, R.; Hernando, A.; Fiorante, S.; Maseda, D.; Matarranz, M.; Costa, J.R.; Alonso, B.; Pulido, F. Características clínico-epidemiológicas de los pacientes inmigrantes con infección por el VIH: Estudio de 371 casos. *Enferm. Infecc. y Microbiol. Clínica* **2012**, *30*, 441–451. [CrossRef] [PubMed]

9. Vázquez, M.L.; Vargas, I.; Aller, M. Reflexiones sobre el impacto de la crisis en la salud y la atención sanitaria de la población inmigrante. Informe SESPAS 2014. *Gac. Sanit.* **2014**, *28*, 142–146. [CrossRef] [PubMed]

10. Porthé, V.; Vargas, I.; Ronda, E.; Malmusi, D.; Bosch, L.; Vázquez, M.L. Has the quality of health care for the immigrant population changed during the economic crisis in Catalonia (Spain)? Opinions of health professionals and immigrant users. *Gac. Sanit.* **2018**, *32*, 425–432. [CrossRef] [PubMed]

11. Petrelli, A.; Di Napoli, A.; Rossi, A.; Costanzo, G.; Mirisola, C.; Gargiulo, L. The variation in the health status of immigrants and Italians during the global crisis and the role of socioeconomic factors. *Int. J. Equity Health* **2017**, *16*, 98. [CrossRef] [PubMed]

12. Sargent, C.; Kotobi, L. Austerity and its implications for immigrant health in France. *Soc. Sci. Med.* **2017**, *187*, 259–267. [CrossRef] [PubMed]

13. Garcia-Subirats, I.; Vargas, I.; Sanz-Barbero, B.; Malmusi, D.; Ronda, E.; Ballesta, M.; Vázquez, M.L. Changes in Access to Health Services of the Immigrant and Native-Born Population in Spain in the Context of Economic Crisis †. *Int. J. Environ. Res. Public Health* **2014**, *11*, 10182–10201. [CrossRef] [PubMed]

14. Llácer, A.; Fernández-Cuenca, R.; Navarro, J.F.M. Crisis económica y patología infecciosa. Informe SESPAS 2014. *Gac. Sanit.* **2014**, *28*, 97–103. [CrossRef] [PubMed]

15. Suhrcke, M.; Stuckler, D.; Suk, J.E.; Desai, M.; Senek, M.; McKee, M.; Tsolova, S.; Basu, S.; Abubakar, I.; Hunter, P.R.; et al. The Impact of Economic Crises on Communicable Disease Transmission and Control: A Systematic Review of the Evidence. *PLoS ONE* **2011**, *6*, e20724. [CrossRef] [PubMed]

16. Martínez García, M.F.; Sánchez, A.; Martínez, J. Crisis económica, salud e intervención psicosocial en España. *Apunt. Psicol.* **2017**, *35*, 5–24.

17. Peralta-Gallego, L.; Badia, J.G.; Gallo, P. Effects of undocumented immigrants exclusion from health care coverage in Spain. *Heal. Policy* **2018**, *122*, 1155–1160. [CrossRef] [PubMed]

18. Instituto Nacional de Estadística. Principales Series de Población Desde 1998. Población por Provincias, Edad (3 Grupos de Edad), Españoles/extranjeros, Sexo y Año. INEbase. Available online: http://www.ine.es/jaxi/Tabla.htm?path=/t20/e245/p08/l0/&file=03001.px&L=0 (accessed on 20 June 2019).

19. Ferrando, J.; Palència, L.; Gotsens, M.; Puig-Barrachina, V.; Marí-Dell'Olmo, M.; Rodríguez-Sanz, M.; Bartoll, X.; Borrell, C. Trends in cancer mortality in Spain: The influence of the financial crisis. *Gac. Sanit.* **2019**, *33*, 229–234. [CrossRef] [PubMed]

20. Valverde, E.E.; DiNenno, E.; Schulden, J.D.; Oster, A.; Painter, T. Sexually transmitted infection diagnoses among Hispanic immigrant and migrant men who have sex with men in the United States. *Int. J. STD AIDS* **2016**, *27*, 1162–1169. [CrossRef] [PubMed]

21. Bermúdez, M.P.; Castro, A.; Madrid, J.; Buela-Casal, G. Análisis de la conducta sexual de adolescentes autóctonos e inmigrantes lationoamericanos en España. *Int. J. Clin. Health Psychol.* **2010**, *10*, 89–103.

22. Cano Murillo, A.M.; Castilla Catalán, J.; Del Amo Valero, J. Epidemiology of HIV infection in immigrants in Spain: Information sources, characteristics, magnitude and tendencies. *Gac. Sanit.* **2010**, *24*, 81–88.

23. Robert, G.; Martínez, J.M.; Garcia, A.M.; Benavides, F.G.; Ronda, E. From the boom to the crisis: Changes in employment conditions of immigrants in Spain and their effects on mental health. *Eur. J. Public Health* **2014**, *24*, 404–409. [CrossRef] [PubMed]

24. Pérez-Morente, M.; Sánchez-Ocón, M.T.; Martínez-García, E.; Martín-Salvador, A.; Hueso-Montoro, C.; Garcia, I.G. Differences in Sexually Transmitted Infections between the Precrisis Period (2000–2007) and the Crisis Period (2008–2014) in Granada, Spain. *J. Clin. Med.* **2019**, *8*, 277. [CrossRef] [PubMed]

25. Ramos, J.M.; Gutiérrez, F.; Padilla, S.; Masiá, M.; Escolano, C. Características clínicas y epidemiológicas de la infección por el VIH en extranjeros en Elche, España (1998–2003). *Enferm. Infecc. y Microbiol. Clínica* **2005**, *23*, 469–473. [CrossRef] [PubMed]

26. Gotsens, M.; Malmusi, D.; Villarroel, N.; Vives-Cases, C.; Garcia-Subirats, I.; Hernando, C.; Borrell, C. Health inequality between immigrants and natives in Spain: The loss of the healthy immigrant effect in times of economic crisis. *Eur. J. Public Health* **2015**, *25*, 923–929. [CrossRef] [PubMed]

27. Gobierno de España. *Real Decreto-ley de medidas urgentes para garantizar la sostenibilidad del Servicio Nacional de Salud (SNS) y mejorar la calidad y seguridad de sus prestaciones*; L.N°16/2012 (20 abril 2012). BOE núm. 98, de 24 de abril de 2012; Gobierno de España: Madrid, Spain, 2012.

28. Folch, C.; Casabona, J.; Sanclemente, C.; Esteve, A.; González, M.V. Tendencias de la prevalencia del VIH y de las conductas de riesgo asociadas en mujeres trabajadoras del sexo en Cataluña. *Gac. Sanit.* **2014**, *28*, 196–202. [CrossRef] [PubMed]

29. Ministerio del Interior. Gobierno de España. Lucha Contra la Inmigración Irregular. Balance. 2014. Available online: http://www.interior.gob.es/documents/10180/3066430/Balance+2014+de+la+lucha+contra+la+inmigraci%C3%B3n+irregular/4a33ce71-3834-44fc-9fbf-7983ace6cec4 (accessed on 4 March 2020).

30. Granada Acoge. Memoria. 2008. Available online: http://www.granadaacoge.org/wp-content/uploads/2012/04/Memoria-2008.pdf (accessed on 4 March 2020).

International Journal of
Environmental Research and Public Health

Article

Understanding the Experiences and Needs of Migrant Women Affected by Female Genital Mutilation Using Maternity Services in Australia

Sabera Turkmani [1,*], Caroline S. E. Homer [2] and Angela J. Dawson [1]

[1]   Australian Centre for Public and Population Health Research, Faculty of Health, University of Technology Sydney, Ultimo NSW 2007, Australia; angela.dawson@uts.edu.au
[2]   Maternal, Child and Adolescent Health Program, Burnet Institute, Melbourne VIC 3004, Australia; caroline.homer@burnet.edu.au
*   Correspondence: sabera.turkmani@uts.edu.au

Received: 5 February 2020; Accepted: 18 February 2020; Published: 26 February 2020

Abstract: Female genital mutilation (FGM) is a cultural practice defined as the partial or total removal of the external female genitalia for non-therapeutic reasons. Changing patterns of migration in Australia and other high-income countries has meant that maternity care providers and health systems are caring for more pregnant women affected by this practice. The aim of the study was to identify strategies to inform culturally safe and quality woman-centred maternity care for women affected by FGM who have migrated to Australia. An Appreciative Inquiry approach was used to engage women with FGM. We conducted 23 semi-structured interviews and three focus group discussions. There were four themes identified: (1) appreciating the best in their experiences; (2) achieving their dreams; (3) planning together; and (4) acting, modifying, improving and sustaining. Women could articulate their health and cultural needs, but they were not engaged in all aspects of their maternity care or considered active partners. Partnering and involving women in the design and delivery of their maternity care would improve quality care. A conceptual model, underpinned by women's cultural values and physical, emotional needs, is presented as a framework to guide maternity services.

Keywords: female genital mutilation (FGM); women's health needs; equality; quality of maternity care; midwifery continuity of care

---

1. Introduction

Female genital mutilation (FGM) is defined as the partial or total removal of external female genitalia for non-therapeutic reasons [1]. This practice is deeply rooted in culture, with social or religious obligation and marriageability considered to be the most important reason for its continuation (UNICEF 2013). FGM is also performed for fear of being excluded from opportunities as a young woman [2]. FGM is traditionally practiced in 30 African and Middle Eastern countries, and some parts of Asia and South America [3]. Changing patterns of migration have led to an increase in the prevalence of women with FGM in many high-income countries [4,5].

It is estimated that globally over 200 million women and girls have undergone FGM and another three million women and girls are at risk annually [6]. There is a lack of reliable and high-quality data in relation to the numbers of women affected by FGM in high-income countries (HICs) [7]. It is, therefore, challenging for countries to develop effective policies, allocate relevant resources and evaluate the results of interventions [8,9]. In Australia, a recent report [10] estimated that in 2017 there were 53,000 women and girls affected by FGM in Australia, which represents a prevalence of 4 per 1000 girls and women.

Women affected by FGM in HICs are usually migrants or refugees and may have complex needs in addition to their clinical care [11,12]. These women are more likely to face socio-economic and cultural challenges due to language barriers, low education levels, and financial difficulties, which can hinder access to health services [13,14]. Migrant and refugee women from low- and middle-income countries, especially those from African countries, are reported to have poorer perinatal outcomes due to a higher rate of complications during pregnancy and childbirth [15,16]. FGM poses an additional burden to affected women and there are potential adverse consequences during pregnancy and childbirth such as an increased risk of caesarean section, post-partum haemorrhage, instrumental birth and prolonged labour [6].

Research has found that health services in some HICs may not be adequately prepared to provide quality care to FGM-affected women [17]. For example, many health professionals lack clinical skills and knowledge about the law in relation to FGM [17,18]. Health professionals have also been found to have a poor understanding of the cultural background of women affected by FGM and find communication challenging [19,20]. These issues, combined with inadequate support services, such as interpreting and counselling services, mean that many women may face difficulties expressing their needs [21].

The World Health Organization (WHO) highlights the importance of improving the quality of maternity care for women with complex needs to minimise further complications and harm [6]. The WHO's quality care standards outline eight domains of quality of care that encompass the provision of care and a woman's experience of care [22]. Quality of health services would improve if women trust and are confident to utilise the services on the basis of their positive and satisfactory experiences [23]. While research has provided insight into what constitutes quality maternal care from a health system perspective, gaps remain concerning the views and needs of women with FGM and what they regard as quality care. This study aimed to identify approaches to achieve culturally safe and high-quality woman-centred care for migrant women who have been personally affected by FGM.

2. Materials and Methods

The study employed Appreciative Inquiry (AI), a qualitative methodology to gain a deep insight into women's experiences of midwifery care in an Australian setting [24]. Appreciative Inquiry is a well-accepted methodology in health research [25]. This methodology has been used to explore patient experience of clinical healthcare [26] and to address the complex needs of families and children in primary health care [27]. McAdam and Mirza [28] used AI to describe the experience of marginalised youth engaged in drug and alcohol misuse and the implications of positive stories on health and social well-being.

We applied the four-stepped processes of AI to elicit examples of positive care interactions and envisage what best quality maternity care might look like in the future [29]. AI is open-ended, allowing a flexible approach to be taken depending on the needs of the participants [30]. The collaborative nature of AI is helpful in the development of confidence and can motivate participants to become actively involved in change [31]. We designed this research not only to identify the changes that are needed to improve maternity care but encourage women to become involved in this change as users and beneficiaries of maternity services.

Ethical considerations for this project were of particular importance as the migrant women with FGM often feel vulnerable, stigmatised and marginalised [32]. Ethical approval (UTS HREC REF NO. ETH17-1525) was obtained from the Human Research Ethics Committee of UTS in August 2017 before the recruitment or data collection process.

This study was conducted in Western Sydney, New South Wales (NSW), the area that has the largest number of non-English speaking women in this state of Australia [33]. The participants were English-speaking migrant and refugee women who were personally affected by FGM and lived in Sydney. The women had given birth in Australia in the last ten years or were currently pregnant. A written information package was given to women to invite them to participate. The research was designed and conducted in direct consultation with experts in this area from government and non-governmental organizations, and an independent activist and advocate who is a survivor of FGM. In addition, a member of the community was involved throughout the study process and guided the development of the research tools, assisted with recruitment of the study population and ensured that the project was conducted in an ethical and culturally appropriate manner. Chain referral sampling was employed to approach potential women [34]. This method of sampling is useful for recruiting participants in research where the topic is sensitive or in populations that are stigmatised and hard to reach [35]. Participants signed the informed consent form prior to the commencement of interviews or group discussions. Participant anonymity was assured by allocating a code to the woman's name.

Data were gathered through in-depth interviews and followed by focus group discussions over five months (October 2017–February 2018). Interviews lasted for one hour and group discussions were conducted with five to eight women for two hours. The interviews and discussions were guided by questions (Supplementary Materials) following concepts of AI that were flexible enough to enable an exploration of ideas and experiences that women raised [36]. Interviews were held at a time and place convenient to women and group discussions were held in community centres. We offered small gift cards as compensation for their time and travel to the interview location.

Braun and Clarke's [37] approach to qualitative data analysis was followed because it offered a way of analysing the data according to the 4Ds representing the discovery, dream, design and deploy phases of AI in the first instance followed by a closer analysis of women's experiences as maternity service users.

The interview transcripts were first transcribed verbatim [38] enabling the researchers to familiarise themselves with the data tomake sense of it and reflect on overall meanings and general ideas [39]. Data were exported into the NVivo qualitative data management software to enable coding of the text according to the four phases of AI. Woman's narratives were coded into themes representing 4Ds and appropriate sub-themes [40]. The final step of data analysis involved the interpretation of data to draw recommendations for future maternity care policy and practice [39].

3. Results

In total, 23 individual interviews and three focus group discussions were conducted. The women were from Sudan ($n = 9$), Somalia ($n = 6$), Sierra Leone ($n = 3$), Egypt ($n = 2$), Indonesia ($n = 2$), and Ethiopia ($n = 1$). The majority of women ($n = 21$) had undergone FGM when they were 0-10 years old. All of the women came to Australia as refugees except for four who entered the country on spousal visas or employment visas. English was the second language for all women (Table 1).

The main findings are presented under four themes in line with the 4D cycle of Appreciative Inquiry: Appreciating the positives in their maternity care (Discovering); Desiring the best in maternity services (Dreaming); Planning together for improved maternity services (Designing); Improving and sustaining (Developing/Deploying). The four themes and their associated sub-themes are further elaborated in the sections below (Figure 1).

Table 1. Demographic information.

Study Code	Age	Age Underwent FGM	Country of Origin	Date of Last Birth in Australia	Education Level	First Language	Employment Status	# Children Born in Australia	# Live Birth	Years Lived in Australia
Astur	30–35	5–10	Somali	2012	Secondary	Somali	Employed	1	1	20–25
Bilan	30–35	5–10	Somali	2005	Primary	Somali	Housewife	5	5	20–25
Calaso	30–35	1–5	Somali	2013	Secondary	Somali	Employed	1	1	10–15
Bilqis	30–35	1–5	Somali	2010	Secondary	Somali	Employed	3	3	20–25
Indah	40–45	<1	Indonesia	2004	Tertiary	Indonesian	Employed	3	3	15–20
Aminata	40–45	10–15	Sierra Leone	2013	Tertiary	Creole Temne	Employed	2	3	10–15
Binta	25–30	5–10	Sierra Leone	2016	Tertiary	Temne	Employed	2	2	15–20
Arifa	30–35	1–5	Sudan	2013	Secondary	Arabic	Employed	3	4	10–15
Fiza	35–40	5–10	Sudan	2009	Tertiary	Arabic	Employed	2	2	15–20
Mariatu	25–30	15–20	Sierra Leone	2017	Secondary	Creole Temne	Housewife	2	2	5–10
Hiba	40–45	1–5	Sudan	2011	Secondary	Arabic	Housewife	3	5	10–15
Nadia	40–45	<1	Sudan	2006	Tertiary	Arabic	Employed	1	1	10–15
Rita	35–40	1–5	Sudan	2015	Tertiary	Arabic	Housewife	3	5	5–10
Yusra	35–40	5–10	Sudan	2017	Tertiary	Arabic	Housewife	4	5	5–10
Faduma	40–45	1–5	Somali	2009	Secondary	Somali	Housewife	5	5	15–20
Kia	35–40	1–5	Ethiopia	2011	Secondary	Arabic	Employed	3	3	15–20
Zara	25–30	5–10	Sudan	2016	Tertiary	Arabic	Housewife	2	2	10–15
Fatma	40–45	1–5	Sudan	2012	Secondary	Arabic	Housewife	2	5	10–15
Nour	30–35	5–10	Egypt	2016	Tertiary	Arabic	Employed	3	3	5–10
Gamal	35–40	1–5	Egypt	2015	Tertiary	Arabic	Employed	3	3	5–10
Asima	30–35	1–5	Sudan	2014	Tertiary	Arabic	Housewife	1	1	5–10
Harum	35–40	<1	Indonesia	2012	Tertiary	Bahasa	Housewife	2	3	5–10
Zaineb	40–45	1–5	Somali	2007	Primary	Somali	Housewife	1	1	10–15

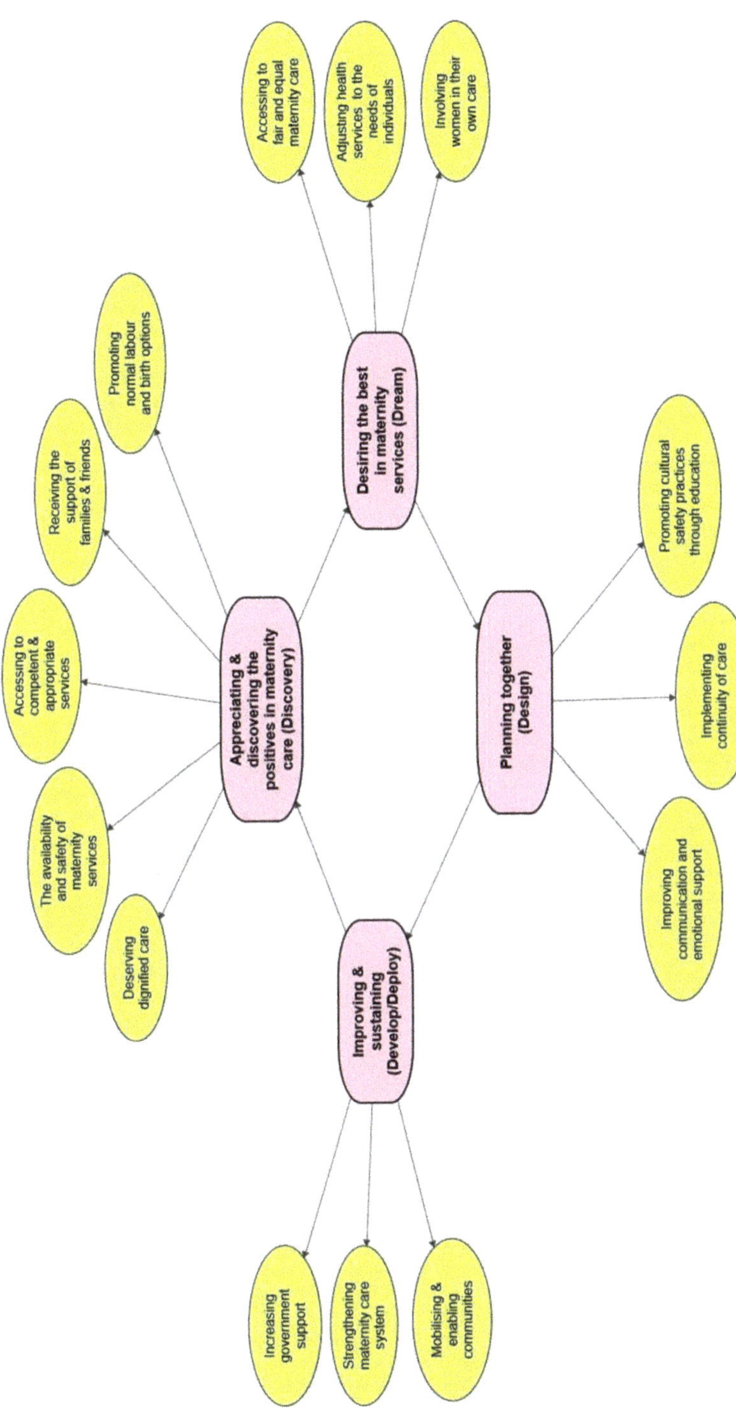

Figure 1. Thematic data analysis based on 4Ds cycle (stages) of Appreciative Inquiry.

3.1. Appreciating and Discovering the Positives in Maternity Care (Discovering)

Appreciating the positives in maternity care concentrated on women's description of events during their maternity care in Australia, and the strategies or approaches that they perceived to be useful, or inappropriate. For the most part, women were appreciative of, and satisfied with, the maternity care they received. This included being provided with respectful care, a feeling of having a safe service, receiving the required information, having access to skilled health care providers, and being able to have advance care planning and family support. Women frequently reported that "Maternity services are really good in Australia compared to where we came from".

Women felt that the maternity services are safe and technologically advanced in Australia and they expected their maternity care providers to have an appropriate level of knowledge about FGM, possess effective communication skills, be sensitive to their cultural needs and involve women in their care. For example:

> *The good thing was always feeling safe, knowing there are all the facilities, medicines and machines and skills you might need available within the hospital. I really felt relaxed in both my deliveries. Overall pregnancy was a happy experience for me and I knew they would help me straightaway compared to my country where nothing is available.* (W18)

Most of the women appreciated maternity care providers who were helpful, sensitive and responsive to their needs, especially when they had no family and relatives around to support them. Women were impressed by the way maternity care providers made them feel cared for, particularly when they followed up to make sure women did not miss their appointments.

A few women thought that while caring for women with FGM was not a common experience for Australian maternity care providers, women expected providers to know how to deal with FGM and how to communicate with women. This woman said:

> *... it is not like that the doctors and midwives in Australia come across a circumcised woman every day, you know. And I don't blame them if they are surprised or ask you millions of questions.* (W13)

Several women were worried about the care they might receive because they did not think maternity care providers were adequately prepared to manage FGM as this woman explained:

> *The medical staff need to understand this issue [FGM] and be knowledgeable about it and if they don't have hands-on experience and skills please do not touch us and make our situation worse. You need to feel safe knowing that they get training before coming to women with FGM.* (W23)

The women acknowledged that developing trust with a maternity care provider was directly associated with the provider's competence. Some felt anxious and lost confidence in the ability of maternity care providers to deliver good care when they saw that their care providers were surprised or shocked when they encountered FGM. One woman in a focus group said:

> *If these midwives and doctors know where to cut (de-infibulation), how to cut and when to cut it will be so helpful for us and for them because we will not have a problem and they will be relaxed and confident in what they do. Now, as soon as they see us they are shaking ... Oh my God. They can get advice from doctors and midwives who worked in our country and have real experience of treatment of women with FGM.* (FGD3)

3.2. Desiring the Best in Maternity Services (Dreaming)

Women expressed their vision for the best maternity care in the future, including how they would wish to be treated within the healthcare system. They described the need for equality and for FGM-affected women to be treated the same way as other women. This included a desire for personalised care to be delivered by a provider from a similar cultural background with services tailored to the needs of the individual woman. In practical terms, women described how individualised care should mean the provision of support services for women following de-infibulation.

Women believed that each pregnancy is an individual experience and expressed a clear understanding of the need for services to facilitate informed choice and shared decision making in a way that involves women with FGM in their own care. They wanted maternity care providers to listen to each woman and adjust care to suit her individual needs, rather than following the same course for every woman. One woman said:

> *They need to listen to women as they know their body better. Not everything is going to be according to the recipe in the book. They have to look at each individual pregnancy separately.*

Women wanted to be treated in the same way as other women without being labelled as different while accessing Australian maternity services. They also wanted to have access to appropriate mental health support that took into consideration their special circumstances due to their FGM, for example:

> *If a woman has undergone FGM they need to look after her even after birth and even if there is not any visible harm there is always a change and she needs that emotional support.* (W20)

Many women struggled with their body image and the emotional impact of de-infibulation, and some wished to see their bodies the way they were used to seeing it since childhood (infibulated). This was exacerbated by the fact that legislation in the state did not provide the option of re-infibulation. They perceived a reluctance of health staff to consider any form of reconstruction of the vulval or perineal area and attributed this to laws prohibiting re-infibulation. Women wanted reconstructive surgery to be part of the services offered to them. Most believed that their de-infibulation had been done 'badly' and their body would be 'in better shape' if they were re-infibulated after birth. They desired varying degrees of re-infibulation and used the term 'closed-back' when describing reconstructive surgery. This comment captures such feelings:

> *After they open you during delivery I wish there is someone who stitches it very very nicely so it doesn't look very open.* (FGD1)

Most women felt embarrassed and uncomfortable with their bodies and described their vulva as 'ugly', 'too open', 'not in good shape', 'hanging skin', and 'horrible'. Some women chose to undergo a caesarean section to avoid de-infibulation or they travelled back to their home country to be re-infibulated as this woman did:

> *... I went overseas and closed it by a midwife in my country. You know last time I [got] closed myself in Sudan it was because it was so big and ugly they left me totally open at least they could have stitched me back to make me look like normal.* (W22)

Women recognised and valued their capacity to experience birth as a normal process without unnecessary intervention. Some wished that their FGM was not considered as a barrier to undergoing normal labour and birth and questioned interventions such as caesarean section. Several women stated, 'We had our baby normally and easier in our country; why not here'.

Some women felt they were vulnerable, disempowered and dominated by maternity care providers and these providers took control of the situation. Some agreed to let their family members make decisions on their behalf, while others expressed their strong desire to be involved in a collaborative way with maternity care providers. Culture, personal attitudes, and emergencies were also identified by most women as factors influencing the degree to which they could be involved in decision making. For example:

> *My husband and mother in law made the decision for me. If it was up to me I would have chosen a caesar straightaway. I did not want all that pain and trauma, but midwife went with my husband and mother in law's decision without listening to me.* (FGD1)

Many women perceived that their lack of health literacy and knowledge about access to certain options or health services led to their exclusion from decision making. One woman explained:

Sometimes you are in a position where you have to follow whatever they say. Maybe because our knowledge is limited and the language also is a big, big problem. (W18)

3.3. Planning Together for Improved Maternity Services (Designing)

The 'design' phase of AI invited women, individually or as part of a group, to develop a plan for what they need to achieve their dream for quality care.

In designing future maternity services, women discussed the need for education initiatives that enabled maternity care providers to provide emotional support, promote cultural safety and communicate in ways that are appropriate for supporting women with FGM. The need for training to involve women themselves to improve provider understanding of and familiarity with the cultural beliefs behind the practice of FGM was highlighted, as these women explained:

If I am a midwife I make you feel good and I need to understand what you believe in so I can understand if you see FGM as a good thing or bad thing. Then I can talk to you and guide you accordingly ... first you need to get a sense of what women believe in, otherwise they may not disclose anything. (W12)

Women also noted that, while maternity care providers need to be respectful and integrate the cultural aspects into their services, they also need to be mindful of harmful cultural practices that may place women at risk. For example, in some cultures, women do not use direct communication to explain their problems related to childbirth, maternity care or FGM. Many women mentioned that they avoided disclosing their FGM as they thought this was culturally inappropriate, as explained here:

I was shy and hide my FGM until birth and I am sure many other would do that. In our culture women won't talk about it believe me or not. There is shame and stigma with those topics'. (W23)

Midwifery continuity of care was one of the models of care or services most appreciated by women who received it. Women understood midwifery continuity of care as being cared for by a known midwife over the entire period of pregnancy and childbirth and after birth. Being with the same midwife and building a relationship based on mutual trust and understanding was perceived to improve women's sense of safety and confidence and increase their involvement in their care. Most women, however, did not have access to this model of care. There were a few women who received midwifery continuity of care during pregnancy, but during labour and birth, their known midwives were not present. They expressed feelings of anxiety and distress with being cared for during labour and birth by midwives they had not met before. Women suggested that maternity services should be designed to enable all women to have access to such a model of care, for example:

It is very important for women because we want to trust someone and by changing midwives and doctors we will be lost. I will also develop my confidence in her competence and make sure she can manage my birth and I am in safe hands. That's a huge support for me knowing that I am safe and someone knows my issues and concerns. (W17)

Women viewed high-quality maternity care in terms of the way that maternity care providers had behaved towards them. Considering the sensitivity of a topic such as FGM, the women believed positive and effective communication was a key component of maintaining a sense of connection, trust, and collaboration with health providers. For example, they wanted to be heard, touched and welcomed. Many women indicated that building trust happened over time as they got to know their maternity care providers through their direct interactions. As this woman explains, this was especially important in addressing the embarrassment that many women felt because of their FGM:

You know little by little each time after I started to visit the doctors and midwives and they didn't make me feel embarrassed [because of FGM] and they asked me so many questions when I went to them. And the way they talked to me was so good. You know, you feel so good when someone listens to you. They were not in a rush to get to the next patient and kick me out of their office. They spend time with you and do what they need to do while they kept privacy. (W13)

Women wanted to receive emotional support to address their trauma including dealing with health issues related to FGM.

Sometimes you just want someone to talk to and ask for nothing else, just someone to ask you what your feelings after birth are or how you are because it is a hard time. … I want a midwife or nurse to provide care for me beyond giving medicines, I want them to talk to me and support me emotionally and mentally. (W17)

Many women felt that there was no transparent, clear, and mutual communication between them and their maternity care providers. As a result, women were often suspicious of the maternity care services they received and were not always willing to accept advice from maternity providers as illustrated with the following quote:

Sometimes they don't even talk about FGM with us and just write everything down and say all is good without giving us the details. I think it is mostly because they don't know anything about FGM and they just look at you and they have no idea. (W17)

3.4. Improving and Sustaining Maternity Services (Developing/Deploying)

The final theme reflected strategies that women regarded as useful to support their plans to improve maternity services. The women's suggestions represented three levels of action: mobilising and enabling communities, strengthening maternity care systems and increasing government support.

Women believed that communities need to be mobilised to create a supportive environment in which pregnant women and new mothers affected by FGM can feel safe and healthy. Advocacy and campaigns for policy, professional practice, and at a community level were considered critical in creating a supportive environment to improve health outcomes for women in the long term. Raising community awareness, through formal and informal education, campaigns in the community and schools, and involving women, men and young people, were considered essential to delivering positive change. One woman said:

Still many people in the community believe it is a good thing to do on their daughters [FGM]. … I will not let my daughter to undergo FGM but we need to remove pressure of the community on families. If no one wants a girl without FGM then everybody forced to do it. We need to end that by educating community and change this culture. (W22)

Women believed the practice of FGM was continuing in their communities, even in Australia, and emphasised the need for a reporting system at the community level. Women stated that the success of community-based interventions, such as education and media campaigns, depends upon the involvement of all members of the community including religious and community leaders in the planning and implementation processes of change. Women emphasised the central role of families in bringing a sustainable change to stop a culture such as FGM as explained here:

Change is dependent on families. In my family, I have already talked to my kids about the stuff like FGM and the even bigger impact of it on society. I think that's how we will spread the word and stop it, otherwise it is never going to be stopped. Now people believe in this society that talking about this issue is wrong or Haram [prohibited by religion]. I don't care; I will talk to my children because I don't want them to grow up blindly. (W18)

Men were regarded as important actors of change, but women thought that they lacked knowledge about the physical and mental health consequence of FGM. Women felt that men believe that FGM is women's business and that their views on cultural obligation enabled the continuation of FGM. The women, therefore, perceived that men's involvement as a crucial part of the solution to end this practice but it might be very challenging as men are not interested in taking part in such a movement. An example such as this was given:

At the moment most of the trainings are for women. We need men to talk to men so we can engage them otherwise you cannot force them to sit in a class. You need to train more men to open up and talk about this issue with other men in the community and engage them at the same level as women. Men are still looking at it as a good thing. (W15)

Women described feeling empowered when they shared their stories and regarded these as an important resource for mutual support and to educate the community and challenge cultural beliefs about FGM. Women also mentioned that they feared being socially ostracised by their families and communities if they expressed dissenting views. This woman explained:

... We need to create an environment where people talk about it. You know it is very hard to disclose such issues at community level, as it is a very private matter. I guess if we bring up stories and how women are suffering this would be effective to change this culture in the future. Imagine you're living for someone else's pleasure and you're getting none. (W13)

Women considered government support as a cross-cutting issue linked to all future actions and approaches. Women used the word government to mean all high-level decisions, policy and funding at local, state and territory and federal levels. They wanted resources for improving the health of affected women, introducing FGM as a topic in the school curriculum and making meaningful linkages with communities. They believed such strategies would ultimately lead to the improvement of the health of women with FGM and society as a whole.

They [policy makers] need to identify women with FGM as a priority at policy level and provide them with things they want. We want services which all women deserve . . . . We are in a developed country and we should have access to standard care from an experienced health provider. (W21)

Women also spoke of the need for mental health support and counselling services, both at facility and community level, for example:

Make sure they [women affected with FGM] are OK, mentally and physically. Do the follow up afterwards. Education and individualised support not only for women who have undergone FGM but also to train staff and the community. It goes both ways. (W16)

Women pointed to cultural taboos that make it challenging to have open discussions about FGM with male members of the family. Several women made suggestions like this:

Facilitating and funding community training such as workshops for men and women we can raise the awareness. It is also helpful to open the discussion around this issue. At the moment it is not culturally appropriate to even talk about it even in the family. (W20)

4. Discussion

This research identified the maternity care experiences of women affected by FGM and their views concerning the care they wished to receive in the future and how this might be achieved. Women in this study acknowledged that the maternity care they received had not always been at the level of quality that they desired or had expected. Women reported that being meaningfully involved in their care design and delivery was a crucial strategy for building trust and improving and validating the quality of maternity services. It has previously been shown that women who are well educated and have adequate information about FGM are more likely to have control over health care, access to shared decision making. Making an informed choice is key to respectful care for women with FGM [41] and they are less likely to perform FGM for their daughter [42]. While most women were motivated to be involved in their care, they struggled with poor communication and a lack of information tailored to their individual needs as reported elsewhere [43,44]. Women wanted to be cared for by skilled and culturally competent providers who treated them as 'special' but also as normal and equal to 'other women'. This has been described by other research where they ensured equality by including Aboriginal and Torres Strait Islander midwives, who can interact holistically and provide culturally sensitive services [45]. Finally, women described the importance of having access to evidence-based models of care such as midwifery continuity of care and available services including, reconstructive surgery, management of trauma, emotional support, psychotherapy services and cultural support.

A conceptual framework (Figure 2) was developed based on the findings of this research that highlights four priority approaches required to achieve quality care for women with FGM: co-production, woman-centred care, equity and equality and evidence-based models of care. These approaches are underpinned by four strategies that facilitate women's engagement and include involvement in developing health information to being an equal partner in decision making and the co-design of maternity services.

Figure 2. Conceptual model of quality improvement within maternity services for women with female genital mutilation (FGM).

4.1. Co-Design of Health Literacy Interventions

Women regularly described the need for information that is tailored to their individual needs and noted that support services, such as counselling, were not always accessible due to language and cultural barriers. Some women stated that these services were not available or integrated into maternity care. This is similar to other studies in high-income countries that have found that women affected by FGM do not always receive or understand the information and resources they required or needed because of social isolation, stigma and a lack of health literacy [46]. When women have lower levels of health literacy, they are less prepared to engage and comply with their care regimes or protocols and as a result, do not receive optimum care [47]. Improving the health literacy of women with FGM may change the attitudes of women towards their own FGM and reduce the likelihood of their daughter's being circumcised [48].

Every woman should feel empowered to build her capacity and skills to use health information effectively and make an informed choice [49]. Many women in our study stated that they were not adequately engaged in their health care because of low levels of health literacy, inadequate information and unfamiliarity with their health rights. Again, these findings concur with other studies [50,51] and confirm that women's participation in the process of health information design leads to more satisfying and positive experiences with enhanced health outcomes [52]. Health literacy programs that involve women designing and delivering programs not only build the capacity of women to facilitate the sharing of stories and experiences but also empowers women to support others in their community [53]. Such approaches are likely to be useful for women affected by FGM.

4.2. Co-Design of Evidence-Based Models of Care

Most women in our study reported different types of FGM-related trauma, which affected their overall quality of life. Women expected health care providers to be responsive to their psychological, emotional and socio-cultural needs as found in other studies [54]. The central philosophy that underpins high-quality maternity care does not only involve a focus on physical health but also emotional well-being and includes quality of life issues [55]. Despite the emotional and mental consequences of FGM, most studies are focused on the physical aspects and implications [56,57]. Laio et al. [58] indicated that women affected by FGM are often silent about their emotional problems due to the stigma associated with FGM and have difficulty communicating with health providers. It is difficult for care providers to recognise or determine the level of psychological trauma that may be caused by FGM, but our study highlights the importance of these considerations.

FGM related trauma is important to note because it can negatively impact on childbirth and sexual relationships highlighting the need for individualised trauma-informed interventions for such vulnerable women. FGM related mental health issues such as PTSD, negative body image and feelings of shame and stigma may also affect women's health-seeking behaviour [59]. A trauma-informed model of care may be an approach to providing safe supportive care to women who have been affected by violence to reduce the consequences of trauma in their life [60].

Women should also be involved in the design of such trauma-informed services so that individual needs, views and experiences can be addressed in a collaborative way [61]. Efforts in the area of trauma-informed care currently focus on strengthening health provider's knowledge and skills based on their interactions with consumers, rather than understanding a women's experiences and needs [62,63]. Implementing participatory interventions, however, requires both the health system and community change [64]. Women need to be supported to become empowered to recognise their potential and utilise their capacity in the design and delivery of services [65]. Creating an environment of collaboration and mutual trust by engaging women and acknowledging their values and lived experiences may ensure that women's needs are understood and their views and culture are taken into account in service design, thereby, improving the quality of culturally safe care.

Correa-Velez and Ryan [66] emphasise the need for specific models of maternity care for marginalised and high-risk women, such as women with FGM, that encompass continuity of care plus educational interventions and the delivery of mental health support. Our study indicates that continuity of care can lead to improved interpersonal communication and can boost women's confidence, the collaboration between a woman and her provider and help facilitate women engagement in the process of care design and delivery [67]. Midwifery continuity of care enables health providers to consider the socio-cultural and emotional needs of marginalised women and, therefore, empowers women to achieve positive outcomes [68]. Such care models ensure the continuous assessment and evaluation of women's experiences, opinions and views that can improve the quality of care for marginalised groups [69].

4.3. Co-Design Approaches to Shared Decision Making

The health system must offer women adequate support to enable them to be empowered to communicate, to ask for help and to question their care [60,70]. Patient participation in the process of service design and delivery is often missing as patients are perceived not to have adequate medical and clinical knowledge [71]. A review of the literature found that consumer involvement in the training of health providers ensures that the health system reflects their needs and desires in the design and delivery of services [72]. Collaborative partnerships have been found to have a positive impact on nursing practice by improving communication and shared decision making [73]. Another example from the field of mental health demonstrates the benefits of sharing the experiences and insights of patients through story-telling and using different aspects of personal experience in the development of a mental health assessment tool [74]. There is limited evidence in maternal health research and further research is needed to determine the best approach to engage women and evaluate the impact of their involvement in the co-design of education and training material, guidelines and health service processes.

4.4. Co-Design of Health Professional Education and Training

Women in our study described the need for health providers to receive special training on the cultural aspects of care for women from diverse backgrounds. This would help to address their need for a model of maternity care that integrates a woman's cultural and individual values with excellent communication and referral paths to promote their well-being and safety as described in other research [75]. The involvement of women in teaching health professionals may be a useful strategy to increase the knowledge of clinicians. One study that investigated the outcomes of learning where consumers delivered classes found that nursing students improved their cultural knowledge and understanding of empathic care [76]. The involvement of mental health consumers in the education of nurses also showed improvements in nurses' communication skills and decreased cultural barriers for consumers as well as reduced discrimination [77]. The integration of cultural safety in practice is challenging as it requires the involvement of service users in the co-design of such services and involving a vulnerable population requires a paradigm shift in power differences between service users and health professionals [51]. Future health services need to be co-produced with women to disrupt the inherent power imbalances.

This study is one of the first of its kind in Australia to analyse this group of women's views and experiences of their maternity care. The use of AI as the methodology was unique and enabled women to focus on their positive experiences and come up with solutions for future action and changes within the health system. The study has highlighted the voices of women providing important knowledge to improve the quality of maternity care for marginalised women.

This study included only women who lived in Sydney, which is generally well resourced in terms of services for migrant populations. Therefore, the results may not be generalisable to the other states across Australia and suggested solutions and recommendations might be specific to the local context.

Sampling bias is a possible limitation. Potential women were recruited through chain referral sampling. Therefore, those who decided to participate in this study might be those who had more interest in this subject area, and this might have led the discussion either more positively or negatively.

5. Conclusions

The engagement of individuals and communities is critical to the process of improving the quality of maternity health services and to address the socio-cultural needs of women affected by FGM. Empowering women and raising their awareness of their health care rights can help to engage women as active partners in the design and delivery of health information, models of care approach to shared decision making and health professional education and training which is based on their needs and context.

Further research is needed to explore the replicability of the suggested framework at policy and practice levels. Research is required to establish the feasibility of the co-production of maternity services and how this improves the quality of care and equitable health outcomes for women affected by FGM.

Supplementary Materials: The following are available online at http://www.mdpi.com/1660-4601/17/5/1491/s1.

Author Contributions: Conceptualization, S.T., A.J.D. and C.S.E.H.; methodology, S.T. and A.J.D.; software, S.T.; validation, S.T., A.J.D. and C.S.E.H.; formal analysis, S.T.; investigation, S.T.; resources, A.J.D. and C.S.E.H.; data curation, S.T., A.J.D. and C.S.E.H.; writing—original draft preparation, S.T.; writing—review and editing, A.J.D. and C.S.E.H.; supervision, A.J.D. and C.S.E.H.; project administration, S.T. All authors have read and agreed to the published version of the manuscript.

Acknowledgments: This research was supported by the Australian Government Research Training Program (RTP). The authors are thankful for the support provided by the Faculty of Health, University of Technology Sydney (UTS). We also thank women who generously shared their perspectives and ideas and this study would not have been possible without them.

Conflicts of Interest: The authors declare no conflict of interest.

References

1. UN. *Eliminating Female Genital Mutilation: A Joint Interagency Statement*; UNAIDS; UNDP; UNECA; UNESCO; UNFPA; UNHCHR; UNICEF; UNIFEM; WHO: Geneva, Switzerland, 2008.

2. UNICEF. *The Dynamics of Social Change-Towards the Abandonment of Female Genital Mutilation/Cutting in Five African Countries*; UNICEF: Florence, Italy, 2010.

3. UNICEF. *Female Genital Mutilation/Cutting: A Global Concern*; UNICEF: New York, NY, USA, 2015.

4. Korfker, D.G.; Reis, R.; Rijnders, M.E.; Meijer-van Asperen, S.; Read, L.; Sanjuan, M.; Herschderfer, K.; Buitendijk, S.E. The lower prevalence of female genital mutilation in the Netherlands: A nationwide study in Dutch midwifery practices. *Int. J. Public Health* **2012**, *57*, 413–420. [CrossRef] [PubMed]

5. Stockdale, J.; Fyle, J. *Royal College of Midwives—Female Genital Mutilation: Report of a Survey on Midwives Views and Knowledge*; The Royal College of Midwives: London, UK, 2012.

6. WHO. WHO Guidelines on the Management of Health Complications from Female Genital Mutilation. 2016. Available online: Http://apps.who.int/iris/bitstream/10665/206437/1/9789241549646_eng.pdf?ua=1 (accessed on 17 October 2018).

7. Blignault, I.; Haghshenas, A. Identification of Australians from culturally and linguistically diverse backgrounds in national health data collections. *Aust. Health Rev.* **2005**, *29*, 455–468. [CrossRef] [PubMed]

8. Teixeira, A.L.; Lisboa, M. Estimating the Prevalence of Female Genital Mutilation in Portugal. *Public Health* **2016**, *139*, 53–60. [CrossRef] [PubMed]

9. Ziyada, M.M.; Norberg-Schulz, M.; Johansen, R.E.B. Estimating the magnitude of female genital mutilation/cutting in Norway: An extrapolation model. *BMC Public Health* **2016**, *16*, 110. [CrossRef]

10. Australian Institute of Health and Welfare. *Towards Estimating the Prevalence of Female Genital Mutilation/Cutting in Australia*; AIHW: Canberra, Australia, 2019.

11. Hulton, L.A.; Matthews, Z.; Stones, R.W. *A framework for the Evaluation of Quality of Care in Maternity Services*; University of Southampton: Southampton, UK, 2000.

12. Van den Broek, N.; Graham, W. Quality of care for maternal and newborn health: The neglected agenda. *BJOG Int. J. Obstet. Gynaecol.* **2009**, *116* (Suppl. 1), 18–21. [CrossRef]

13. Burnett, A.; Ndovi, T. The health of forced migrants. *BMJ* **2018**, *363*, k4200. [CrossRef]

14. Varol, N.; Hall, J.J.; Black, K.; Turkmani, S.; Dawson, A. Evidence-based policy responses to strengthen health, community and legislative systems that care for women in Australia with female genital mutilation/cutting. *Reproduct. Health* **2017**, *14*, 63. [CrossRef]

15. Gibson-Helm, M.; Teede, H.; Block, A.; Knight, M.; East, C.; Wallace, E.M.; Boyle, J. Maternal health and pregnancy outcomes among women of refugee background from African countries: A retrospective, observational study in Australia. *BMC Pregnancy Childbirth* **2014**, *14*, 392–403. [CrossRef]

16. Belihu, F.B.; Davey, M.-A.; Small, R. Perinatal health outcomes of East African immigrant populations in Victoria, Australia: A population based study. *BMC Pregnancy Childbirth* **2016**, *16*, 86. [CrossRef]

17. Dawson, A.J.; Turkmani, S.; Varol, N.; Nanayakkara, S.; Sullivan, E.; Homer, C.S.E. Midwives' experiences of caring for women with female genital mutilation: Insights and ways forward for practice in Australia. *Women Birth* **2015**, *28*, 207–214. [CrossRef]

18. Dawson, A.; Homer, C.S.; Turkmani, S.; Black, K.; Varol, N. A systematic review of doctors' experiences and needs to support the care of women with female genital mutilation. *Int. J. Gynecol. Obstet.* **2015**, *131*, 35–40. [CrossRef]

19. Dixon, S.; Agha, K.; Ali, F.; El-Hindi, L.; Kelly, B.; Locock, L.; Otoo-Oyortey, N.; Penny, S.; Plugge, E.; Hinton, L. Female genital mutilation in the UK—Where are we, where do we go next? Involving communities in setting the research agenda. *Res. Involv. Engagem.* **2018**, *4*, 1–8. [CrossRef] [PubMed]

20. Connelly, E.; Murray, N.; Baillot, H.; Howard, N. Missing from the debate? A qualitative study exploring the role of communities within interventions to address female genital mutilation in Europe. *BMJ Open* **2018**, *8*, e021430. [CrossRef] [PubMed]

21. Ogunsiji, O. Female Genital Mutilation (FGM): Australian Midwives' Knowledge and Attitudes. *Health Care Women Int.* **2014**, *36*, 179–193. [CrossRef] [PubMed]

22. WHO. Standards for Improving Quality of Maternal and Newborn Care in Health Facilities. Available online: http://apps.who.int/iris/bitstream/10665/249155/1/9789241511216-eng.pdf?ua=1 (accessed on 16 December 2018).

23. Kruk, M.E.; Gage, A.D.; Arsenault, C.; Jordan, K.; Leslie, H.H.; Roder-DeWan, S.; Adeyi, O.; Barker, P.; Daelmans, B.; Doubova, S.V.; et al. High-quality health systems in the Sustainable Development Goals era: Time for a revolution. *Lancet Glob. Health* **2018**, *6*, e1196–e1252. [CrossRef]

24. Bazely, P. *Qualitative Data Analysis: Practical Strategies*; Sage: London, UK, 2013.

25. Trajkovski, S.; Schmied, V.; Vickers, M.; Jackson, D. Using appreciative inquiry to transform health care. *Contemp. Nurse* **2013**, *45*, 95–100. [CrossRef]

26. Chenail, R.J. How to conduct clinical qualitative research on the patient's experience. *Qual. Rep.* **2011**, *16*, 1173–1189.

27. Carter, C.A.; Ruhe, M.C.; Weyer, S.; Litaker, D.; Fry, R.E.; Stange, K.C. An appreciative inquiry approach to practice improvement and transformative change in health care settings. *Qual. Manag. Healthc.* **2007**, *16*, 194–204. [CrossRef]

28. McAdam, E.; Mirza, K.A.H. Drugs, hopes and dreams: Appreciative inquiry with marginalized young people using drugs and alcohol. *J. Fam. Ther.* **2009**, *31*, 175–193. [CrossRef]

29. Michael, S. The promise of appreciative inquiry as an interview tool for field research. *Dev. Pract.* **2005**, *15*, 222–230. [CrossRef]

30. Reed, J. *Appreciative Inquiry*; Sage Publications: Thousand Oaks, CA, USA, 2006.

31. Edwin, T.C. *Appreciative Inquiry: A Positive Approach to Change*; Institute for Public Service and Policy Research, University of South Carolina: Columbia, SC, USA, 2013.

32. Berggren, V.; Bergstrom, S.; Edberg, A.-K. Being different and vulnerable: Experiences of immigrant African women who have been circumcised and sought maternity care in Sweden. *J. Transcult. Nurs.* **2006**, *17*, 50–57. [CrossRef] [PubMed]

33. Centre for Epidemiology and Evidence. *New South Wales Mothers and Babies*; Ministry of Health: Sydney, NSW, Australia, 2017.

34. Heckathorn, D.D. Respondent-driven sampling: A new approach to the study of hidden populations. *Soc. Probl.* **1997**, *44*, 174–199. [CrossRef]

35. Penrod, J.; Preston, D.B.; Cain, R.E.; Starks, M.T. A discussion of chain referral as a method of sampling hard-to-reach populations. *J. Transcult. Nurs.* **2003**, *14*, 100–107. [CrossRef] [PubMed]

36. Gill, P.; Stewart, K.; Treasure, E.; Chadwick, B. Methods of data collection in qualitative research: Interviews and focus groups. *Br. Dent. J.* **2008**, *204*, 291–295. [CrossRef]
37. Braun, V.; Clarke, V. Using thematic analysis in psychology. *Qual. Res. Psychol.* **2006**, *3*, 77–101. [CrossRef]
38. Stuckey, H. The first step in Data Analysis: Transcribing and managing qualitative research data. *J. Soc. Health Diabet.* **2014**, *2*, 6–8. [CrossRef]
39. Creswell, J.W. *Research Design: Qualitative, Quantitative, and Mixed Methods Approaches*, 2nd ed.; Sage Publications: Thousand Oaks, CA, USA, 2009.
40. Woods, L. Qualitative Data Analysis—Explorations with NVivo. *Nurse Res.* **2002**, *9*, 86–87. [CrossRef]
41. WHO. *Care of Girls and Women Living with Female Genital Mutilation: A Clinical Handbook*; World Health Organization: Geneva, Switzerland, 2018.
42. Van Rossem, R.; Meekers, D.; Gage, A.J. Women's position and attitudes towards female genital mutilation in Egypt: A secondary analysis of the Egypt demographic and health surveys, 1995–2014. *BMC Public Health* **2015**, *15*, 1. [CrossRef]
43. Carolan, M.; Cassar, L. Antenatal care perceptions of pregnant African women attending maternity services in Melbourne, Australia. *Midwifery* **2010**, *26*, 189–201. [CrossRef]
44. Lundberg, P.C.; Gerezgiher, A. Experiences from pregnancy and childbirth related to female genital mutilation among Eritrean immigrant women in Sweden. *Midwifery* **2008**, *24*, 214–225. [CrossRef]
45. Homer, C.S.; Foureur, M.J.; Allende, T.; Pekin, F.; Caplice, S.; Catling-Paull, C. 'It's more than just having a baby' women's experiences of a maternity service for Australian Aboriginal and Torres Strait Islander families. *Midwifery* **2012**, *28*, e509–e515. [CrossRef] [PubMed]
46. Turkmani, S.; Homer, C.S.E.; Dawson, A. Maternity care experiences and health needs of migrant women from female genital mutilation–practicing countries in high-income contexts: A systematic review and meta-synthesis. *Birth* **2019**, *46*, 3–14. [CrossRef]
47. Gustavsson, S.; Gremyr, I.; Kenne, S.E. Designing quality of care—Contributions from parents; Parents' experiences of care processes in paediatric care and their contribution to improvements of the care process in collaboration with healthcare professionals. *J. Clin. Nurs.* **2015**, *25*, 742–751. [CrossRef]
48. Alo, O.A.; Gbadebo, B. Intergenerational attitude changes regarding female genital cutting in Nigeria. *J. Women Health* **2011**, *20*, 1655–1661. [CrossRef] [PubMed]
49. Asekun-Olarinmoye, E.O.; Amusan, O.A. The impact of health education on attitudes towards female genital mutilation (FGM) in a rural Nigerian community. *Eur. J. Contracept. Reproduct. Health Care* **2008**, *13*, 289–297. [CrossRef]
50. Origlia, P.; Jevitt, C.; Sayn-Wittgenstein, F.Z.; Cignacco, E. Experiences of Antenatal Care Among Women Who Are Socioeconomically Deprived in High-Income Industrialized Countries: An Integrative Review. *J. Midwifery Women Health* **2017**, *62*, 589–598. [CrossRef] [PubMed]
51. Ebert, L.; Bellchambers, H.; Ferguson, A.; Browne, J. Socially disadvantaged women's views of barriers to feeling safe to engage in decision-making in maternity care. *Women Birth* **2014**, *27*, 132–137. [CrossRef]
52. Svensson, P.; Carlzén, K.; Agardh, A. Exposure to culturally sensitive sexual health information and impact on health literacy: A qualitative study among newly arrived refugee women in Sweden. *Cult. Health Sex.* **2017**, *19*, 752–766. [CrossRef]
53. Manderson, L.; Allotey, P. Storytelling, marginality, and community in Australia: How immigrants position their difference in health care settings. *Med. Anthropol.* **2003**, *22*, 1–21. [CrossRef]
54. Mattern, E.; Lohmann, S.; Ayerle, G.M. Experiences and wishes of women regarding systemic aspects of midwifery care in Germany: A qualitative study with focus groups. *BMC Pregnancy Childbirth* **2017**, *17*, 389. [CrossRef]
55. Renfrew, M.J.; McFadden, A.; Bastos, M.H.; Campbell, J.; Channon, A.A.; Cheung, N.F.; Silva, D.R.A.D.; Downe, S.; Kennedy, H.P.; Malata, A.; et al. Midwifery and quality care: Findings from a new evidence-informed framework for maternal and newborn care. *Lancet* **2014**, *384*, 1129–1145. [CrossRef]
56. Andro, A.; Cambois, E.; Lesclingand, M. Long-term consequences of female genital mutilation in a European context: Self perceived health of FGM women compared to non-FGM women. *Soc. Sc. Med.* **2014**, *106*, 177–184. [CrossRef]
57. Vloeberghs, E.; Van der Kwaak, A.; Knipscheer, J.; van den Muijsenbergh, M. Coping and chronic psychosocial consequences of female genital mutilation in the Netherlands. *Ethn. Health* **2012**, *17*, 677–695. [CrossRef] [PubMed]

58. Liao, L.M.; Elliott, C.; Ahmed, F.; Creighton, S.M. Adult recall of childhood female genital cutting and perceptions of its effects: A pilot study for service improvement and research feasibility. *J. Obstet. Gynaecol.* **2013**, *33*, 292–295. [CrossRef]

59. Caillet, M.; O'Neill, S.; Minsart, A.F.; Richard, F. Addressing FGM with Multidisciplinary Care. The Experience of the Belgian Reference Center CeMAViE. *Curr. Sex. Health Rep.* **2018**, *10*, 44–49. [CrossRef]

60. Shaw, D.; Guise, J.M.; Shah, N.; Gemzell-Danielsson, K.; Joseph, K.S.; Levy, B.; Wong, F.; Woodd, S.; Main, E.K. Drivers of maternity care in high-income countries: Can health systems support woman-centred care? *Lancet* **2016**, *388*, 2282–2295. [CrossRef]

61. Asma, H.; Trister, G.K.; Nicole, W. A Meta-Synthesis of the Birth Experiences of African Immigrant Women Affected by Female Genital Cutting. *J. Midwifery Women Health* **2018**, *63*, 185–195.

62. Wilson, J.M.; Fauci, J.E.; Goodman, L.A. Bringing trauma-informed practice to domestic violence programs: A qualitative analysis of current approaches. *Am. J. Orthopsychiatry* **2015**, *85*, 586–599. [CrossRef] [PubMed]

63. Kulkarni, S. Intersectional Trauma-Informed Intimate Partner Violence (IPV) Services: Narrowing the Gap between IPV Service Delivery and Survivor Needs. *J. Fam. Violence* **2019**, *34*, 55–64. [CrossRef]

64. Mihelicova, M.; Brown, M.; Shuman, V. Trauma-Informed Care for Individuals with Serious Mental Illness: An Avenue for Community Psychology's Involvement in Community Mental Health. *Am. J. Orthopsychiatry* **2018**, *61*, 141–152. [CrossRef]

65. Castro, E.M.; Malfait, S.; Van Regenmortel, T.; Van Hecke, A.; Sermeus, W.; Vanhaecht, 5. Co-design for implementing patient participation in hospital services: A discussion paper. *Patient Educ. Couns.* **2018**, *101*, 1302–1305. [CrossRef] [PubMed]

66. Correa-Velez, I.; Ryan, J. Developing a best practice model of refugee maternity care. *Women Birth* **2012**, *25*, 13–22. [CrossRef] [PubMed]

67. Dahlberg, U.; Aune, I. The woman's birth experience—The effect of interpersonal relationships and continuity of care. *Midwifery* **2013**, *29*, 407–415. [CrossRef] [PubMed]

68. WHO. *Recommendations: Intrapartum Care for a Positive Childbirth Experience*; World Health Organization: Geneva, Switzerland, 2018.

69. Hackett, C.L.; Mulvale, G.; Miatello, A. Co-designing for quality: Creating a user-driven tool to improve quality in youth mental health services. *Health Expect.* **2018**, *21*, 1013–1023. [CrossRef] [PubMed]

70. Higginbottom, G.M.A.; Hadziabdic, E.; Yohani, S.; Paton, P. Immigrant women's experience of maternity services in Canada: A meta-ethnography. *Midwifery* **2014**, *30*, 544–559. [CrossRef]

71. Longtin, Y.; Sax, H.; Leape, L.L.; Sheridan, S.E.; Donaldson, L.; Pittet, D. Patient participation: Current knowledge and applicability to patient safety. *Mayo Clin. Proc.* **2010**, *85*, 53–62. [CrossRef]

72. Repper, J.; Breeze, J. User and carer involvement in the training and education of health professionals: A review of the literature. *Int. J. Nurs. Stud.* **2007**, *44*, 511–519. [CrossRef]

73. Kieft, R.A.; de Brouwer, B.B.; Francke, A.L.; Delnoij, D.M. How nurses and their work environment affect patient experiences of the quality of care: A qualitative study. *BMC Health Serv. Res.* **2014**, *14*, 249–258. [CrossRef]

74. Coupland, K.; Davis, E.; Gregory, K. Learning from life. *Mental Health Care* **2001**, *4*, 166–169.

75. Beckett, K.; Earthy, S.; Sleney, J.; Barnes, J.; Kellezi, B.; Barker, M.; Clarkson, J.; Coffey, F.; Elder, G.; Kendrick, D.; et al. Providing effective trauma care: The potential for service provider views to enhance the quality of care (qualitative study nested within a multicentre longitudinal quantitative study). *BMJ Open* **2014**, *4*, e005668. [CrossRef]

76. Stickley, T.; Rush, B.; Shaw, R.; Smith, A.; Collier, R.; Cook, J.; Shaw, T.; Gow, D.; Felton, A.; Roberts, S. Participation in nurse education: The PINE project. *J. Mental Health Train. Educ. Pract.* **2009**, *4*, 11–18. [CrossRef]

77. Simpson, A.; Reynolds, L.; Light, I.; Attenborough, J. Talking with the experts: Evaluation of an online discussion forum involving mental health service users in the education of mental health nursing students. *Nurse Educ. Today* **2008**, *28*, 633–640. [CrossRef] [PubMed]

International Journal of
***Environmental Research
and Public Health***

Review

Prevalence of Intimate Partner Violence in Pregnancy: An Umbrella Review

Rosario M. Román-Gálvez [1,2], **Sandra Martín-Peláez** [3,4,*], **Juan Miguel Martínez-Galiano** [5,6], **Khalid Saeed Khan** [3,6] and **Aurora Bueno-Cavanillas** [3,4,6]

[1] Departamento de Enfermería, Facultad de Ciencias de la Salud, Universidad de Granada, 18071 Granada, Spain; galvezmaro@gmail.com
[2] Unidad Asistencial Churriana de la Vega, Servicio Andaluz de Salud, Churriana de la Vega, 18194 Granada, Spain
[3] Departamento de Medicina Preventiva y Salud Pública, Facultad de Medicina, Universidad de Granada, 18071 Granada, Spain; profkkhan@ugr.es (K.S.K.); abueno@ugr.es (A.B.-C.)
[4] Instituto de Investigación Biosanitaria de Granada IBS, 18014 Granada, Spain
[5] Department of Nursing, University of Jaén, 23071 Jaén, Spain; jgaliano@ujaen.es
[6] Consortium for Biomedical Research in Epidemiology and Public Health (CIBERESP), 28029 Madrid, Spain
* Correspondence: sandramartin@ugr.es; Tel.: +34-958-24-20-57

Received: 1 December 2020; Accepted: 13 January 2021; Published: 15 January 2021

Abstract: Background: Intimate partner violence (IPV) is a public health concern, especially during pregnancy, and needs to be urgently addressed. In order to establish effective actions for the prevention of IPV during pregnancy, authorities must be aware of the real burden of IPV. This review aimed to summarize the existing evidence about IPV prevalence during pregnancy worldwide. Methods: A review of reviews was carried out. All published systematic reviews and meta-analyses published until October 2020 were identified through PubMed, Scopus, and Web of Science. The main outcome was the IPV prevalence during pregnancy. Results: A total of 12 systematic reviews were included in the review, 5 of them including meta-analysis. The quality of the reviews was variable. Physical IPV during pregnancy showed a wide range (1.6–78%), as did psychological IPV (1.8–67.4%). Conclusions: Available data about IPV prevalence during pregnancy were of low quality and showed high figures for physical and psychological IPV. The existing evidence syntheses do not capture the totality of the worldwide disease burden of IPV in pregnancy.

Keywords: intimate partner violence; pregnancy; prevalence; umbrella review

---

1. Introduction

Intimate partner violence (IPV), defined as physical violence, sexual violence, harassment, and psychological assault (including coercive tactics) by a current or former intimate partner [1], is a public health concern that urgently needs to be addressed. During pregnancy, the woman experiences a situation of special dependence, both physical and emotional. In this period, exposure to violence affects not only the mother but also the fetus, which is at greater risk than in other stages of life [2]. In fact, IPV has been associated with adverse pregnancy outcomes including increased risk of human immunodeficiency virus infection [3], perinatal depression [4], insufficient weight gain during pregnancy [5], uterine rupture, hemorrhage, maternal death [6], prematurity, low birth weight, newborns small for gestational age [7], stillbirth [8], and reduced levels of breastfeeding [9]. At the same time, routine contacts with the health system offer an excellent detection window to identify it and establish protective measures. Despite this, IPV during pregnancy is a neglected condition, even though it is more common than many maternal health conditions like preeclampsia and gestational diabetes [10].

IPV during pregnancy should be an avoidable global public health problem. However, in order to establish effective actions for the prevention of IPV during pregnancy, such as the performance of systematic screening and diagnosis of IPV in the antenatal visits, authorities must be aware of the real worldwide burden of IPV. However, information about IPV prevalence is not consistent. Whereas some studies indicate higher prevalence of IPV during pregnancy than before [11] or after [12,13] the pregnancy, other studies report a smaller prevalence [14,15]. Furthermore, the prevalence of IPV during pregnancy is reported to vary depending on the definition used [1], the screening strategy [16,17], and the development status of the population studied [10,18]. These factors make comparison between individually reported rates difficult.

This review aims to summarize the existing evidence about IPV prevalence during pregnancy worldwide through a synthesis of systematic reviews and meta-analysis. Prevalence studies provide a snapshot of a situation in a specific context, so it is important to bring together different existing studies for a global understanding. This work analyzes the existing reviews, identifying their strengths and limitations and laying the foundations for future reviews that clarify the situation of IPV during pregnancy in a complete and realistic way.

2. Materials and Methods

This umbrella review was written according to the Preferred Reporting Items for Systematic Reviews and Meta-Analyses (PRISMA) statement for systematic reviews and meta-analyses [19] and the Aromataris' guidelines for performing umbrella reviews [20].

2.1. Inclusion and Exclusion Criteria

Systematic reviews and meta-analyses including observational studies reporting IVP prevalence suffered by women during pregnancy were considered in this review. The types of IPV were classified as physical, sexual, psychological and any type of IPV. Studies not following a systematic review approach, narrative reviews, and primary studies were excluded. No language restrictions were applied in this review.

2.2. Literature Search and Selection of Studies

Relevant systematic reviews and meta-analyses according to inclusion criteria were identified through systematic searches of the following electronic databases: PubMed, Web of Science and Scopus, CYNAHL, PsycINFO, Social Science Database, and Sociological Abstracts. The full search strings can be found in Table 1. When the search engine used only allowed selecting systematic reviews or meta-analyses, the terms "systematic reviews" OR "meta-analyses" were not included in the search string; otherwise, those terms were included. Studies published from inception until October 2020 were included. Reference lists of identified studies were checked.

Table 1. Search strings.

Database	Searching String
PUBMED	("Intimate Partner Violence" [Mesh]) AND (("Pregnancy" [Mesh]) OR ("Pregnant Women" [Mesh]) OR ("Prenatal Care" [Mesh])) AND ("Prevalence" [Mesh])
Rest of databases	("Intimate Partner Violence") AND (("Pregnancy") OR ("Pregnant Women") OR ("Prenatal Care")) AND ("Prevalence")

2.3. Data Collection and Analysis

Eligible studies were selected through a multistep approach (elimination of duplicates, title reading, abstract and full-text assessment). Two researchers (S.M.-P. and R.M.R.-G.) independently examined titles and abstracts, evaluating afterwards full texts according to the inclusion criteria described above. Any disagreement between the reviewers was resolved by means of a consensus session with a third reviewer (A.B.-C.). In case of ambiguity in reporting or lack of data, primary authors were contacted for clarification.

2.4. Data Extraction and Management

Data were independently extracted by two researchers (S.M.-P. and R.M.R.-G.), and the following information was considered for each article: (1) first author and year of publication; (2) interval of time covered by the review; (3) countries (of studies included in the systematic review or meta-analysis); (4) number of studies included; (5) study design (of studies included in the systematic review or meta-analysis); (6) sample characteristics; (7) IPV as main outcome; (8) type of IPV investigated; (9) meta-analysis performance; (10) IPV outcome.

2.5. Quality Assessment Tools

The updated AMSTAR 2 version for systematic reviews and meta-analyses was used to evaluate the methodological quality and risk of bias of studies included in the systematic review [21]. The overall final rating of each systematic review was judged as high, moderate, low, or critically low. In case of disagreements, a consensus session with the third reviewer (A.B.-C.) was held.

3. Results

The electronic search initially resulted in 199 citations. A total of 80 studies were excluded after elimination of duplicates. From the 119 remaining, 61 were excluded after title and abstract screening and 58 full-text articles were selected and read. From those, a total of 12 systematic reviews were included in this umbrella review, of which 5 were meta-analyses. The reasons for exclusion were the lack of data about IPV prevalence during pregnancy (20), the use of violence other than IPV or the indistinct report of IPV or domestic violence (10), investigations on populations with a specific risk, supposedly different from the general population (12), and not being a systematic review or meta-analysis (4). The list of the excluded articles is presented in Supplementary Materials Table S1. Figure 1 shows the PRISMA flowchart and the study selection process.

Characteristics of included systematic reviews and meta-analyses are shown in Table 2. Only two reviews included global data [22,23], most of which were limited to a country or a group of countries, mainly from Asia [4,23–28] and Africa [4,26,27,29,30], followed by America [31,32], Europe [32], and Australia [32]. The number of studies included in the reviews giving information about IPV prevalence during pregnancy ranged from 2 [24] to 73 [23].

Most of the studies included in the selected reviews were cross-over studies, in which there was only a single evaluation of the women sometime during pregnancy. Some of the reviews also included cohort studies [4,24,25,30–32]. Reviews included studies giving information about IPV only at pregnancy [22–24, 27,29,30], both during pregnancy or at postpartum [4,30–32], during pregnancy, or having a child 2 years old or younger [25] or at current pregnancy or any pregnancy [26].

Wide differences were also observed regarding the type of IPV violence investigated. Of the selected reviews, nine investigated prevalence of physical violence [4,22,24–28,30,31], nine psychological violence [4, 22,24–28,30,31], ten sexual violence [4,22–28,30,31], and three any type of violence [26,29,32]. From the selected studies, four did not report IPV pregnancy during pregnancy as main outcome [24–26,32]. Five studies showed a summarized estimate of IPV during pregnancy [23,24,28–30]. In most of the reviews ranges of IPV prevalence are given [4,22,24,26,27,31,32].

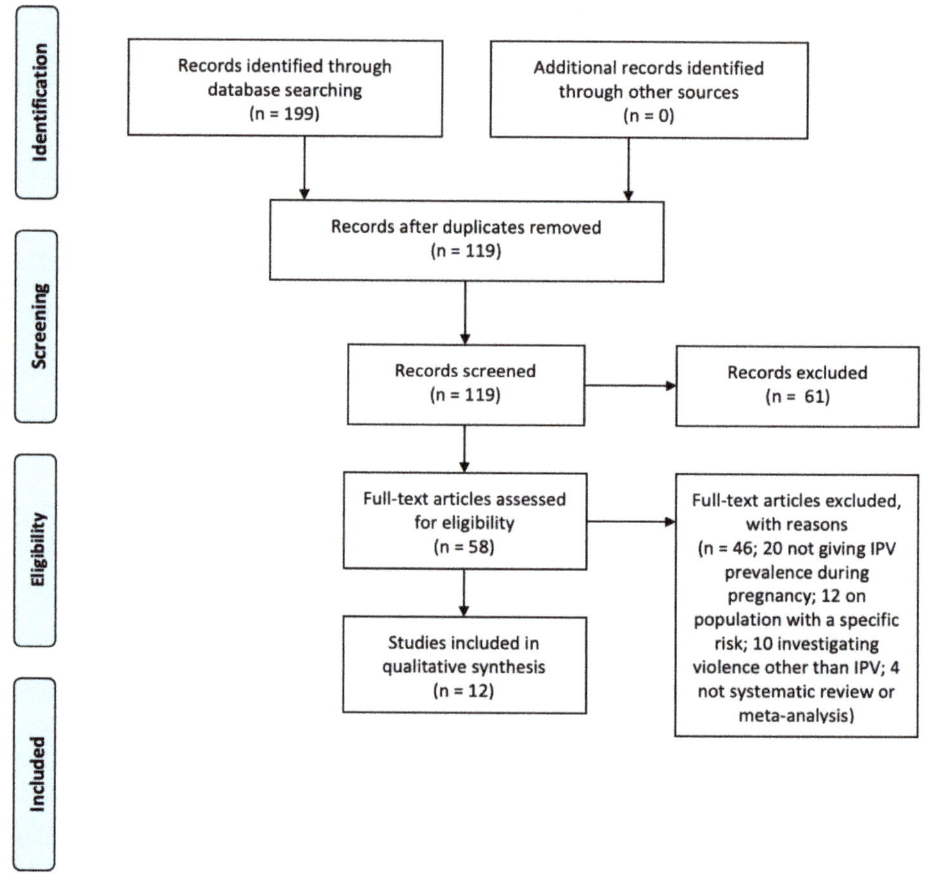

Figure 1. PRISMA flowchart of the study selection process.

From the reviews included in the study, many showed data about any IPV prevalence worldwide. The data about prevalence of any kind of IPV during pregnancy were obtained from different countries [26,29,32].

The highest range of any IPV prevalence was obtained in Portugal, USA, and Australia [32] (15.4–40%), followed by Ethiopia [29] (26.1% (95% CI: 20–32.3)) and countries from the Arab League [26] (40.9–44.1%).

Regarding physical IPV during pregnancy, China [28] and Vietnam [25] showed the lowest ranges; (3.6% (95% CI: 1.6–6.2%)) and (3–8.5%) respectively. Higher ranges were found in countries from Latin America [31] (2.5–38.7%), Africa and Asia [4] (2–35%) followed by low- and middle-income countries [27] (5–52.8%), countries from the Arab League [26] (10.4–34.6%) and African countries [30] (22.5–40%), being the widest range the one found in Saudi Arabia [24] (21–78%). James and colleagues [22] showed a prevalence of 13.8% in all over the world.

Table 2. Characteristics of the studies selected for the umbrella review of reviews of worldwide prevalence of IPV in pregnancy.

First Author, Year	Interval of Time	Country	Studies Included (n)	Study Design Included	Population Characteristics	IPV in Pregnancy Main Outcome Y/N	Type of Violence	Meta-Analysis Y/N	IPV Outcome
Alebel et al., 2018 [29]	Inception–February 2018	Ethiopia	8	Observational	Pregnant women	Y	Any	Y	26.1% (95% CI: 20, 32.3)
Alhalal et al., 2019 [24]	Inceptio–March 2018	Saudi Arabia	2 *	Cross-sectional Case-control Cohort	Pregnant women	N	Ph, Ps, S	N	Ph, 21–78% (other kind of violence figures provides only by an author)
Bazyar et al., 2018 [23]	1997–2015	World and Iran	73	Any	Pregnant women	Y	S	Y	World 17% (95% CI:15–18%) Iran 28% (95% CI: 23–32%).
Do et al., 2019 [25]	1970–2018	Vietnam	8	Cross-Sectional Cohort	Pregnant women or having a 2-year or younger child	N	Ph, Ps, S	Y	Ph, 3–8.5% Ps, 6–32.5% S, 3.4–10%
Elghossain, 2019 [26]	January 2000–January 2016	Arab league countries	7 *	Facility-based studies	Current pregnancy; Any pregnancy/any perpretator	N	Ph, Ps, S, Any	N	Ph, 10.4–34.6% Ps, 23.4–32.6% S, 5.7–15.0% Any, 40.9–44.1%

Table 2. *Cont.*

First Author, Year	Interval of Time	Country	Studies Included (n)	Study Design Included	Population Characteristics	IPV in Pregnancy Main Outcome Y/N	Type of Violence	Meta-Analysis Y/N	IPV Outcome
Halim et al., 2018 [4]	February 1990–May 2007	Bangladesh, India, Nepal, Pakistan, Vietnam, Tanzania, Ethiopia, Malawi, Zimbabwe and Egypt	24	Cross-sectional Cohort	Women during pregnancy or at postpartum	Y	Ph, Ps, S	N	Ph, 2–35% Ps, 22–65% S, 9–40%
James et al., 2013 [22]	NR	All over the world		NR	Pregnant women	Y	Ph, Ps, S	N	Ph, 13.8% Ps, 28.4% S, 8.0%
Méndez-Figueroa et al., 2013 [32]	January 1990–May 2012	Australia Portugal USA	5	Case-control Cohort	Women during pregnancy and postpartum	N	Any	N	15.4–40%
Shamu et al., 2011 [30]	2000–2010	Nigeria, South Africa, Zimbabwe, Uganda, Rwanda	19	Cross-section Cohort Case-control RCT	Pregnant women or within two months of giving birth	Y	Ph, Ps, S	Y	Ph, 22.5–40%. Ps, 24.8–49% S, 2.7–26.5%
Udmuangpia et al., 2020 [27]	January 2009–May 2018	Bangladesh, Colombia, Ethiopia, Kenya, India, Iran, Nepal, Rwanda, and Uganda	12	Cross-sectional Qualitative studies	Pregnant women 15–24 year old	Y	Ph, Ps, S	N	Ph, 5–52.8% Ps, 17–67.4% S, 2.8–21%

Table 2. *Cont.*

First Author, Year	Interval of Time	Country	Studies Included (n)	Study Design Included	Population Characteristics	IPV in Pregnancy Main Outcome Y/N	Type of Violence	Meta-Analysis Y/N	IPV Outcome
Wang et al., 2017 [28]	Inception– January 2016	China	12	Cross-sectional	Pregnant women	Y	Ph, Ps, S	Y	Overall, 7.7% (95% CI: 5.6–10.1%) Ph, 3.6% (95% CI: 1.6–6.2%) Ps, 4.2% (95% CI: 1.8–7.5%) S, 1.3% (95% CI: 0.6–2.5%)

NR, not reported; Ph, Physical violence; Ps, Psychological violence; S, Sexual violence. * Number of studies included in the review related to IPV during pregnancy, when in the review reports IPV during pregnancy and others.

As for any and physical IPV, China [28] showed the lowest and smallest ranges of psychological IPV prevalence during pregnancy (4.2% (95% CI: 1.8–7.5%)). Higher ranges were found in Vietnam [25] (6–32.5%) and countries from the Arab League [26] (23.4–32.6%), Latin America [31] (13–44%), and African countries [30] (24.8–49%). The widest ranges were found in low- and middle-income countries [27] (17–67.4%) and countries from Asia and Africa [4] (22–65%). James and colleagues [22] showed a prevalence of 28.4% throughout the world.

In general, sexual violence showed lower ranges of prevalence than the other types of IPV violence during pregnancy, being the lowest in China [28] (1.3% (95% CI: 0.6–2.5%)), followed by Vietnam [25] (3.4–10%), countries from the Arab League [26] (5.7–15.0%), low- and middle-income countries [27] (2.8–21%), Africa [30] (2.7–26.5%) and Latin America [31] (3–34.4%). The highest and widest ranges of sexual IPV prevalence during pregnancy were found in countries from Asia and Africa in the study of Halim and colleagues [4] (9–40%). Worldwide, prevalence of sexual IPV during pregnancy remained lower than 18% [22,23].

Quality assessment is reported in Table 3. Although all of the studies used a comprehensive literature search strategy, only two of the selected reviews did not include the components of PICO in their research questions and inclusion criteria [22,24], the reviews described the included studies in adequate detail, with the exception of three studies [22,26,32], and in only three of the reviews [22,25,27], authors did not report any statement about potential sources of conflict of interest. For some other aspects of the AMSTAR2 checklist, the quality remained low. Thus, none of the reviews included an explicit statement that the review methods were established prior to the conduct of the review nor the sources of funding for the studies included. Only one review provided a list of excluded studies and justified the exclusions [28]. Only three of the reviews explained their selection of the study design for inclusion in the review [24–26]. Half of the reviews performed study selection in duplicate [4,23–26,28], whereas only three did not perform data extraction in duplicate [4,22,31]. Half of the studies included did not use a satisfactory technique for assessing the risk of bias in individuals included in the review [22,23,27,29,31,32], and more than half did not account for risk of bias in individual studies when interpreting/discussing the results of the review [22–25,31–33]. Only two of the reviews provided a satisfactory explanation for, and discussion of, any heterogeneity observed in the results of the review [22,26]. In the reviews where meta-analysis was performed, the authors used appropriate methods for statistical combination of results [23,25,28–30], but in only three of them, the review authors assessed the potential impact of risk of bias in individual studies [25,28,30] and only two [28,29] carried out an adequate investigation of publication bias.

Table 3. Evaluation of selected IPV during pregnancy reviews based on AMSTAR 2 guidelines.

Author, Year	AMSTAR 2 Checklist Items *															
	1	2	3	4	5	6	7	8	9	10	11	12	13	14	15	16
Alebel et al., 2018 [29]	Y	N	N	Y	N	Y	N	Y	N	N	Y	N	Y	N	Y	Y
Alhalal et al., 2019 [24]	N	N	Y	Y	Y	Y	N	Y	Y	N	NA	NA	N	N	NA	Y
Bazyar et al., 2018 [23]	Y	N	N	Y	Y	Y	N	Y	N	N	Y	N	N	N	N	Y
Do et al., 2019 [25]	Y	N	Y	Y	Y	Y	N	Y	Y	N	Y	Y	N	N	N	N
Elghossain, 2019 [26]	Y	N	Y	Y	Y	Y	N	N	Y	N	NA	NA	Y	Y	NA	Y
Han and Stewart, 2014 [31]	Y	N	N	Y	N	N	N	Y	N	N	NA	NA	N	N	NA	Y
Halim et al., 2018 [4]	Y	N	N	Y	Y	N	N	Y	Y	N	NA	NA	Y	N	NA	Y
James et al., 2013 [22]	N	N	N	Y	N	N	N	N	N	N	NA	NA	N	Y	NA	N
Méndez-Figueroa et al., 2013 [32]	Y	N	N	Y	N	Y	N	N	N	N	NA	NA	N	N	NA	Y
Shamu et al., 2011 [30]	Y	N	N	Y	N	Y	N	Y	Y	N	Y	Y	N	N	N	Y
Udmuangpia et al., 2020 [27]	Y	N	N	Y	N	Y	N	Y	N	N	NA	NA	Y	N	NA	N
Wang et al., 2017 [28]	Y	N	N	Y	Y	Y	Y	Y	Y	N	Y	Y	Y	N	Y	Y

* Each number corresponds with an AMSTAR 2 checklist item as follows: 1. Did the research questions and inclusion criteria for the review include the components of PICO? 2. Did the report of the review contain an explicit statement that the review methods were established prior to the conduct of the review and did the report justify any significant deviations from the protocol? 3. Did the review authors explain their selection of the study designs for inclusion in the review? 4. Did the review authors use a comprehensive literature search strategy? 5. Did the review authors perform study selection in duplicate? 6. Did the review authors perform data extraction in duplicate? 7. Did the review authors provide a list of excluded studies and justify the exclusions? 8. Did the review authors describe the included studies in adequate detail? 9. Did the review authors use a satisfactory technique for assessing the risk of bias (RoB) in individual studies that were included in the review? 9. Did the review authors report on the sources of funding for the studies included in the review? 10. If meta-analysis was performed did the review authors use appropriate methods for statistical combination of results? 11. If meta-analysis was performed, did the review authors assess the potential impact of RoB in individual studies on the results of the meta-analysis or other evidence synthesis? 12. Did the review authors account for RoB in individual studies when interpreting/discussing the results of the review? 13. Did the review authors provide a satisfactory explanation for, and discussion of, any heterogeneity observed in the results of the review? 14. If they performed quantitative synthesis did the review authors carry out an adequate investigation of publication bias (small study bias) and discuss its likely impact on the results of the review? 15. Did the review authors report any potential sources of conflict of interest, including any funding they received for conducting the review? NA, Not applicable.

4. Discussion

The aim of this umbrella review was to provide a summary of the evidence currently available on global IPV prevalence in women during pregnancy. Despite the fact that the selected reviews were recent, they are of low quality as assessed against most of the AMSTAR2 recommendations. There were only two reviews giving worldwide IPV prevalence during pregnancy [22,23], both of them complying with less than half of the AMSTAR2 criteria.

4.1. Limitations

We selected systematic reviews for prevalence of IPV in pregnancy, yet we obtained very diverse data. In the reviews, sometimes the concepts of IPV and domestic violence were mixed together. Most of the included studies were cross-sectional self-report surveys, which may have been associated with inaccurate recall [24,28,31]. They did not always specify the gestational time point of IPV evaluation. It was common to find a mix among studies assessing IPV at any time of pregnancy or even after pregnancy. This is important since data can vary depending on the gestational age when IPV is measured, antenatally or after delivery. One review [4] also included studies that assessed violence for a period that is inclusive of, but not exclusive to, pregnancy.

Some of the included reviews, in spite of being systematic reviews, showed possible bias in studies included for evidence synthesis. They generally failed to adequately address the heterogeneity of results [22,28,32]. Others had a very narrow geographical coverage [24–26,29]. In addition, sample sizes of the included studies were generally small [29–31], and the use of standardized and validated IPV instruments was low. Geographical coverage of the reviews selected was mainly focused on low-income countries, a fact that invites readers to infer that IPV is a problem exclusive of those countries, which is far from the reality [34].

The main strength is that we have conducted an umbrella review following up the PRISMA and Aromataris' guidelines. Our search has been exhaustive, collecting all kind of IPV.

4.2. Implications

Whereas the consequences of IPV during pregnancy on the mother and on the newborn are widely known [3–9], the frequency and types of IPV in that period are not fully characterized. WHO recommendations on antenatal care for a positive pregnancy experience advise considering clinical inquiry about the possibility of IPV at antenatal care visits when assessing conditions that may be caused or complicated by IPV [35]. Other prenatal care guidelines affirm that clinical practitioners should be aware of the possibility of IPV, but do not include any specific recommendation related to the screening [36]. It is well known that IPV is associated with adverse mental health and obstetrical health consequences for the mother, fetus, and child, but women are reluctant to speak about this topic without a previous inquiry [37]. The American College of Obstetricians and Gynecologists guidelines recommend screening for IPV at the first prenatal visit, at least once per trimester, and at the postpartum checkup [38]. However, the overall rate of screening asymptomatic women is distressing [39]. Due to the high prevalence of this serious problem, estimated violence during pregnancy ranges from 15 to 40.5% for any type of violence, figures higher than those previously reported by Perttu et al. [40]; it is vital to have a correct estimation of its magnitude. These evaluations are necessary to underscore the importance of systematic screening: only when health staff are aware of the right prevalence and repercussion of IPV will they be able to cope with the screening barriers and to identify the most vulnerable populations by introducing screening programs in antenatal care.

Isolated prevalence studies may underestimate the true IPV prevalence due to barriers to open disclosure. These barriers could vary in different cultures and religions; e.g., widespread social norms in some regions support husbands' right to physically discipline wives. Abused women often face high social, economic, and legal barriers to divorce, a situation that is made worse by unresponsive law enforcement and health care institutions. In this social context, women are often reluctant to report violence to authorities and may hesitate to disclose violence to survey interviewers. In many societies, women are also reluctant to report violence because they are ashamed of living under this kind of relationship.

Summarizing the figures of IPV prevalence in pregnancy is needed to highlight the public health importance of this problem, with rates in some studies reported to be over 50%. These figures point towards the need of systematic screening in pregnancy. However, the analysis of published IPV reviews showed weaknesses in the research available on this topic. This umbrella review allows us to identify some methodological aspects that should be addressed in future reviews, related to geographic scope, study selection, and bias assessment.

5. Conclusions

Available data about IPV prevalence during pregnancy are of low quality. The existing evidence syntheses do not capture the totality of the disease burden in IPV in pregnancy. Despite there being wide variability in existing prevalence figures, it is worth noting that no less than 1 out of 50, and as many as 1 out of 2 women, could be suffering physical IPV in pregnancy. Psychological IPV violence is reported to be even more frequent in the published reviews. The existing evidence syntheses do not capture the totality of the worldwide disease burden of IPV in pregnancy.

Supplementary Materials: The following are available online at http://www.mdpi.com/1660-4601/18/2/707/s1, Table S1: List of excluded articles.

Author Contributions: R.M.R.-G. and S.M.-P. were responsible for data collection and analysis, data extraction and management, and quality assessment. A.B.-C. resolved any disagreement between R.M.R.-G. and S.M.-P. R.M.R.-G., S.M.-P., K.S.K., A.B.-C., and J.M.M.-G. participated in study design, analysis, and data interpretation. All authors have read and agreed to the published version of the manuscript.

Funding: This research received no external funding.

Acknowledgments: K.S.K. is a Distinguished Investigator funded by the Beatriz Galindo (senior modality) Program grant given to the University of Granada by the Ministry of Science, Innovation, and Universities of the Spanish Government.

Conflicts of Interest: The authors declare no conflict of interest.

References

1. Dicola, D.; Spaar, E. Intimate partner violence. *Am. Fam. Physician* **2016**, *94*, 646–651.
2. Committee Opinion No. 518: Intimate Partner Violence. *Obstet. Gynecol.* **2012**, *119*, 412–417. Available online: https://pubmed.ncbi.nlm.nih.gov/22270317/ (accessed on 4 January 2021).
3. WHO. *Global and Regional Estimates of Violence against Women: Prevalence and Health Effects of Intimate Partner Violence and Non-Partner Sexual Violence*; WHO: Geneva, Switzerland, 2013; p. 368.
4. Halim, N.; Beard, J.; Mesic, A.; Patel, A.; Henderson, D.; Hibberd, P. Intimate partner violence during pregnancy and perinatal mental disorders in low and lower middle income countries: A systematic review of literature, 1990–2017. *Clin. Psychol. Rev.* **2018**, *66*, 117–135. [CrossRef]
5. Moraes, C.L.; Amorim, A.R.; Reichenheim, M.E. Gestational weight gain differentials in the presence of intimate partner violence. *Int. J. Gynecol. Obstet.* **2006**, *95*, 254–260. [CrossRef]
6. El Kady, D.; Gilbert, W.M.; Xing, G.; Smith, L.H. Maternal and neonatal outcomes of assaults during pregnancy. *Obstet. Gynecol.* **2005**, *105*, 357–363. [CrossRef]
7. Donovan, B.; Spracklen, C.; Schweizer, M.; Ryckman, K.; Saftlas, A. Intimate partner violence during pregnancy and the risk for adverse infant outcomes: A systematic review and meta-analysis. *BJOG Int. J. Obstet. Gynaecol.* **2016**, *123*, 1289–1299. [CrossRef]
8. Pastor-Moreno, G.; Ruiz-Pérez, I.; Henares-Montiel, J.; Petrova, D. Intimate partner violence during pregnancy and risk of fetal and neonatal death: A meta-analysis with socioeconomic context indicators. *Am. J. Obstet. Gynecol.* **2020**, *222*, 123–133.e5. [CrossRef]
9. Lau, Y.; Chan, K.S. Influence of Intimate Partner Violence During Pregnancy and Early Postpartum Depressive Symptoms on Breastfeeding among Chinese Women in Hong Kong. *J. Midwifery Women Health* **2007**, *52*, e15–e20. [CrossRef]
10. Devries, K.M.; Kishor, S.; Johnson, H.; Stöckl, H.; Bacchus, L.J.; Garcia-Moreno, C.; Watts, C. Intimate partner violence during pregnancy: Analysis of prevalence data from 19 countries. *Reprod. Health Matters* **2010**, *18*, 158–170. [CrossRef]
11. Das, S.; Bapat, U.; Shah More, N.; Alcock, G.; Joshi, W.; Pantvaidya, S.; Osrin, D. Intimate partner violence against women during and after pregnancy: A cross-sectional study in Mumbai slums. *BMC Public Health* **2013**, *13*, 817. Available online: http://bmcpublichealth.biomedcentral.com/articles/10.1186/1471-2458-13-817 (accessed on 4 January 2021). [CrossRef]
12. Mahenge, B.; Stöckl, H.; Abubakari, A.; Mbwambo, J.; Jahn, A. Physical, sexual, emotional and economic intimate partner violence and controlling behaviors during pregnancy and postpartum among women in Dar es Salaam, Tanzania. *PLoS ONE* **2016**, *11*, e0164376. Available online: https://pubmed.ncbi.nlm.nih.gov/27755559/ (accessed on 4 January 2021). [CrossRef] [PubMed]
13. Martin, S.L.; Mackie, L.; Kupper, L.L.; Buescher, P.A.; Moracco, K.E. Physical abuse of women before, during, and after pregnancy. *J. Am. Med. Assoc.* **2001**, *285*, 1581–1584. Available online: https://jamanetwork.com/ (accessed on 4 January 2021). [CrossRef] [PubMed]
14. Ezeudu, C.C.; Akpa, O.; Waziri, N.E.; Oladimeji, A.; Adedire, E.; Saude, I.; Nguku, P.; Nsubuga, P.; Fawole, O.I. Prevalence and correlates of intimate partner violence, before and during pregnancy among attendees of maternal and child health services, Enugu, Nigeria: Mixed method approach, January 2015. *Pan Afr. Med. J.* **2019**, *32*, 14. Available online: https://pubmed.ncbi.nlm.nih.gov/30949288/ (accessed on 4 January 2021). [CrossRef]
15. Gomez-Beloz, A.; Williams, M.A.; Sanchez, S.E.; Lam, N. Intimate partner violence and risk for depression among postpartum women in Lima, Peru. *Violence Vict.* **2009**, *24*, 380–398. Available online: https://pubmed.ncbi.nlm.nih.gov/19634363/ (accessed on 4 January 2021). [CrossRef] [PubMed]

16. Velasco, C.; Luna, J.D.; Martin, A.; Caño, A.; Martin-De-las-heras, S. Intimate partner violence against Spanish pregnant women: Application of two screening instruments to assess prevalence and associated factors. *Acta Obstet. Gynecol. Scand.* **2014**, *93*, 1050–1058. [CrossRef] [PubMed]

17. Canterino, J.C.; VanHorn, L.G.; Harrigan, J.T.; Ananth, C.V.; Vintzileos, A.M. Domestic abuse in pregnancy: A comparison of a self-completed domestic abuse questionnaire with a directed interview. *Am. J. Obstet. Gynecol.* **1999**, *181*, 1049–1051. [CrossRef]

18. Garcia-Moreno, C.; Jansen, H.A.; Ellsberg, M.; Heise, L.; Watts, C.H. Prevalence of intimate partner violence: Findings from the WHO multi-country study on women's health and domestic violence. *Lancet* **2006**, *368*, 1260–1269. [CrossRef]

19. Liberati, A.; Altman, D.G.; Tetzlaff, J.; Mulrow, C.; Gøtzsche, P.C.; Ioannidis, J.P.A.A.; Clarke, M.; Devereaux, P.J.; Kleijnen, J.; Moher, D.; et al. The PRISMA statement for reporting systematic reviews and meta-analyses of studies that evaluate healthcare interventions: Explanation and elaboration. *BMJ* **2009**, *339*, b2700. Available online: https://www.bmj.com/content/339/bmj.b2700 (accessed on 19 May 2020). [CrossRef]

20. Aromataris, E.; Fernandez, R.; Godfrey, C.M.; Holly, C.; Khalil, H.; Tungpunkom, P. Summarizing systematic reviews: Methodological development, conduct and reporting of an umbrella review approach. *Int. J. Evid. Based. Healthc.* **2015**, *13*, 132–140. Available online: https://pubmed.ncbi.nlm.nih.gov/26360830/ (accessed on 20 October 2020). [CrossRef]

21. Shea, B.J.; Reeves, B.C.; Wells, G.; Thuku, M.; Hamel, C.; Moran, J.; Moher, D.; Tugwell, P.; Welch, V.; Kristjansson, E.; et al. AMSTAR 2: A critical appraisal tool for systematic reviews that include randomised or non-randomised studies of healthcare interventions, or both. *BMJ* **2017**, *358*. Available online: https://pubmed.ncbi.nlm.nih.gov/28935701/ (accessed on 20 October 2020). [CrossRef]

22. James, L.; Brody, D.; Hamilton, Z. Risk factors for domestic violence during pregnancy: A meta-analytic review. *Violence Vict.* **2013**, *28*, 359–380. [CrossRef] [PubMed]

23. Bazyar, J.; Safarpour, H.; Daliri, S.; Karimi, A.; Safi Keykaleh, M.; Bazyar, M. The prevalence of sexual violence during pregnancy in Iran and the world: A systematic review and meta-analysis. *J. Inj. Violence Res.* **2018**, *10*, 63–74. Available online: http://www.ncbi.nlm.nih.gov/pubmed/29500334 (accessed on 23 October 2020). [PubMed]

24. Alhalal, E.; Ta'an, W.; Alhalal, H. Intimate Partner Violence in Saudi Arabia: A Systematic Review. In *Trauma, Violence, and Abuse*; SAGE Publications Ltd.: Thousand Oaks, CA, USA, 2019.

25. Do, H.P.; Tran, B.X.; Nguyen, C.T.; Van Vo, T.; Baker, P.R.A.; Dunne, M.P. Inter-partner violence during pregnancy, maternal mental health and birth outcomes in Vietnam: A systematic review. *Child. Youth Serv. Rev.* **2019**, *96*, 255–265. [CrossRef]

26. Elghossain, T.; Bott, S.; Akik, C.; Obermeyer, C.M. Prevalence of intimate partner violence against women in the Arab world: A systematic review. *BMC Int. Health Hum. Rights* **2019**, *19*, 1–16. [CrossRef]

27. Udmuangpia, T.; Yu, M.; Laughon, K.; Saywat, T.; Bloom, T. Prevalence, risks, and health consequences of intimate partner violence during pregnancy among young women: A systematic review. *Pacific Rim Int. J. Nurs. Res.* **2020**, *24*, 412–429.

28. Wang, T.; Liu, Y.; Li, Z.; Liu, K.; Xu, Y.; Shi, W.; Chen, L. Prevalence of intimate partner violence (IPV) during pregnancy in China: A systematic review and meta-analysis. *PLoS ONE* **2017**, *12*, e0175108. [CrossRef]

29. Alebel, A.; Kibret, G.D.; Wagnew, F.; Tesema, C.; Ferede, A.; Petrucka, P.; Bobo, F.T.; Birhanu, M.Y.; Tadesse, A.A.; Eshetie, S. Intimate partner violence and associated factors among pregnant women in Ethiopia: A systematic review and meta-analysis. *Reprod. Health* **2018**, *15*, 1–12. [CrossRef]

30. Shamu, S.; Abrahams, N.; Temmerman, M.; Musekiwa, A.; Zarowsky, C. A systematic review of African studies on intimate partner violence against pregnant women: Prevalence and risk factors. *PLoS ONE* **2011**, *6*, e17591. [CrossRef]

31. Han, A.; Stewart, D.E. Maternal and fetal outcomes of intimate partner violence associated with pregnancy in the Latin American and Caribbean region. *Int. J. Gynecol. Obstet.* **2014**, *124*, 6–11. Available online: https://pubmed.ncbi.nlm.nih.gov/24182684/ (accessed on 23 September 2020). [CrossRef]

32. Mendez-Figueroa, H.; Dahlke, J.D.; Vrees, R.A.; Rouse, D.J. Trauma in pregnancy: An updated systematic review. *Am. J. Obstet. Gynecol.* **2013**, *209*, 1–10. [CrossRef]

33. Shamu, S.; Zarowsky, C.; Roelens, K.; Temmerman, M.; Abrahams, N. High-frequency intimate partner violence during pregnancy, postnatal depression and suicidal tendencies in Harare, Zimbabwe. *Gen. Hosp. Psychiatry* **2016**, *38*, 109–114. Available online: http://www.ncbi.nlm.nih.gov/pubmed/26607330 (accessed on 21 May 2020). [CrossRef] [PubMed]

34. Sumner, S.A.; Mercy, J.A.; Dahlberg, L.L.; Hillis, S.D.; Klevens, J.; Houry, D. Violence in the United States: Status, challenges, and opportunities. *JAMA J. Am. Med. Assoc.* **2015**, *314*, 478–488. Available online: https://pubmed.ncbi.nlm.nih.gov/26241599/ (accessed on 30 November 2020). [CrossRef] [PubMed]

35. World Health Organization. WHO Recommendations on Antenatal Care for a Positive Pregnancy Experience. 2016. Available online: http://library1.nida.ac.th/termpaper6/sd/2554/19755.pdf (accessed on 15 December 2020).

36. Ministerio de Sanidad, Servicios Sociales e Igualdad. Guía de práctica clínica de atención en el embarazo y puerperio. 2014. Available online: https://www.mscbs.gob.es/organizacion/sns/planCalidadSNS/pdf/Guia_practica_AEP.pdf (accessed on 15 December 2020).

37. Halpern-Meekin, S.; Costanzo, M.; Ehrenthal, D.; Rhoades, G. Intimate Partner Violence Screening in the Prenatal Period: Variation by State, Insurance, and Patient Characteristics. *Matern. Child Health J.* **2019**, *23*, 756–767. Available online: https://pubmed.ncbi.nlm.nih.gov/30600519/ (accessed on 7 January 2021). [CrossRef] [PubMed]

38. The American College of Obstetricians and Gynecologists, American Academy of Pediatrics. Guidelines for Perinatal Care, 8th ed.; 2017; pp. 1–712. Available online: https://www.acog.org/clinical-information/physician-faqs/--media/3a22e153b67446a6b31fb051e469187c.ashx (accessed on 16 December 2020).

39. Alvarez, C.; Fedock, G.; Grace, K.T.; Campbell, J. Provider Screening and Counseling for Intimate Partner Violence: A Systematic Review of Practices and Influencing Factors. *Trauma Violence Abus.* **2017**, *18*, 479–495. Available online: https://pubmed.ncbi.nlm.nih.gov/27036407/ (accessed on 4 January 2021). [CrossRef] [PubMed]

40. Perttu, S.; Kaselitz, V. *Addressing Intimate Partner Violence Guidelines for Health Professionals in Maternity and Child Health Care*; University of Helsinki: Helsinki, Finland, 2006.

International Journal of
Environmental Research
and Public Health

Review

Relationship Between Prolonged Second Stage of Labor and Short-Term Neonatal Morbidity: A Systematic Review and Meta-Analysis

Nuria Infante-Torres [1], Milagros Molina-Alarcón [2], Angel Arias-Arias [1], Julián Rodríguez-Almagro [3,*] and Antonio Hernández-Martínez [3]

[1] Mancha Centro Hospital, Av. Constitución, 3, Alcázar de San Juan, 13600 Ciudad Real, Spain; nuria.i.t@hotmail.com (N.I.-T.); angel_arias_arias81@hotmail.com (A.A.-A.)

[2] Department of Nursing, Physiotherapy and Occupational Therapy, Faculty of Nursing, University of Castilla-La Mancha, Av. de España, s/n, 02001 Albacete, Spain; milagros.molina@uclm.es

[3] Department of Nursing, Physiotherapy and Occupational Therapy, Faculty of Nursing, University of Castilla-La Mancha, Camilo José Cela, 14, 13071 Ciudad Real, Spain; antonio.hmartinez@uclm.es

* Correspondence: julianj.rodriguez@uclm.es; Tel.: +346-7668-3843

Received: 2 September 2020; Accepted: 21 October 2020; Published: 23 October 2020

Abstract: To evaluate the association between prolonged second stage of labor and the risk of adverse neonatal outcomes with a systematic review and meta-analysis. PubMed, Scopus and EMBASE were searched using the search strategy "Labor Stage, Second" AND (length OR duration OR prolonged OR abnormal OR excessive). Observational studies that examine the relationship between prolonged second stage of labor and neonatal outcomes were selected. Prolonged second stage of labor was defined as 4 h or more in nulliparous women and 3 h or more in multiparous women. The main neonatal outcomes were 5 min Apgar score <7, admission to the Neonatal Intensive Care Unit, neonatal sepsis and neonatal death. Data collection and quality assessment were carried out independently by the three reviewers. Twelve studies were selected including 266,479 women. In nulliparous women, a second stage duration greater than 4 h increased the risk of 5 min Apgar score <7, admission to the Neonatal Intensive Care Unit and neonatal sepsis and intubation. In multiparous women, a second stage of labor greater than 3 h was related to 5 min Apgar score <7, admission to the Neonatal Intensive Care Unit, meconium staining and composite neonatal morbidity. Prolonged second stage of labor increased the risk of 5 min Apgar score <7 and admission to the Neonatal Intensive Care Unit in nulliparous and multiparous women, without increasing the risk of neonatal death. This review demonstrates that prolonged second stage of labor increases the risk of neonatal complications in nulliparous and multiparous women.

Keywords: Apgar score; meta-analysis; Neonatal Intensive Care Unit; neonatal morbidity; newborn care; labor stage; second; systematic review

---

1. Introduction

The second stage of labor is the period of time between full cervical dilatation and birth of the baby, during which the woman has an involuntary urge to bear down, as a result of expulsive uterine contractions [1].

The description of the onset of the second stage of labor in clinical practice is often not precisely known. If complete dilatation is found on vaginal examination, it remains uncertain how long this cervical status has been present [2].

Multiple observational studies [2–4] have observed an increase in maternal complications associated with a prolonged second stage of labor, such as operative vaginal delivery, third-/fourth-degree perineal lacerations, caesarean delivery, urinary retention, postpartum hemorrhage and chorioamnionitis, as well as an increase in neonatal complications like seizures, hypoxic-ischemic encephalopathy, sepsis and increased mortality. However, the criteria these studies used to define the second stage of labor are heterogenous.

Thus, diagnosis and management of prolonged second stage of labor and its complications are difficult and often pose a dilemma to the obstetrician regarding timing and type of intervention [5]. Additionally, evidence on the duration of the second stage of labor is of very low certainty [1] and it is unclear whether there is a point of time from which the risk of perinatal complications increases and at which health professionals should intervene to prevent adverse events [3,6].

Nevertheless, there are professionals involved in childbirth care that try to reduce the duration of the second stage by obstetric interventionism in order to avoid neonatal complications. Paradoxically, these interventions, such as immediate pushing (initiated as soon as complete dilation is identified) [7], instrumental birth [8] or fundal pressure [9], may themselves increase the risk of neonatal morbidity.

In the past, a prolonged second stage of labor had been defined as a period of time that lasted beyond 2 h with epidural analgesia or 1 h without epidural analgesia for multiparous women. For nulliparous women, a prolonged second stage is defined as a period of time that lasted beyond 3 h with epidural analgesia or 2 h without epidural analgesia [10]. Recently, though, the American College of Obstetricians and Gynecologists (ACOG) [11] and the National Institute for Health and Care Excellence (NICE) [12] have allowed longer durations in specific cases. In spite of this, the correct management of the second stage of labor should be individualized according to birth progress, fetal malposition or the use of epidural analgesia [11,12]. For example, the Eunice Kennedy Shriver National Institute of Child Health and Human Development document suggested allowing one additional hour for the use of epidural analgesia. Thus, at least 3 h in multiparous women and 4 h in nulliparous women would be considered to diagnose a prolonged second stage of labor [11].

Thus, our objective was to evaluate the evidence on the association between prolonged second stage of labor (defined as 4 h in nulliparous women and 3 h in multiparous women) and the risk of adverse neonatal outcomes.

2. Materials and Methods

This systematic review with a meta-analysis was done according to PRISMA (Preferred Reporting Items for Systematic Review and Meta-Analyses) [13,14].

2.1. Data Sources and Searches

The adopted search strategy was: "Labor Stage, Second" (Mesh) AND (length OR duration OR prolonged OR abnormal OR excessive). Studies were identified in three main databases: PubMed [15], Scopus [16] and EMBASE [17], from 1 January 1990 to 1 November 2019. As well as published studies, we included non-published studies which had been included in the conference proceedings of the main scientific associations and indexed in the databases consulted. All languages were included. The search results for each database are provided in detail in Table A1.

All members of the research team had prior training in the methodology of systematic reviews, literature reviews and critical reading. AAA and AHM are also experts in meta-analysis.

Studies were included according to four criteria: (I) duration of second stage of labor greater than 4 h in nulliparous women; (II) duration of second stage of labor greater than 3 h in multiparous women; (III) studies reporting neonatal outcomes in relation to duration of second stage of labor; (IV) studies that stratified results by parity. Reference lists from the selected studies were also examined to locate further studies not identified using the search strategy. Two authors (NIT and AAA) independently performed the literature search and excluded any articles that did not meet the established inclusion criteria. A third author (MMA) was consulted to resolve any disagreements or uncertainty regarding inclusion.

2.2. Main Outcomes

The primary outcomes were 5 min Apgar score < 7, admission to the Neonatal Intensive Care Unit, neonatal sepsis and neonatal death. All neonatal outcomes examined by the available studies were included in this review. The definitions of some of the variables included in our study are shown in Table 1.

Table 1. Definition of variables.

Definitions	Authors						
	1995; Menticoglou [18]	2007; Cheng [19]	2009; Allen [4]	2009; Rouse [20]	2012; Bleich [21]	2017; Sandström [22]	2019; Infante [23]
Acidosis	NR	NR	NR	NR	NR	A pH value <7.05 and base excess <−12 in the umbilical artery.	NR
Birth depression	NR	NR	Delay in initiating and maintaining respirations after birth requiring resuscitation by mask or endotracheal tube for at least 3 min, a 5 min Apgar score of 3 or less, or neonatal seizures due to hypoxic-ischemic encephalopathy.	NR	NR	NR	NR
Intubation	NR	NR	NR	Intubation in delivery room.	NR	NR	NR
Heart compressions	NR	NR	NR	NR	NR	Resuscitation in delivery room with heart compressions and/or intubation.	NR
Advanced neonatal resuscitation	NR	NR	NR	NR	NR	NR	Type III: Oxygen therapy with positive intermittent pressure. Type IV: Endotracheal intubation, Type V: Cardiac massage and/or using drugs

Table 1. *Cont.*

Definitions	Authors						
	1995; Menticoglou [18]	2007; Cheng [19]	2009; Allen [4]	2009; Rouse [20]	2012; Bleich [21]	2017; Sandström [22]	2019; Infante [23]
Admission to Neonatal Intensive Care Unit	Need for admission to the Neonatal Intensive Care Unit for any reason at all or with a 5 min Apgar score < 7 or arterial cord pH < 7.20.	NR	Neonatal intensive care unit admission with duration of stay longer than 24 h.	Admission to a neonatal intensive care unit for >48 h.	NR	NR	NR
Prolonged neonatal stay	NR	Neonatal stay >2 d for vaginal delivery and >4 d for caesarean delivery.	NR	NR	NR	NR	NR
Neonatal seizures	NR	NR	NR	NR	Seizures in the first 24 h of life.	NR	NR
Sepsis	NR	NR	Positive blood culture, septicemia or systematic infection.	NR	Positive blood culture.	NR	NR
Minor trauma	NR	NR	One or more of the following neonatal traumas: linear skull fracture, other fractures (clavicle, ribs, numerus, or femur), facial palsy, or cephalohematoma.	NR	NR	NR	NR
Major trauma	NR	NR	One or more of the following neonatal traumas: depressed skull fracture, intracranial hemorrhage, or brachial plexus palsy.	NR	NR	NR	NR

Table 1. *Cont.*

Definitions	Authors						
	1995; Menticoglou [18]	2007; Cheng [19]	2009; Allen [4]	2009; Rouse [20]	2012; Bleich [21]	2017; Sandström [22]	2019; Infante [23]
Composite neonatal morbidity	NR	Composite variable for 5 min Apgar <7, UA pH <7.0, UA base excess ≥12, shoulder dystocia, NICU stay, and birth trauma (which includes brachial plexus injury, facial nerve palsy, clavicular fracture, skull fracture, head laceration, and cephalohematoma defined and diagnosed by the attending pediatrician).	Composite of any of the other neonatal outcomes.	Any of the following occurrences: a 5 min Apgar score <4, an umbilical artery pH <7.0, seizures, intubation, stillbirth, neonatal death, or admission to a NICU.	NR	NR	Composite of any of the other neonatal outcomes.
Neonatal death	Death during the second stage of labor or in the first 28 d of life	NR	NR	NR	NR	NR	NR

NR: not reported.

2.3. Data Extraction and Quality Assessment

Data collection and quality assessment were carried out independently by the three reviewers (NIT, AHM and JRA). We tried to contact the authors of several studies to provide us with data that did not appear in their manuscripts.

We used the Joanna Briggs Institute Critical Appraisal tools for use in JBI Systematic Reviews to assess the risk of bias in each included study [24]. Eleven domains were assessed to appraise the methodological quality of a study and to determine the extent to which a study had addressed the possibility of bias in its design, conduct and analysis.

2.4. Data Synthesis

For the categorical results, the odds ratio (OR) was used along with its 95% confidence intervals (95% CI). To calculate the OR, either the Mantel–Haenszel fixed-effects or Der Simonian–Laird random-effects models were used, depending on whether there was heterogeneity between the studies. Heterogeneity was assessed using the I^2 and the statistical Cochran's Q tests. I^2 values of < 25%, 25–50 and >50% normally correspond to small, medium and large heterogeneity, respectively [14,25,26]. Publication bias was also evaluated using the Egger asymmetry test and funnel plots [14,27]. Statistical significance was defined at the ≤0.05 level.

All calculations were done with the StatsDirect statistical software, version 2.7.9. (Stats Direct Ltd., Cheshire, England) [14].

3. Results

3.1. Study Selection

A total of 1868 studies were selected from the literature search. After removing any duplicated articles, 267 were selected by title and abstract. After applying the inclusion/exclusion criteria, twelve articles were selected for the qualitative and quantitative analyses (meta-analysis) (Figure 1).

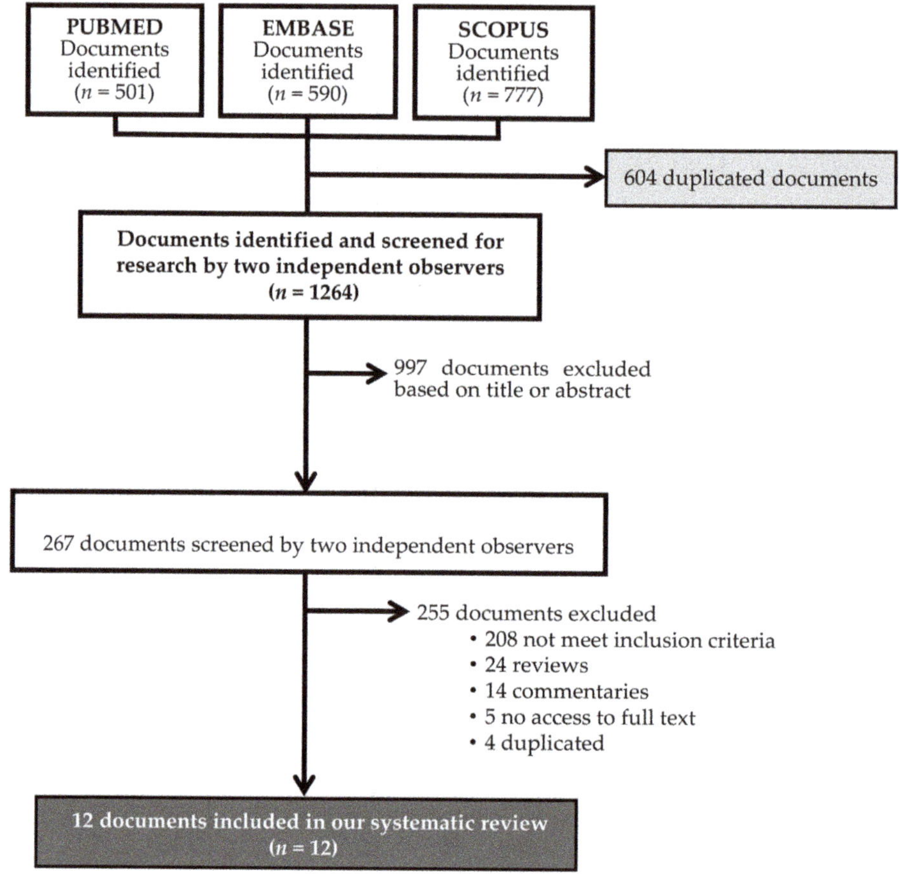

Figure 1. PRISMA flow diagram of the literature reviewing process.

3.2. Study Characteristics

The description of the studies included in this systematic review are shown in Table 2. The sample included 268,624 women. The selected studies were conducted in Canada [4,18,28], United States [19–21, 29,30], China [31], Sweden [22,32] and Spain [23]. The sample size of these studies ranged from 307 [31] to 121,490 [4]. All studies were restricted to singleton infants with cephalic presentation. Eight of these articles studied nulliparous women [18,20–22,28,30–32], two studied multiparous women [19,23] and two studied both (nulliparous and multiparous women) [4,29]

Table 2. Characteristics of the studies analyzed.

	YEAR OF PUBLICATION AUTOR	COUNTRY	STUDY DESIGN	POPULATION UNDER STUDY	DURATION OF SECOND STAGE OF LABOUR n (%)					DELIVERY MODE N (%)			USE OF EA N (%)	INCLUSION/EXCLUSION CRITERIA
					0–1 h	1–2 h	2–3 h	3–4 h	>4 h	Spontaneous Vaginal Delivery	Operative Vaginal Delivery	Caesarean Section		
NULLIPAROUS	**1995/ Menticoglou** [18]	Canada	Cohort study	6041	2622 (43.4)	1805 (29.9)	927 (15.3)	379 (6.3)	308 (5.1)	4942 (81.8)	932 (15.5)	167 (2.7)	NR	• January 1988 to December 1992. Singleton babies in cephalic presentation. • BW ≥ 2500 g • Fetal death diagnosed before labor and caesarean section before labor or during the first stage of labor were excluded.
	2009/ Rouse [20]	United States	Secondary analysis of a clinical trial	4126	1901 (46.1)	1251 (30.3)	614 (14.9)	217 (5.2)	143 (3.5)	3054 (74.0)	765 (18.5)	307 (7.5)	3916 (95.0)	• Nulliparous women with a singleton vertex fetus who labored spontaneously or were induced at ≥ 36 WG and who reached the second stage of labor. • Exclusion criteria included maternal fever and serious medical conditions.
	2011/ Li [31]	China	Case-control study	307	206 (67.1)		29 (9.4)	60 (19.5)	12 (4.0)	NR	NR	NR	NR	• NR
	2012/ Bleich [21]	United States	Cohort study	21,991	13,736 (62.5)	4933 (22.4)	1833 (8.3)	1062 (4.8)	427 (2.0)	19,326 (87.9)	1367 (6.2)	1298 (5.9)	13,676 (62.2)	• Nulliparous women who reached the second stage of labor. • Singleton live-born infants at ≥ 37 WG and cephalic presentation. • Between January 2003 to December 2008. • Fetal malformations, placenta previa and multiple gestation were excluded.

Table 2. *Cont.*

YEAR OF PUBLICATION AUTOR	COUNTRY	STUDY DESIGN	POPULATION UNDER STUDY	DURATION OF SECOND STAGE OF LABOUR *n* (%)					DELIVERY MODE N (%)			USE OF EA N (%)	INCLUSION/EXCLUSION CRITERIA
				0–1 h	1–2 h	2–3 h	3–4 h	>4 h	Spontaneous Vaginal Delivery	Operative Vaginal Delivery	Caesarean Section		
2015/ Altman [32]	Sweden	Cohort study	32,796	10,731 (32.7)	9491 (29.0)	5856 (17.8)	3898 (11.9)	2820 (8.6)	NR	6728 (20.5)	NR	19,417 (59.2)	• First live singleton infant in cephalic presentation at ≥ 37 WG. • From January 2008 to December 2012. • Caesarean and induced deliveries and deliveries with incomplete data were excluded.
2015/ Hunt [28]	Canada	Cohort study	1515	NC	NC	NC	629 (41.5)	886 (58.5)	615 (40.6)	662 (43.7)	238 (15.7)	NR	• Nulliparous women who delivered non-anomalous, term (≥ 36 WG), cephalic, live singleton neonatal weight ≥ 2500 g and who had a prolonged second stage of labor. • Between January 1993 and April 2006.
2017/ Sandström [22]	Sweden	Cohort study	42,539	13,558 (31.9)	12,225 (28.7)	7710 (18.1)	5238 (12.3)	3808 (9.0)	NR	NR	NR	NR	• Between January 2008 to December 2013. • Nulliparous women with cephalic presentation at 37 WG or later. • Elective caesarean deliveries, emergency caesareans during first stage of labor and deliveries with incomplete data were excluded (without labor partograph or notation on complete dilation of the cervix).
2018/ Souter [30]	United States	Cohort study in a poster session	20,029	16,682 (83.3)			3347 (16.7)		14,942 (74.6)	3015 (15.0)	2072 (10.4)	20,029 (100)	• Singleton deliveries at 37 + 0 to 42 + 6 WG between January 2012 and December 2016

Table 2. *Cont.*

	YEAR OF PUBLICATION AUTOR	COUNTRY	STUDY DESIGN	POPULATION UNDER STUDY	DURATION OF SECOND STAGE OF LABOUR n (%)					DELIVERY MODE N (%)			USE OF EA N (%)	INCLUSION/EXCLUSION CRITERIA
					0–1 h	1–2 h	2–3 h	3–4 h	>4 h	Spontaneous Vaginal Delivery	Operative Vaginal Delivery	Caesarean Section		
MULTIPAROUS	2007/ Cheng [19]	United States	Cohort study	5158	4112 (79.7)	550 (10.7)	239 (4.6)	257 (5.0)		4480 (86.8)	414 (8.1)	263 (5.1)	2274 (44.1)	• Between 1991 and 2001. • All term and post-term, cephalic, live singleton births to multiparous women who had spontaneous onset of labor. • Caesarean delivery before the completion of the first stage of labor, placenta previa, intrauterine fetal demise or known lethal congenital anomalies were excluded.
MULTIPAROUS	2019/ Infante [23]	Spain	Cohort study	2145	1589 (74.1)	327 (15.2)	165 (7.7)	64 (3.0)		2070 (96.5)	75 (3.5)	NR	1675 (78.1)	• Women who had given birth vaginally, with cephalic presentation and singleton babies between 2013 and 2016. • Births with < 35 WG and antepartum fetal death were excluded.
NULLIPAROUS AND MULTIPAROUS	2009/ Allen [4]	Canada	Cohort study	55,936 nulliparous	38,790 (69.3)		7832 (14.0)	4406 (7.9)	4908 (8.8)	101,897 (83.8)	15,865 (13.1)	3734 (3.1)	61,077 (50.3)	• Between 1988 and 2006. • Liveborn singleton at or after 37 WG reaching full cervical dilatation. • Deliveries that occurred before onset of labor, a major congenital anomaly, at least one previous caesarean delivery, severe pregnancy-related medical disorders or missing outcome data were excluded
NULLIPAROUS AND MULTIPAROUS	2009/ Allen [4]	Canada	Cohort study	65,554 multiparous	59,227 (90.3)	4171 (6.3)	1188 (1.8)	968 (1.6)						
NULLIPAROUS AND MULTIPAROUS	2017/ Ogunyemi [29]	United States	Poster session	10,487 *	NC	NC	NC	NC	NC	NR	NR	NR	NR	• Singleton at term.

NC: not calculated, NR: not reported, EA: epidural analgesia, BW: birthweight, WG: weeks gestation, *: no data on nulliparous/multiparous.

3.3. Study and Data Quality

The included studies had a low risk of bias, except for three studies that did not identify confounding factors [18,28,31] and four studies that did not include strategies to deal with confounding factors [18,28,30,31] (Table A2).

With regard to the selection of subjects, all studies except one [31] specified inclusion and exclusion criteria, selecting all women (nulliparas and/or multiparas) with singleton cephalic presentation that reached second stage of labor within a specific period of time.

Seven of the studies included in the meta-analysis [4,18,20,21,23,28,32] correctly defined prolonged second stage of labor (in this case, second stage of labor longer than 4 h in nulliparas and longer than 3 h in multiparas). Conversely, only three of them [18,21,28] established the maneuver used once prolonged second stage of labor was diagnosed (instrumental birth, continuing maternal pushing, caesarean, etc.).

As for data and information collection, five studies [4,22,23,30,32] included missing or incomplete data as exclusion criteria, so they were not included in the analysis.

3.4. Main Outcomes and Meta-Analysis

3.4.1. Nulliparous Women

3.4.1.1. min Apgar score <7

To determine the relation between prolonged second stage of labor in nulliparous women (Table A3) and risk of low 5 min Apgar score (<7), six studies were included (n = 116,624) [4,18,28,30–32]. A significant increase in low 5 min Apgar score was observed when the second stage of labor lasted more than 4 h with respect to when the second stage of labor was ≤ 4 h. (OR = 1.65; 95% CI: 1.20–2.27). For this analysis, a random-effects model was used since heterogeneity was observed (Cochran's Q p-value = 0.0041; I2 = 71.0) (Figure 2a; Table 3).

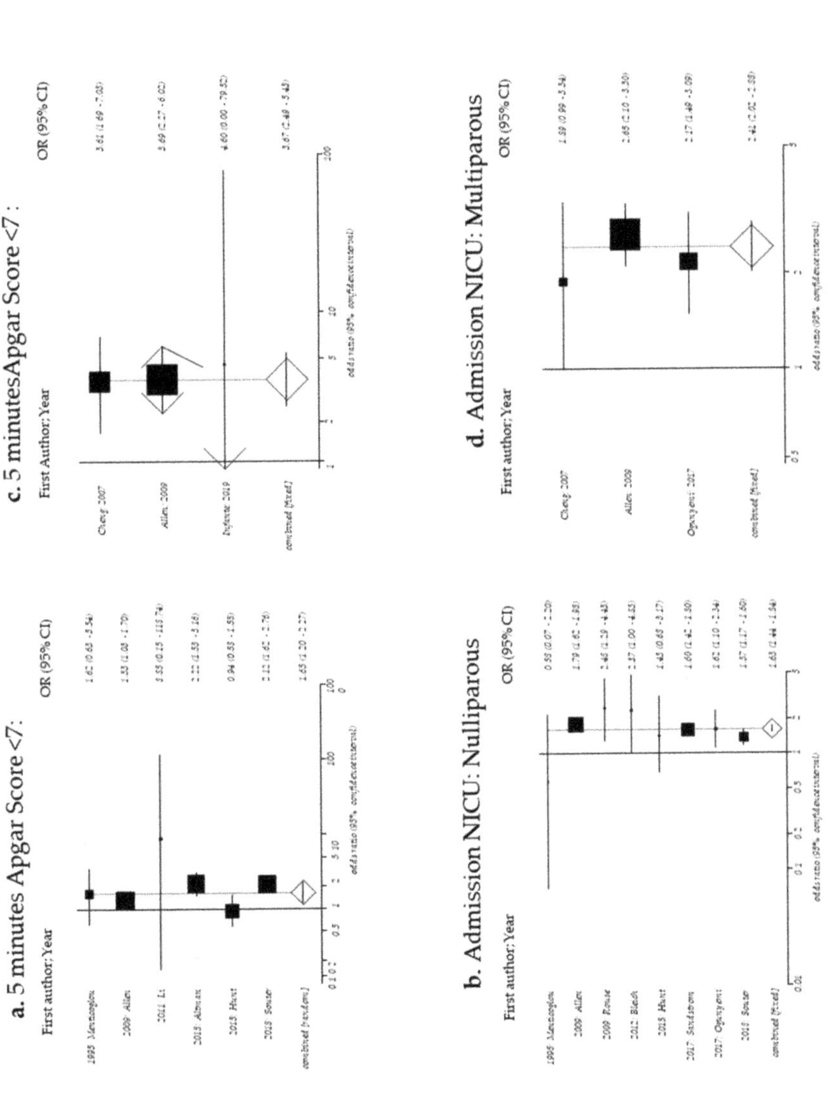

Figure 2. Forest plot for 5 min Apgar Score < 7 in nulliparous women (**a**), admission to NICU in nulliparous women (**b**), 5 min Apgar Score < 7 in multiparous women (**c**) and (**d**) admission to NICU in multiparous women.

Table 3. Summary of results obtained following meta-analysis of all variables studied in nulliparous women. Summary of results obtained following meta-analysis of all variables studied in multiparous women.

Variable	Number of Studies	Number of Subjects	Egger Bias (*p*-Value)	I^2 95% CI	Cochran's Q (*p*-Value)	OR 95% CI
1 min Apgar Score <7	1	307	NC	NC	NC	NC
5 min Apgar score <7	6	116,624	0.7861	71.0 (2,1–85,6)	0.0041	1.65 (1.20–2.27)
5 min Apgar score <4	2	36,922	NC	NC	0.7026	2.27 (1.08–4.74)
5 min Apgar Score <3	1	21,991	NC	NC	NC	NC
Umbilical artery pH <7	2	29,117	0.8132	NC	0.8132	2.30 (0.94–5.69)
Umbilical artery pH <7.10	0	0	NR	NR	NR	NR
Umbilical artery base excess >−12	0	0	NR	NR	NR	NR
Acidosis	1	33,429	NC	NC	NC	NC
Birth depression	1	55,936	NC	NC	NC	NC
Resuscitation at delivery	2	42,020	NC	NC	<0.001	2.60 (0.81–8.63)
Intubation	2	46,665	NC	NC	0.681	2.19 (1.23–3.90)
Heart compressions	1	42,539	NC	NC	NC	NC
ANR	0	0	NR	NR	NR	NR
Meconium aspiration	1	42,539	NC	NC	NC	NC
Meconium-stained amniotic fluid	1	4487	NC	NC	NC	NC
Admission to Neonatal Intensive Care Unit	8	156,650	0.8326	48.8 (0.0–75.4)	0.0573	1.63 (1.44–1.84)
Prolonged neonatal stay	0	0	NR	NR	NR	NR
Neonatal seizures	3	70,571	NC	92.3 (78.6–95.8)	<0.001	4.67 (0.78–27.78)
Neonatal sepsis	3	82,053	NC	0.0 (0–72.9)	0.7962	1.57 (1.07–2.29)
Birth trauma	1	4064	NC	NC	NC	NC
Minor trauma	1	55,936	NC	NC	NC	NC
Major trauma	1	55,936	NC	NC	NC	NC

Table 3. *Cont.*

Variable	Number of Studies	Number of Subjects	Egger Bias (*p*-Value)	I² 95% CI	Cochran's Q (*p*-Value)	OR 95% CI
Shoulder dystocia	1	20,029	NC	NC	NC	NC
Brachial plexus injury	1	4126	NC	NC	NC	NC
Erb's palsy	1	21,991	NC	NC	NC	NC
Hypoxic ischemic encephalopathy	1	42,539	NC	NC	NC	NC
Hypothermia treatment	1	42,539	NC	NC	NC	NC
Composite neonatal morbidity	1	4126	NR	NR	NR	NR
Any perinatal morbidity	0	0	NR	NR	NR	NR
Neonatal death	2	28,032	NC	NC	NC	7.21 (0.37–139.71)
Variable	**Number of Studies**	**Number of Subjects**	**Egger Bias (*p*-Value)**	**I² 95% CI**	**Cochran's Q (*p*-Value)**	**OR 95% CI**
1 min Apgar Score < 7	0	0	NR	NR	NR	NR
5 min Apgar score < 7	3	72,857	NC	0.0 (0.0–72.9)	0.987	3.67 (2.48–5.43)
5 min Apgar score < 4	0	0	NR	NR	NR	NR
5 min Apgar Score ≤ 3	0	0	NR	NR	NR	NR
Umbilical artery pH < 7	1	5158	NC	NC	NC	NC
Umbilical artery pH < 7.10	1	1912	NC	NC	NC	NC
Umbilical artery base excess > −12	1	5158	NC	NC	NC	NC
Acidosis	0	0	NR	NR	NR	NR
Birth depression	1	65,554	NC	NC	NC	NC
Resuscitation at delivery	0	0	NR	NR	NR	NR
Intubation	0	0	NR	NR	NR	NR
Heart compressions	0	0	NR	NR	NR	NR
ANR	1	2145	NC	NC	NC	NC
Meconium aspiration	0	0	NR	NR	NR	NR

Table 3. *Cont.*

Variable	Number of Studies	Number of Subjects	Egger Bias (*p*-Value)	I^2 95% CI	Cochran's Q (*p*-Value)	OR 95% CI
Meconium-stained amniotic fluid	2	11,193	NC	NC	0.121	1.29 (1.01–1.66)
Admission to Neonatal Intensive Care Unit	3	76,692	NC	0.0 (0.0–72.9)	0.417	2.41 (2.02–2.88)
Prolonged neonatal stay	1	5158	NC	NC	NC	NC
Neonatal seizures	0	0	NR	NR	NR	NR
Neonatal sepsis	0	0	NR	NR	NR	NR
Birth trauma	0	0	NR	NR	NR	NR
Minor trauma	1	65,554	NC	NC	NC	NC
Major trauma	1	65,554	NC	NC	NC	NC
Shoulder dystocia	1	5158	NC	NC	NC	NC
Brachial plexus injury	0	0	NR	NR	NR	NR
Erb's palsy	0	0	NR	NR	NR	NR
Hypoxic ischemic encephalopathy	0	0	NR	NR	NR	NR
Hypothermia treatment	0	0	NR	NR	NR	NR
Composite neonatal morbidity	2	7303	NC	NC	0.330	1.97 (1.39–2.80)
Any perinatal morbidity	1	65,554	NC	NC	NC	NC
Neonatal death	0	0	NR	NR	NR	NR

ANR: advanced neonatal resuscitation; NC: not calculated; NR: not reported.

Admission to Neonatal Intensive Care Unit

To assess the risk of admission to the Neonatal Intensive Care Unit, eight studies were employed (n = 156,650) [4,18,20–22,28–30].

The risk significantly increased when the second stage of labor lasted more than 4 h with respect to when the second stage of labor was ≤ 4 h (OR, 1.63; 95%CI 1.44–1.84). For this analysis, a random-effects model was used since medium heterogeneity was observed (Cochran's Q p-value = 0.057; I2 = 48.8) (Figure 2B; Table 3).

Neonatal Sepsis

By combining three studies (n = 82,053) [4,20,21], we found that the risk of neonatal sepsis increased when the duration of the second stage of labor was longer than 4 h with respect to when the second stage of labor was ≤ 4 h (OR, 1.57; 95% CI 1.07–2.29). For this analysis, a fixed-effects model was used since no heterogeneity was observed (Cochran's Q p-value = 0.7962; I2 = 0.0) (Table 3).

Neonatal Death

Two studies (n = 28,032) [18,21] were employed to determine the relationship between prolonged second stage of labor and risk of neonatal death, and no differences were found (OR, 7.21; 95% CI 0.37–139.71) (Table 3).

Other Neonatal Outcomes

No significant associations were reported between prolonged second stage in nulliparous women and 1 min Apgar score < 1, 5 min Apgar score ≤ 3, umbilical artery pH < 7, acidosis, meconium-stained amniotic fluid, meconium aspiration, birth depression, minor or major trauma, birth trauma, shoulder dystocia, brachial plexus injury, Erb's palsy, resuscitation at birth, heart compressions, hypoxic ischemic encephalopathy, hypothermia treatment or composite neonatal morbidity. When the results of two studies were combined [20,22], only an increased risk of neonatal intubation in women with a second stage of labor > 4 h was observed (OR, 2.19; 95% CI 1.23–3.90) (Table 3).

3.4.2. Multiparous Women

3.4.2.1. min Apgar Score < 7

To determine the relation between prolonged second stage of labor in multiparous women (Table A4) and risk of low 5 min Apgar score (< 7), three studies were included (n = 72,857) [4,19,23]. A significant increase in low 5 min Apgar score was observed when the second stage of labor lasted more than 3 h with respect to when the second stage of labor was ≤ 3 h (OR, 3.67; 95% CI 2.49–5.43). For this analysis, a fixed-effects model was used since no heterogeneity was observed (Cochran's Q p-value = 0.987; I2 = 0.0) (Figure 2C; Table 3).

Admission to the Neonatal Intensive Care Unit

To assess the risk of admission to the Neonatal Intensive Care Unit, three studies were employed (n = 76,692) [4,19,29]. The risk significantly increased when the second stage of labor lasted more than 3 h with respect to when the second stage of labor was ≤ 3 h (OR, 2.41; 95% CI 2.02–2.88). For this analysis, a fixed-effects model was used since no heterogeneity was observed (Cochran's Q p-value = 0.417; I2 = 0.0) (Figure 2D; Table 3).

Neonatal Sepsis

None of the studies that analyzed multiparous women considered this variable when assessing neonatal morbidity in relation to the duration of the second stage of childbirth (Table 3).

Neonatal Death

None of the studies that analyzed multiparous women considered this variable when assessing neonatal morbidity in relation to the duration of the second stage of childbirth (Table 3).

Other Neonatal Outcomes

No significant associations were reported between prolonged second stage in multiparous women and umbilical artery pH < 7.0, umbilical artery pH < 7.10, umbilical artery base excess ≥12, meconium aspiration, shoulder dystocia, prolonged neonatal stay, advanced neonatal resuscitation, birth depression, minor or major trauma or any perinatal morbidity. After combining two studies [19,29], only an increase in the risk of meconium staining was observed (OR, 1.29; 95%CI, 1.01–1.66), and an increase in composite neonatal morbidity (OR,1.97; 95% CI, 1.39–2.80) was observed after another two studies were combined [19,23] (Table 3).

3.4.3. Publication Bias

We did not observe publication bias for the study in any of the variables studied (Tables A3 and A4).

We can observe a summary of results obtained following meta-analysis of all variables studied in nulliparous and multiparous women in Table 3.

4. Discussion

4.1. Main Findings

Our meta-analysis results suggested that duration of second stage of labor of more than 4 h in nulliparous women increased the risk of low 5 min Apgar score < 7, admission to the Neonatal Intensive Care Unit, neonatal sepsis and neonatal intubation. In multiparous women, when the second stage of labor was longer than 3 h, the risk of 5 min Apgar score < 7, admission to Neonatal Intensive Care Unit, meconium staining and composite neonatal morbidity increased.

However, a prolonged second stage of labor did not increase the risk of any of the other variables studied, such as umbilical artery pH < 7, birth depression, neonatal death meconium aspiration or shoulder dystocia.

4.2. Comparison with Existing Literature

The literature has very limited data on neonatal outcomes of women with duration of second stage of labor of more than 4 h in nulliparas and of more than 3 h in multiparas. We were only able to locate 12 articles with these durations for this review.

An example of this is a recent systematic review by Gimovksy et al., which evaluated the maternal and fetal morbidities associated with prolonged second stage of labor in nulliparous women with epidurals, in which the authors defined prolonged second stage as greater than three hours [33]. Only two papers were included in this systematic review, and very discordant neonatal outcomes were analyzed, which did not allow the results to be combined in order to establish conclusions that would be useful for decision-making in clinical practice.

Another systematic review studied the influence of prolonged second stage of labor on the risk of adverse maternal and neonatal outcomes from 1980 until 2005 [34]. It did not report associations between prolonged second stage and adverse neonatal outcomes, but most of the studies analyzed in this review defined the prolongation of the second stage as more than 2 h, without differentiating according to parity. In addition, it did not conform to the new recommendations of allowing longer durations.

Only one randomized controlled trial [35] specifically addressed the effect of this change in obstetric practice on maternal and neonatal outcomes. In that trial, a policy of extending the second stage of labor for at least 1 h in nulliparous women with epidural anesthesia with respect to "usual labor" (3 h) decreased the incidence of caesarean birth by more than half compared with the common practice (19.5%, 8 of 41, vs. 43.2%, 16 of 37; RR, 0.45; 95% CI, 0.22–0.93). Maternal or neonatal morbidity were not statistically different between the groups. Unfortunately, the trial was underpowered to detect significant differences in the frequency of adverse maternal or neonatal outcomes between groups because the sample studied was very small (only 78 nulliparous women) (35).

However, Zipori et al. [36] recently published another study comparing maternal and neonatal outcomes over two distinct time periods. In period I, the duration of the second stage of labor was considered prolonged according to ACOG limits, and it was called a "classic labor curve" (10). The "new labor curve" of period II allowed nulliparous and multiparous women to continue the second stage of labor for an additional 1 h before diagnosing second-stage arrest. Primary caesarean deliveries decreased with the new policy of labor management, with a small rise in instrumental deliveries, but it also increased other immediate maternal and neonatal complications, such as higher rate of lower umbilical artery cord pH.

4.3. Strengths and Limitations

One of the strengths of this study is that it is the first systematic review to define prolonged second stage of labor according the most recent recommendations (11), that is, 4 h for nulliparous women and 3 h for multiparous women. Most of the studies had large sample sizes with sufficient numbers of participants in each group to lend power to the findings, and the majority of them used methods to control for potential confounding factors.

Among the limitations of our systematic review is that neonatal outcome measures were discordant in the included studies, meaning it was difficult to combine data to summarize important clinical findings, and that the definition of two variables (admission to NICU and composite neonatal morbidity) differed among included studies. None of the studies considered the pushing duration or pushing techniques employed (delayed pushing or immediate pushing). Finally, since they were observational studies, there is a risk of confounding bias even though many of the studies included techniques to control confounding.

5. Conclusions

In nulliparous women, a prolonged second stage of labor is not related with an increased risk of neonatal death. However, it is related with an increased risk of 5 min Apgar score < 7, admission to the Neonatal Intensive Care Unit, neonatal sepsis or intubation. In multiparous women, a prolonged second stage of labor is related with an increased risk of 5 min Apgar score < 7, admission to the Neonatal Intensive Care Unit, meconium staining and composite neonatal morbidity.

These potential risks associated with a prolonged second stage of labor in both nulliparous and multiparous women should serve as an incentive for professionals involved in childbirth care to increase supervision of mothers who exceed these durations.

More studies are needed, especially clinical studies, to guarantee the safety of newborns when the second stage of labor exceeds 4 h in nulliparous women and 3 h in multiparous women.

Author Contributions: Conceptualization, N.I.-T. and J.R.-A.; Methodology, M.M.-A. and A.A.-A.; Formal Analysis, A.H.-M. and J.R.-A.; Writing—Original Draft Preparation, N.I.-T. and M.M.-A.; Writing—Review & Editing, N.I.-T. and A.A.-A.; Supervision, J.R.-A. and A.H.-M.; Project Administration, A.H.-M. All authors have read and agreed to the published version of the manuscript.

Funding: This research received no external funding.

Conflicts of Interest: The authors declare no conflict of interest.

Appendix A

Table A1. Search strategies.

Search Strategies		
Search Strategy	**Database**	**Hits**
"Labor Stage, Second" (Mesh) AND (length OR duration OR prolonged OR abnormal OR excessive)	PubMed	501
	Scopus	590
	Embase	777
Search Strategy "PICO"		
Population	Nulliparous and Multiparous women	
Intervention	Second stage labor > 4 h in nulliparous Second stage labor > 3 h in multiparous	
Comparison	Second stage labor ≤ 4 h in nulliparous Second stage labor ≤ 3 h in multiparous	
Outcome	Neonatal morbidity: 5 min Apgar score < 7, admission to the Neonatal Intensive Care Unit, neonatal sepsis and neonatal death.	

Table A2. Checklist for Cohort Studies.

.	1995; Menticoglou	2007; Cheng	2009; Allen	2009; Rouse	2011; Li	2012; Bleich	2015; Altman	2015; Hunt	2017; Sandström	2017; Ogunyemi	2018; Souter	2019; Infante
1. Were the two groups similar and recruited from the same population?	Unclear *	No *	No *	Unclear *	Unclear *	Unclear *	Unclear *	Unclear *	Unclear *	Unclear *	Unclear *	Unclear *
2. Were the exposures measured similarly to assign people to both exposed and unexposed groups?	Yes	Yes	Yes	Yes	Yes	Yes	Yes	Yes	Yes	Yes	Yes	Yes
3. Was the exposure measured in a valid and reliable way?	Yes	Yes	Yes	Yes	No	Yes	Yes	Yes	Yes	Yes	Yes	Yes
4. Were confounding factors identified?	No	Yes	Yes	Yes	No	Yes	Yes	No	Yes	Yes	Yes	Yes
5. Were strategies to deal with confounding factors stated?	No	Yes	Yes	Yes	No	Yes	Yes	No	Yes	Yes	No	Yes
6. Were the groups/participants free of the outcome at the start of the study (or at the moment of exposure)?	Yes	Yes	Yes	Yes	Yes	Yes	Yes	Yes	Yes	Yes	Yes	Yes
7. Were the outcomes measured in a valid and reliable way?	Yes	Yes	Yes	Yes	Yes	Yes	Yes	Yes	Yes	Yes	Yes	Yes
8. Was the follow up time reported and sufficient to be long enough for outcomes to occur?	Yes	Yes	Yes	Yes	Yes	Yes	Yes	Yes	Yes	Yes	Yes	Yes
9. Was follow up complete, and if not, were the reasons to loss to follow up described and explored?	Yes	Yes	Yes	Yes	Yes	Yes	Yes	Yes	Yes	Yes	Yes	Yes
10. Were strategies to address incomplete follow up utilized?	NA	NA	Yes	NA	NA	NA	Yes	NA	Yes	NA	NA	Yes
11. Was appropriate statistical analysis used?	Unclear	Yes	Yes	Yes	Unclear	Yes	Yes	Unclear	Yes	Yes	Unclear	Yes

NA: not applicable, *: groups recruited from the same population.

Table A3. Neonatal morbidity outcomes in nulliparous women (≤4 h versus >4 h second-stage labor).

	1 min Apgar Score <7		5 min Apgar Score <7		5 min Apgar Score <4		5 min Apgar Score ≤3		Umbilical Artery pH <7		Acidosis		Birth Depression		Resuscitation at Delivery		Intubation		Heart Compressions	
Author	≤4 h	>4 h	≤4 h	>4 h	≤4 h	>4 h	≤4 h	>4 h	≤4 h	>4 h	≤4 h	>4 h	≤4 h	>4 h	≤4 h	>4 h	≤4 h	>4 h	≤4 h	>4 h
1995; Menticoglou	NR	NR	81/5733	7/308	NR	NR	NR	NR	NR	NR	NR	NR	NR	NR	NR	NR	NR	NR	NR	NR
2009; Allen	NR	NR	589/51,028	75/4908	NR	NR	NR	NR	NR	NR	NR	NR	862/51,028	138/4908	NR	NR	NR	NR	NR	NR
2009; Rouse	NR	NR	NR	NR	3/3983	0/143	NR	NR	15/3983	1/143	NR	NR	NR	NR	NR	NR	19/3983	1/143	NR	NR
2011; Li	17/295	8/12	3/295	1/12	NR	NR	NR	NR	NR	NR	NR	NR	NR	NR	NR	NR	NR	NR	NR	NR
2012; Bleich	NR	NR	NR	NR	NR	NR	17/21,564	2/427	83/21,564	4/427	NR	NR	NR	NR	138/21,564	13/427	NR	NR	NR	NR
2015; Altman	NR	NR	188/29,976	39/2820	32/29,976	8/2820	NR	NR	NR	NR	NR	NR	NR	NR	NR	NR	NR	NR	NR	NR
2015; Hunt	NR	NR	33/629	44/886	NR	NR	NR	NR	NR	NR	NR	NR	NR	NR	NR	NR	NR	NR	NR	NR
2017; Sandström	NR	NR	NR	NR	NR	NR	NR	NR	NR	NR	330/30,453	31/2976	NR	NR	NR	NR	58/38,731	13/3808	36/38,731	10/3808
2017; Ogunyemi	NR	NR	NR	NR	NR	NR	NR	NR	NR	NR	NR	NR	NR	NR	NR	NR	NR	NR	NR	NR
2018; Souter	NR	NR	200/16,682	84/3347	NR	NR	NR	NR	NR	NR	NR	NR	NR	NR	851/16,682	248/3347	NR	NR	NR	NR
Egger Bias (p-value)			0.7861		NC				0.8132						NC		NC			
I² 95% CI			71.0 (2.1–85.6)		NC				NC						NC		NC			
Q Cochran (p-value)			0.0041		0.7026				0.8132						<0.001		0.681			
OR 95% CI			1.65 (1,20–2.27) *		2.27 (1.08–4.74) *				2.30 (0.94–5.69)						2.60 (0.81–8.63) *		2.19 (1.23–3.90)			

	Meconium Aspiration		Meconium-Stained Amniotic Fluid		Admission to NICU		Neonatal Seizures		Sepsis		Birth Trauma		Minor Trauma		Major Trauma		Shoulder Dystocia	
Author	≤4 h	>4 h	≤4 h	>4 h	≤4 h	>4 h	≤4 h	>4 h	≤4 h	>4 h	≤4 h	>4 h	≤4 h	>4 h	≤4 h	>4 h	≤4 h	>4 h
1995; Menticoglou	NR	NR	NR	NR	64/5733	2/308	5/5733	0/308	NR	NR	NR	NR	NR	NR	NR	NR	NR	NR
2009; Allen	NR	NR	NR	NR	3071/51,028	505/4908	NR	NR	195/51,028	30/4908	NR	NR	1311/51,028	162/4908	78/51,028	16/4908	NR	NR
2009; Rouse	NR	NR	NR	NR	167/3983	14/143	NR	NR	6/3983	0/143	NR	NR	NR	NR	NR	NR	NR	NR
2011; Li	NR	NR	NR	NR	NR	NR	NR	NR	NR	NR	NR	NR	NR	NR	NR	NR	NR	NR
2012; Bleich	NR	NR	NR	NR	172/21,564	8/427	41/21,564	13/427	39/21,564	0/427	NR	NR	NR	NR	NR	NR	NR	NR
2015; Altman	NR	NR	NR	NR	NR	NR	NR	NR	NR	NR	NR	NR	NR	NR	NR	NR	NR	NR
2015; Hunt	NR	NR	NR	NR	12/629	24/886	NR	NR	NR	NR	NR	NR	NR	NR	NR	NR	NR	NR
2017; Sandström	58/38,731	10/3808	NR	NR	2373/38,731	360/3808	78/38,731	17/3808	NR	NR	NR	NR	NR	NR	NR	NR	NR	NR

Table A3. *Cont.*

2017; Ogunyemi	NR	NR	309/4215	37/272	372/4201	37/272	NR	NR	NR	NR	33/3814	1/250	NR	NR	NR	NR	NR	NR
2018; Souter	NR	NR	NR	NR	817/16,682	221/3347	NR	NR	NR	NR	NR	NR	NR	NR	NR	NR	367/16,682	74/3347
Egger Bias (*p*-value)					0.8326		NC		NC									
I² 95% CI					48.8 (0.0–75,4)		92.3 (78.6–95,8)		0.0 (0–72.9)									
Q Cochran (*p*-value)					0.0573		<0.001		0.7962									
OR 95% CI					**1.63 (1.53–1.74)**		4.67 (0,78–27.78) *		**1.57 (1.07–2.29)**									

Author	Brachial Plexus Injury		Erb's Palsy		Hypoxic Ischemic Encephalopathy		Hypothermia Treatment		Composite Neonatal Morbidity		Neonatal Death	
	≤4 h	>4 h	≤4 h	>4 h	≤4 h	>4 h	≤4 h	>4 h	≤4 h	>4 h	≤4 h	>4 h
1995; Menticoglou	NR	NR	NR	NR	NR	NR	NR	NR	NR	NR	0/5733	0/308
2009; Allen	NR	NR	NR	NR	NR	NR	NR	NR	NR	NR	NR	NR
2009; Rouse	10/3983	1/143	NR	NR	NR	NR	NR	NR	98/3983	6/143	NR	NR
2011; Li	NR	NR	NR	NR	NR	NR	NR	NR	NR	NR	NR	NR
2012; Bleich	NR	NR	82/21,564	2/427	NR	NR	NR	NR	NR	NR	3/21,564	0/427
2015; Altman	NR	NR	NR	NR	NR	NR	NR	NR	NR	NR	NR	NR
2015; Hunt	NR	NR	NR	NR	NR	NR	NR	NR	NR	NR	NR	NR
2017; Sandström	NR	NR	NR	NR	75/38,731	22/3808	16/38,731	7/3808	NR	NR	NR	NR
2017; Ogunyemi	NR	NR	NR	NR	NR	NR	NR	NR	NR	NR	NR	NR
2018; Souter	NR	NR	NR	NR	NR	NR	NR	NR	NR	NR	NR	NR
Egger Bias (*p*-value)											NC	
I² 95% CI											NC	
Q Cochran (*p*-value)											NC	
OR 95% CI											7.21 (0.37–139.71) *	

NR: not reported; NR: not calculated; CI: confidence interval; * Random effects (DerSimonian–Laird); Bold: Significant results are highlighted.

Table A4. Neonatal Morbidity Outcomes in Multiparous Women (≤3 h versus >3 h second stage of labour).

	5 min Apgar Score <7		Umbilical Artery pH <7.0		Umbilical Artery pH <7.10		Umbilical Artery Base Excess>−12		Birth Depression		Advanced Neonatal Resuscitation		Meconium Amniotic Fluid or Meconium Staining		NICU Admission		Prolonged Neonatal Stay	
Author	≤3 h	>3 h	≤3 h	>3 h	≤3 h	>3 h	≤3 h	>3 h	>3 h	≤3 h	≤3 h	>3 h	≤3 h	>3 h	≤3 h	>3 h	≤3 h	>3 h
2007; Cheng	60/4901	11/257	17/4901	1/257	NR	NR	31/4901	3/257	NR	NR	NR	NR	1061/4901	73/257	145/4901	14/257	445/4901	34/257
2009; Allen	311/64,586	17/968	NR	NR	NR	NR	NR	NR	553/64,586	27/968	NR	NR	NR	NR	2409/64,586	90/968	NR	NR
2017; Ogunyemi	NR	NR	NR	NR	NR	NR	NR	NR	NR	NR	NR	NR	290/5759	12/276	410/5700	40/278	NR	NR
2019; Infante	3/2081	0/64	NR	NR	37/1858	3/54	NR	NR	NR	NR	39/2081	3/64	NR	NR	NR	NR	NR	NR
Egger Bias (*p*-value)	NC												NC		NC			
I[2] 95% CI	0% (0.0%–72.9%)												NC		0% (0.0%–72.9%)			
Q Cochran (*p*-value)	0.987												0.121		0.417			
OR 95% CI	**3.67 (2.49–5.43)**												**1.29 (1.01–1.66)**		**2.41 (2.02–2.88)**			

	Minor Trauma		Major Trauma		Shoulder Dystocia		Composite Neonatal Morbidity		Any Perinatal Morbidity	
Author	>3 h	>3 h	≤3 h	>3 h	≤3 h	>3 h	≤3 h	>3 h	≤3 h	>3 h
2007; Cheng	NR	NR	NR	NR	117/4901	10/257	361/4901	33/257	NR	NR
2009; Allen	586/64,586	27/968	104/64,586	2/968	NR	NR	NR	NR	3662/64,586	128/968
2017; Ogunyemi	NR	NR	NR	NR	NR	NR	NR	NR	NR	NR
2019; Infante	NR	NR	NR	NR	NR	NR	70/2081	6/64	NR	NR
Egger Bias (*p*-value)			NC							
I[2] 95% CI			NC							
Q Cochran (*p*-value)			0.330							
OR 95% CI			**1.97 (1.39–2.80)**							

NR: not reported; NR: not calculated; CI: confidence interval; Random effects (DerSimonian–Laird); Bold: Significant results are highlighted.

References

1. WHO Recommendations: Intrapartum Care for a Positive Childbirth Experience. Available online: https://www.who.int/reproductivehealth/publications/intrapartum-care-guidelines/en/ (accessed on 2 October 2020).
2. Myles, T.D.; Santolaya, J. Maternal and neonatal outcomes in patients with a prolonged second stage of labor. In *Obstet Gynecol. 2003, 102.*. Available online: http://www.ncbi.nlm.nih.gov/pubmed/12850607 (accessed on 30 January 2020).
3. Cheng, Y.W.; Hopkins, L.M.; Caughey, A.B. How long is too long: Does a prolonged second stage of labor in nulliparous women affect maternal and neonatal outcomes? *Am. J. Obstet. Gynecol.* **2004**, *191*, 933–938. [CrossRef] [PubMed]
4. Allen, V.M.; Baskett, T.F.; O'Connell, C.M.; McKeen, D.; Allen, A.C. Maternal and Perinatal Outcomes With Increasing Duration of the Second Stage of Labor. *Obstet. Gynecol.* **2009**, *113*, 1248–1258. [CrossRef] [PubMed]
5. Singh, S.; Kohli, U.A.; Vardhan, S. Management of prolonged second stage of labor. *Int. J. Reprod. Contracept. Obstet. Gynecol.* **2018**, *7*, 2527–2531. [CrossRef]
6. Le Ray, C.; Audibert, F.; Goffinet, F.; Fraser, W. When to stop pushing: Effects of duration of second-stage expulsion efforts on maternal and neonatal outcomes in nulliparous women with epidural analgesia. *Am. J. Obstet. Gynecol.* **2009**, *201*, 361.e1-361.e7. [CrossRef] [PubMed]
7. Simpson, K.R.; James, D.C. Effects of Immediate Versus Delayed Pushing During Second-Stage Labor on Fetal Well-Being. *Nurs. Res.* **2005**, *54*, 149–157. [CrossRef]
8. Thavarajah, H.; Flatley, C.; Kumar, S. The relationship between the five minute Apgar score, mode of birth and neonatal outcomes. *J. Matern. Neonatal Med.* **2017**, *31*, 1335–1341. [CrossRef]
9. Moiety, F.; Azzam, A.Z. Fundal pressure during the second stage of labor in a tertiary obstetric center: A prospective analysis. *J. Obstet. Gynaecol. Res.* **2014**, *40*, 946–953. [CrossRef]
10. Dystocia and the augmentation of labor. *Int. J. Gynecol. Obstet.* **1996**, *53*, 73–80. [CrossRef]
11. American College of Obstetricians and Gynecologists Obstetric Care Consensus No. 1. *Obstet. Gynecol.* **2014**, *123*, 693–711. [CrossRef]
12. NICE. Intrapartum Care for Healthy Women and Babies—NICE Guidelines. Available online: https://www.nice.org.uk/guidance/cg190/resources/intrapartum-care-for-healthy-women-and-babies-pdf-35109866447557 (accessed on 2 October 2020).
13. Liberati, A.; Altman, D.G.; Tetzlaff, J.; Mulrow, C.; Gøtzsche, P.C.; A Ioannidis, J.P.; Clarke, M.; Devereaux, P.J.; Kleijnen, J.; Moher, D. The PRISMA statement for reporting systematic reviews and meta-analyses of studies that evaluate healthcare interventions: Explanation and elaboration. *BMJ* **2009**, *339*, b2700. [CrossRef]
14. Hernández-Martínez, A.; Arias-Arias, A.; Morandeira-Rivas, A.; Pascual-Pedreño, A.I.; Ortiz-Molina, E.J.; Rodríguez-Almagro, J. Oxytocin discontinuation after the active phase of induced labor: A systematic review. *Women Birth* **2019**, *32*, 112–118. [CrossRef] [PubMed]
15. PubMed. Available online: https://pubmed.ncbi.nlm.nih.gov/ (accessed on 2 October 2020).
16. Scopus. Elsevier. Available online: https://www.elsevier.com/es-es/solutions/scopus (accessed on 2 October 2020).
17. Investigación Biomédica—Embase. Available online: https://www.elsevier.com/es-es/solutions/embase-biomedical-research (accessed on 2 October 2020).
18. Menticoglou, S.M.; Manning, F.; Harman, C.; Morrison, I. Perinatal outcome in relation to second-stage duration. *Am. J. Obstet. Gynecol.* **1995**, *173*, 906–912. [CrossRef]
19. Cheng, Y.W.; Hopkins, L.M.; Laros, R.K.; Caughey, A.B. Duration of the second stage of labor in multiparous women: Maternal and neonatal outcomes. *Am. J. Obstet. Gynecol.* **2007**, *196*, 585.e1-585.e6. [CrossRef] [PubMed]
20. Rouse, D.J.; Weiner, S.J.; Bloom, S.L.; Varner, M.W.; Spong, C.Y.; Ramin, S.M.; Caritis, S.N.; Peaceman, A.M.; Sorokin, Y.; Sciscione, A.; et al. Second-stage labor duration in nulliparous women: Relationship to maternal and perinatal outcomes. *Am. J. Obstet. Gynecol.* **2009**, *201*, 357.e1-357.e7. [CrossRef] [PubMed]
21. Bleich, A.T.; Alexander, J.M.; McIntire, D.D.; Leveno, K.J. An Analysis of Second-Stage Labor beyond 3 Hours in Nulliparous Women. *Am. J. Perinatol.* **2012**, *29*, 717–722. [CrossRef]

22. Sandström, A.; Altman, M.; Cnattingius, S.; Johansson, S.; Ahlberg, M.; Stephansson, O. Durations of second stage of labor and pushing, and adverse neonatal outcomes: A population-based cohort study. *J. Perinatol.* **2016**, *37*, 236–242. [CrossRef] [PubMed]

23. Infante-Torres, N.; Alarcón, M.M.; Gómez-Salgado, J.; Almagro, J.J.R.; Rubio-Álvarez, A.; Martínez, A.H. Relationship between the Duration of the Second Stage of Labour and Neonatal Morbidity. *J. Clin. Med.* **2019**, *8*, 376. [CrossRef]

24. The Joanna Briggs Institute Critical Appraisal for Use in JBI Systematic Reviews. Checklist for Cohort Studies. Available online: https://joannabriggs.org/sites/default/files/2019-05/JBI_Critical_Appraisal-Checklist_for_Cohort_Studies2017_0.pdf (accessed on 2 October 2020).

25. Higgins, J.P.T.; Thompson, S.G. Quantifying heterogeneity in a meta-analysis. *Stat. Med.* **2002**, *21*, 1539–1558. [CrossRef]

26. Higgins, J.P.T.; Thompson, S.G.; Deeks, J.J.; Altman, D.G. Measuring inconsistency in meta-analyses [Internet]. *Br. Med. J.* **2003**, *327*, 557–560. Available online: https://pubmed.ncbi.nlm.nih.gov/12958120/ (accessed on 2 October 2020). [CrossRef]

27. Egger, M.; Smith, G.D. meta-analysis bias in location and selection of studies. *BMJ* **1998**, *316*, 61–66. [CrossRef]

28. Hunt, J.C.; Menticoglou, S.M. Perinatal Outcome in 1515 Cases of Prolonged Second Stage of Labour in Nulliparous Women. *J. Obstet. Gynaecol. Can.* **2015**, *37*, 508–516. [CrossRef]

29. Ogunyemi, D.; Friedman, P.; Jovanovski, A.; Shah, I.; Hage, N.; Whitten, A. 890: Maternal and neonatal outcomes by parity and second-stage labor duration. *Am. J. Obstet. Gynecol.* **2017**, *216*, S508–S509. [CrossRef]

30. Souter, V.; Chien, A.J.; Kauffman, E.; Sitcov, K.; Caughey, A.B. 365: Outcomes associated with prolonged second stage of labor in nulliparas. *Am. J. Obstet. Gynecol.* **2018**, *218*, S226. [CrossRef]

31. Li, W.-H.; Zhang, H.-Y.; Ling, Y.; Jin, S. Effect of prolonged second stage of labor on maternal and neonatal outcomes. *Asian Pac. J. Trop. Med.* **2011**, *4*, 409–411. [CrossRef]

32. Altman, M.; Sandström, A.; Petersson, G.; Frisell, T.; Cnattingius, S.; Stephansson, O. Prolonged second stage of labor is associated with low Apgar score. *Eur. J. Epidemiol.* **2015**, *30*, 1209–1215. [CrossRef] [PubMed]

33. Gimovsky, A.C.; Guarente, J.; Berghella, V. Prolonged Second Stage in Nulliparous with Epidurals. *Obstet. Gynecol. Surv.* **2017**, *72*, 399–401. [CrossRef]

34. Altman, M.R.; Lydon-Rochelle, M.T. Prolonged Second Stage of Labor and Risk of Adverse Maternal and Perinatal Outcomes: A Systematic Review. *Birth* **2006**, *33*, 315–322. [CrossRef] [PubMed]

35. Gimovsky, A.C.; Berghella, V. Randomized Controlled Trial of Prolonged Second Stage. *Obstet. Gynecol. Surv.* **2016**, *71*, 385–387. [CrossRef]

36. Zipori, Y.; Grunwald, O.; Ginsberg, Y.; Beloosesky, R.; Weiner, Z. The impact of extending the second stage of labor to prevent primary cesarean delivery on maternal and neonatal outcomes. *Am. J. Obstet. Gynecol* **2019**, *220*, 191.e1-191.e7. [CrossRef]

International Journal of
Environmental Research and Public Health

Commentary

Patient and Public Involvement in Sexual and Reproductive Health: Time to Properly Integrate Citizen's Input into Science

Miguel García-Martín [1,2,3], **Carmen Amezcua-Prieto** [1,2,3,*], **Bassel H Al Wattar** [4,5], **Jan Stener Jørgensen** [6], **Aurora Bueno-Cavanillas** [1,2,3] and **Khalid Saeed Khan** [1,2]

[1] Department of Preventive Medicine and Public Health, Faculty of Medicine, University of Granada, 18016 Granada, Spain; mgar@ugr.es (G.-M.M.); abueno@ugr.es (B.-C.A.); profkkhan@gmail.com (K.K.S.)
[2] Consortium for Biomedical Research in Epidemiology and Public Health (CIBERESP), 28029 Madrid, Spain
[3] Instituto de Investigación Biosanitaria (ibs.Granada), 18014 Granada, Spain
[4] Reproductive Medicine Unit, Institute for Women's Health, University College London Hospitals, London WC1E 6BT, UK; dr.basselwa@gmail.com
[5] Warwick Medical School, University of Warwick, Coventry CV4 7 AL, UK
[6] Department of Obstetrics and Gynaecology CIMT-Centre for Innovative Medical Technologies Odense University Hospital, 5000 Odense C, Denmark; jan.stener@rsyd.dk
[*] Correspondence: carmezcua@ugr.es; Tel.: +34-95824100 (ext. 20287)

Received: 21 September 2020; Accepted: 28 October 2020; Published: 31 October 2020

Abstract: Evidence-based sexual and reproductive health is a global endeavor without borders. Inter-sectorial collaboration is essential for identifying and addressing gaps in evidence. Health research funders and regulators are promoting patient and public involvement in research, but there is a lack of quality tools for involving patients. Partnerships with patients are necessary to produce and promote robust, relevant and timely research. Without the active participation of women as stakeholders, not just as research subjects, the societal benefits of research cannot be realized. Creating and developing platforms and opportunities for public involvement in sexual and reproductive health research should be a key international objective. Cooperation between healthcare professionals, academic institutions and the community is essential to promote quality research and significant developments in women's health. This cooperation will be improved when involvement of citizens in the research process becomes standard.

Keywords: patient and public involvement; women; health; research; international collaboration

---

1. Introduction

Evidence-based sexual and reproductive health requires a global effort, with political and societal actions. A comprehensive paradigm shift is needed, where not only health professionals and researchers are ready to implement citizen participation, but also where political and administrative control of the health system allows and supports it [1]. Public and patient involvement is a field where policy has tended to outpace evidence. Exploring the impact of citizens in research is limited to investigating researchers' and public and patient contributors' reports of their views and experiences. Objective techniques for evaluating the impact and its influences remain hard to reach in a process that is inherently relational, subjective and socially constructed [2]. Public and patient involvement is as an expression of a democratization of healthcare, and a political and managerial tool to ensure quality, documentation, and equal treatment, thereby governing and controlling a public healthcare system. For example, the Danish public healthcare system is increasingly influenced by politicians. The use of standardized schemes and checklists used for documentation and quality assurance underlines this development, and patient and public involvement might be seen as adding to this movement [1].

The future worldwide vision is focused on building healthier lives, driven by the evidence-based provision of evidence-based healthcare to improve citizen's quality of life [3] However, evidence-based medicine faces challenges due to: lack of proper patient and public involvement or public and participant involvement (PPI) in research, leading to poor buy-in [4]; lack of incorporation of patients' lived experiences in research outcomes, reducing the relevance of evidence [5]; and lack of recommendations for shared decision-making in guidelines leading to ineffective communication between health workers and patients [6].

In the last decade, the incorporation of public participation initiatives has shown the potential to improve the quality and the object of research, allow a wide dissemination of the findings found, and promote a better integration of the results in clinical practice [7–9]. Public participation also allows an approach to sensitive phenomena such as sexual and reproductive health, or ethical aspects that arise during the development of the study [7].

Significant development of women's health requires international collaborations and participation of women in the research process. To improve sexual and reproductive health research, our objective should be to establish the importance of creating and developing the platforms and tools for public involvement. In this sense, creation of new tools and platforms involving a combination of traditional outreach and online strategies. Face-to-face invitations, training meetings, local civil society organizations (CSO) and family medicine health center support can be combined with online tools, such as online announcements, use of google meet, hangout or zoom, and social media platforms (Facebook, Twitter). These tools offer opportunities for community members to engage in interactive ways and can bring new input from the patient and participant involvement process.

2. Evidence-Based Sexual and Reproductive Health

Reproductive health (RH) was defined in 1994 at the International Conference on Population and Development as a "state of complete physical, mental and social well-being and not merely the absence of disease or infirmity, in all matters relating to the reproductive system, and to its functions and processes" [7]. Sexual and reproductive health (SRH) includes a number of health problems, such as (1) family planning; (2) maternal and newborn health care; (3) prevention, diagnosis and treatment of sexually transmitted infections (STIs) [8]; (4) adolescent sexual and reproductive health; (5) cervical cancer detection; and (6) prevention and management of infertility [9]. The World Health Organization include the Sexual and reproductive health in the health topics and highlighted that 'Sexual and reproductive health is a very personal subject, so people may have trouble finding or asking for accurate information about it. This may also help explain why these issues are still not addressed openly, and services are inadequate, fragmented and unfriendly in some countries in the European Region' [10]. Sexual and reproductive health services aim to prevent poor sexual and reproductive health, such as complications of pregnancy and childbirth, unwanted pregnancies, unsafe abortions, complications caused by STIs, sexual violence, and women dying from preventable cancer [10].

Reproductive health, including sexual health, and reproductive rights, as well as gender equality and women's empowerment, are relevant to improving the quality of life for everyone [7] as affirmed at various international forums, such as the International Conference on Population and Development (ICPD) and The Fourth World Conference on Women: Action for Equality, Development and Peace [11,12]. The Millennium Development Goals related to maternal and child health have still not been achieved in many countries; considering 35 countries, less than 30 percent of women of reproductive age use modern contraceptive methods. The method choice is still limited in many countries, due to lack of access or provider biases. Although there are good options for safe abortion, these services remain unavailable in many countries due to legal barriers, lack of training, and stigma. However, significant progress has been made in improving reproductive health. For example, family planning has expanded around the world, spurring a broader coverage of services with greater emphasis on quality and human rights; adolescent sexual health has been addressed with effective messages and services and new approaches try to reduce gender-based violence and clinical and policy guidelines have been incorporated [11].

The 2020 Sustainable Development Goals Report in goal five (gender equality) highlighting that there is still a lack of decision-making power for women, extending to their own reproductive health and that slightly more than half of all women (55%) make their own decisions when it comes to sexual and reproductive health and rights, based on 2007–2018 data from 57 countries on women aged 15 to 49, who are either married or in union. The analysis also found that only three in four women are making their own decisions regarding health care or on whether or not to have sex [12]. United Nations Member States have moved on to confirm their commitment to Sustainable Development Goals (SDG) including its targets designed to "ensure universal access to sexual and reproductive health (SRH) care services, including for family planning, information and education, and the integration of reproductive health into national strategies and programs by 2030" and to "ensure universal access to sexual and reproductive health and reproductive rights" [9].

Patient and public involvement in research could be an approach to effectively address poor sexual and reproductive health outcomes among women [13]. This enables researchers to identify better strategies, focusing, for example, on culturally and contextually appropriate research and prevention, equitable access to effective sexual health information, or quality education and training for public health professionals. The health, demographic change and well-being programs [14] are committed to the implementation of research to improve maternal and child health. Public participation and involvement in the investigation of pregnancy and childbirth differ from those under chronic diseases, especially because motherhood is a transitory health experience in the life of women and their families [4]. Women can have a variety of experiences in their pregnancy and delivery, ranging from low-risk, home birth without intervention to prolonged hospitalization with long-lasting morbidity [15]. Their own experience will influence your attitudes and inevitably your contribution to research. Their participation should guide the priority areas for users of health services. Moreover, the Faculty of Sexual and Reproductive Healthcare of the Royal College of Obstetrician and Gynaecologists values patient and public involvement in the development of clinical guidelines. It actively considers patient and public involvement across all stages of the guideline development process, from consulting with individuals on the proposed scope of the clinical guideline through to getting feedback on patient summaries. In order to facilitate patient and public involvement, a variety of methodologies (e.g., questionnaires, focus groups, interviews) across different settings (e.g., clinics, online, community) are considered [16].

3. Patient and Public Involvement

Effective patient and public involvement can be defined as where research is "being carried out "with" or "by" members of the public" not just "to", "about" or "for" them" [17]. The involvement process needs to change from an ad-hoc informal consultation into an established partnership underpinned by dedicated systems and infrastructure [18]. It must be implemented throughout the research lifecycle from idea generation, prioritization and commissioning into research conduct, publication and implementation in clinical practice [4,19].

It has been known for some time now that patient and public involvement has an important role in the identification of specific research questions [20,21], and that subsequent requests for research funding are more successful. Furthermore, it may be that public participation reverses in the reformulation of research questions [22], or even in the abandonment of a research project request if the involvement suggests that the research questions are not meaningful to the patient, public or citizens [23]. Additionally, patient and public involvement improves trial design [24], by ensuring that the design is acceptable to potential participants [25], facilitates recruitment for studies [26] and allows access to underserved and rarely heard populations, and patient and public involvement helps to disseminate the research results [27]. In short, patient and public involvement increases researchers' confidence in their studies [24] and promotes a greater understanding of the patient's perspective [28].

Internationally, there is increasing recognition of the importance of involving patients and the public in health-care governance and research [29]. Policymakers give the relevance to patient and public involvement as characterized by the phrase 'nothing about me without me' [30]. Thus, local [31] and international organizations, such as 'The European Patient's Forum' or 'The International Association for Public Participation' [32,33] are beginning to harness the power of participation by health service users, promoting public participation in research planning and the development of resources and information to support citizen participation and patient engagement. Promoting the role of patients in health research lifecycle has been at the heart of various international efforts including the European Network for Health Technology Assessment [34] and the Innovative Medicines Initiative which have called for better patient and public involvement in research synthesis projects across Europe [35]. Recently, the European Patients' Academy (EUPATI), a pan-European project was established as a collaborative public–private multi-stakeholder partnership of pharmaceutical industry, academia, and not-for-profit patient organizations [36]. These initiatives have developed various tools, platforms, and training resources for patient advocates to improve their participation in research with focus on four areas: pharmaceutical industry-led medical research, ethics committees, regulatory authorities, and health technology assessment. In turn, prestigious journals such as the British Medical Journal (BMJ) have also emphasized the use of patients and citizens as editors and emphasize the need for patients to participate as co-authors in the articles that are published [37].

Involving lay volunteers for problem-solving provided insights enhanced research design and served to identify weaknesses and barriers. Tensions and barriers have been generated with patient and public involvement in research [38]. On one hand, shared tensions usually are due to unclear roles, absent reporting guidelines, exclusion, framework limitations, resource allocation, and administrative boundaries. For example: active public involvement in the decision-making process of designing trials is less common than consultation on what has already been decided. Volunteers report needing early involvement to propose constructive changes. Researchers' worries about aggressive patients and those without respect for rules of confidentiality or data protection held up the research. Researchers may feel overwhelmed when they cannot fulfill the expectations they have over patients and public involvement [39]. On the other hand, shared barriers include those imposed by cultures, values, and power hierarchies. Limited involvement of the health community may occur through coalitions, collaborations, and partnerships [38].

Because research literacy could be an inconvenience in the process of patients and public involvement, some suggestions for getting the best from public involvement have been published [38]: ongoing support and implementation; training/capacity building; inclusion process; building trust and community and reinforcing patients' value and validity. In the second suggestion, capacity building, it is necessary to (1) provide training in research literacy and ethics, drawing on the many training programs that are available, (2) at every meeting have a jargon bin, when an unfamiliar term comes up, define it and use this to build glossaries, (3) promote a reciprocal learning relationship, letting volunteers know that researchers have made a long-term commitment to the patient and public partnership in research, and (4) encourage realistic expectations in volunteers and researchers and manage relationships with respect. Moreover, the use of patient and public involvement in dissemination planning, design, implementation, and distribution could increase public involvement, contribute to health literacy, and expand knowledge for patient values and preferences. The addition of patient reviewers by journals may contribute to health literacy and provide insights for future participatory research practice [38].

4. Public Involvement in Research to Address Gaps in Evidence

There is growing recognition of the importance of experiential knowledge being approached alongside scientific knowledge [40]. Patient and public involvement throughout the research lifecycle and in research syntheses, such as systematic reviews and clinical practice guidelines, is being demanded by funders, but there is often an intrinsic resistance in the field of biomedical research [41].

For example, the chief investigator who expressed skepticism about patient and public involvement, focused mainly on using public engagement to meet funding requirements or by including them on trial steering committees, whereas those who valued patient and public involvement often described in detail how it was of benefit within their trials [2]. Translation and effective implementation of evidence into practice through requires careful planning in consultation with patients as key stakeholders. When involving patients, it is imperative to take the perspective of the patient and their way of life as a starting point. Patient participation is achieved by inviting them to participate in their treatment and care, and in the research that surrounds them. It requires health professionals to be involved in the daily life of their patients, but it also requires organization and shared decisions that promote patient participation in daily clinical practice [1]. Then, establishing patients as a stakeholder group in a true partnership with researchers is the key to addressing these challenges [42,43]. Health research funders and regulators are promoting patient and public involvement in research and research syntheses [29]. However, many clinicians, researchers, systematic reviewers and guideline makers still do not have the tools required to involve patients. It is not always recognized that through their lived experiences and social background, patients can positively be involved in research with their unique perspectives, increasing its impact on society [44].

Another aspect of interest that must be taken into account is that, in the past, the choice of results in research studies was driven by researchers who did not consider the participation of patients or citizens [6]. This practice has often led to heterogeneity, making it difficult to synthesize research, replicating the same studies, and increasing research waste [5]. The implementation of core (standardized) sets of important outcomes for each health condition has been suggested as a possible novel solution to this problem [45]. A core outcome set (COS) is a minimum consensus-based set of results that should be measured and reported in all clinical trials for a specific health condition or intervention [46]. Issues related to outcome assessments, such as selective reporting and inconsistency between studies, can be addressed by developing a core set. Development requires reaching a consensus on: (1) core outcome domains and (2) core outcome measurement instruments. Besides, methods used to reach consensus include systematic reviews of the literature to inform the process, qualitative research with physicians and patients, group discussions (for example, using the nominal group technique), and structured surveys (for example, using the Delphi technique). Multiple stakeholders must be involved in the process, and particular attention must be paid to patients [47]. Traditionally, the guidelines promoted by leading health experts have generated subjective recommendations that correspond to the provision of local, regional or national clinical care [48]. This has produced a limited effect on health outcomes and has promoted high and unjustified variation in the provision of care and services to users.

We have to adopt a novel approach by involving a wide range of key stakeholders of different disciplines with direct involvement in healthcare planning and provision for women and children. A strong level of cooperation is necessary via effective communication channels between healthcare professionals, academic institutions and the community [49]. Stakeholders and activities for participant involvement in research are shown in Figure 1. We suggest that these developments include clinicians (e.g., gynaecologists and paediatricians), methodologists (e.g., statisticians and systematic reviewers) and lay consumers (e.g., patients with lived experiences and civil society organization (CSO) representatives and related to sexual and reproductive health). Clinicians and professional bodies should promote patient and public participation in research helping to develop know-how in participants engagement. Methodologists—researchers, guideline makers and systematic reviewers included—will help in the scientific training of lay research enablers, to develop the local know–how on participant involvement in research. Lay consumers will provide the voice of lay sexual and reproductive health participants of all backgrounds to engage in the research process and to be involved in the dissemination of the research. The patient perspective is valuable to ensure quality in healthcare. However, the patient perspective is only one perspective, similar to the ones of other important groups (e.g., health professionals, leaders, administrators, politicians). A selected coordinator for patient and public involvement is key to facilitating interactions. We agree with other authors, the coordinator should be experienced in the potential challenges of public and patient involvement, including power and control issues, and the consequences of public and patient contributors lacking knowledge of research processes, terminology, and ethical constraints [39].

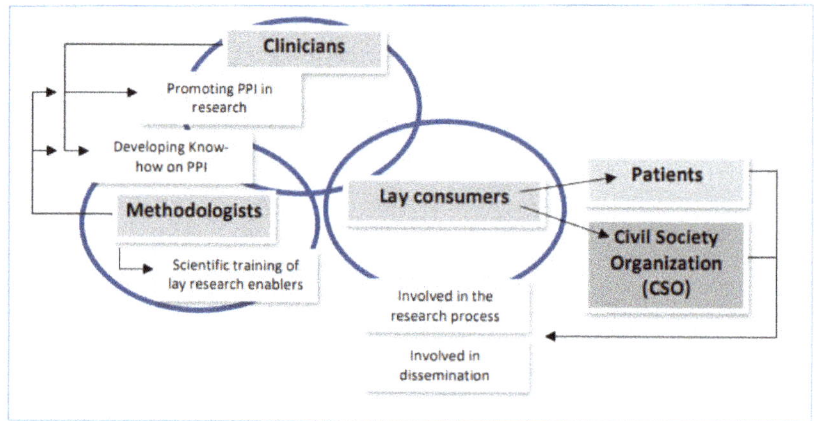

Figure 1. Key stakeholders and activities in public and participant involvement (PPI).

Furthermore, patient and public involvement recruitment strategies are necessary, ensuring that we apply fair and transparent processes, that there is a shared understanding of the roles, offering support and learning for public participation in the research. For example: (1) women participating in actives randomized controlled trials; (2) women and men contacted from midwives who collaborate closely with the clinical institutions; (3) women and men with a disability or belonging to disadvantaged social classes; (4) social media, web sites and banners in health centers and nurseries to recruit participants and (5) civil society organizations (CSO).

The patient and public involvement should be made up of citizens and the public. They should receive advanced training on citizen participation in research, such as has been carried out within individual projects [4]. This training should include some important topics: (1) Introduction to health research; (2) Different types of research and study designs; (3) Perspective of the research from the citizen's point of view; (4) Research life cycle; (5) Different roles for the audience (participation, engagement); (6) Why involve the public in research (including diversity and inclusion in research); (7) Case studies: examples and public impact on research; (8) How to do research with the commitment of citizens; (9) How can the public do "citizen science"?; (10) How to maximize the impact of "citizen science"; (11) Solution of practical problems in the field; (12) Research dissemination and beyond; and (13) Measurement of the value and impact of citizen research. Developing these initiatives further, more citizens should receive basic training in citizen participation and the relevance of the investigation through dedicated training sessions. Patient and public involvement aims are varied depending on the research subject. Therefore, it is less common to involve patients and the public in the research process, than in other areas such as genetics, biobank research, cancer research, clinical trial research or decision-making [50]. The methodology mainly used for patient and public involvement in researching is information sharing, questionnaires, interviews or focus groups. Importantly, conducting a focus group at the beginning of the study and another at the end, you can identify the needs, barriers and feedback of the patient and public participations, trying to respond to how diversity impacts innovation, how to act to promote the inclusion of people with disabilities or from more disadvantaged communities, relationships between citizens, researchers and institutions, the type of language used to communicate and whether there have been unexpected opportunities. On the other hand, the research team is focused on the public and patient involvement of consultation, deliberation and participation [50].

To achieve quality results in the process of patient and public involvement in research, the Guidance for Reporting Involvement of Patients and the Public (GRIPP2) can be useful, the first international evidence-based, consensus informed guidance for reporting patient and public involvement in research [51]. There is a short and a long version of GRIPP2, that aims to improve the quality, transparency of the international patient and public involvement. The short form has five sections or topics: (1) Aim of patient and public involvement in the study; (2) Methods used in the public involvement in the study; (3) Study results: positive and negative; (4) Discussion, describing the positive and negative effects on the extent to which patient and public involvement influenced the study, and conclusions; and (5) Reflections/critical perspective, comment critically the things that went well and those that did not. On the other hand, the long form includes eight sections: (1) Abstract of the paper; (2) Background to the paper; (3) Aims of the paper; (4) Methods of the paper; (5) Measurement of the patient and public involvement: methods used to qualitatively and quantitatively measure or assess the impact of public involvement and methods used to capture or measure its impact; (6) Economic assessment of public involvement in the study; (7) Study results; and (8) Discussion and conclusions. This last section incorporates the reflections and critical perspective of the study. In order to improve current public involvement practice in the sexual health field, a recently published study carried out audits on patients and public involvement plans in the local sexual health service ($n = 18$), and a refined audit tool was completed in research projects ($n = 5$). The responses of the tools used in the audit showed a wide variability in practice. Problems included the combination of patient participation work and qualitative research, lack of goals in patient participation in research and lack of responsiveness around patient needs, and insufficient resources to work with patient and public involvement. Specific research problems included late participation of patients after key decisions had already been made [52].

At present, there are still some limitations about the evidence of the patient and public involvement research outcomes. In a systematic review, mapping the impact of the patient and public involvement in health and social care research [42], authors aimed to examine the conceptualization, measurement, impact and outcomes of patient and public involvement in health and social care research. This study provides the first international evidence of patient and public involvement impact that has emerged in the research process, however, much of the evidence concerning the impact remains weak.

It is important to remark that the "ReseArch with Patient and Public invOlvement: a RealisT evaluation" study (RAPPORT study) [53] aimed to determine the types of patient and public involvement in funded research, analyze the contextual and dynamics of patient and public involvement and explore the experience of patient and public involvement in research. They found that six striking actions were required for effective patient and public involvement: (1) A clear purpose, role and structure for patient and public involvement; (2) a key individual co-ordinating patient and public involvement ensuring diversity; (3) whole research team engagement with patient and public involvement, this is a research team positive about patient and public involvement input and engaged with it; (4) mutual understanding and trust between the researchers and lay representatives; (5) ensuring opportunities for patient and public involvement throughout the research process; and (6) and evaluating patient and public involvement in a proactive and systematic approach. Future work is required, exploring the impact of virtual patient and public involvement, economic evaluation of patient and public involvement, and a longer-term follow-up study of the outcomes of patient and public involvement on research findings and impact on services and clinical practice.

5. Clinical Relevance and Goals of Patient and Public Involvement in Research

The most documented benefits of patient and public involvement in research are the involvement of the patient and public groups at all stages of the trial [39], utility in randomized controlled trials [2], contribution in the design of clinical trials [38], and perceptions of patient involvement in research and clinical practice [1].

In a qualitative study of patients and researchers from a cohort of randomized trials [2], 21 chief investigators, 10 trial managers and 17 patient and public involvement contributors were interviewed from 28 trials. The conclusions were that to enhance patient and public involvement in trials is crucial to develop goals for patient and public involvement at an early stage that fits the needs of the trial, plan patient and public involvement implementation in accordance with these goals, invest in developing good relationships between patient and public involvement contributors and researchers, and favor responsive and managerial roles for contributors. A systematic overview of public and patient involvement in clinical trials design included 27 reviews [38]. Public involvement roles were primarily based on agenda setting, steering committees, ethical review, protocol development, and piloting. Public involvement was reported to have increased the quantity and quality of patient relevant priorities and outcomes, enrolment, funding, design, implementation, and dissemination.

We would like to mention a couple of examples were patients and the public were involved in research. The first is a 3D study [39], where fourteen people living with multiple long-term conditions (multimorbidity) were included as patient and public involvement contributors to a randomized controlled trial to improve care for people with multimorbidity. Meetings took place approximately four times a year throughout the trial, beginning at the grant application phase. Their experience also supports the findings of the RAPPORT study [53] in four of the six conditions most influential in establishing effective patient and public involvement: (1) a research team positive about patient and public involvement, (2) a key individual to co-ordinate, ensuring diversity, and relationships that are established and maintained over time, (3) a shared understanding of moral and methodological purposes of patient and public involvement and (4) proactive and systematic evaluation of patient and public involvement. The second example is a secondary analysis of women's perceptions and experiences of egg aspiration in fertility treatment [1]. This secondary analysis was initially inserted in a randomized controlled study where the aim was to gain insight into perceptions and experiences within a group of women undergoing fertility treatment through two focus group interviews. They pointed out that knowledge and awareness of the difference in perspectives is important when healthcare professionals seek to involve patients both in clinical practice and in research.

Secondly, patient and public involvement in research has doubled the impact in trials: focused and diffuse [2]. Focused impact comprises patient and public contributors' input that, from the perspective of the informant, changed or influenced an aspect of the trial, whereas diffuse impacts comprise patient and public contributions that influenced the way researchers thought or felt about the trial. Focused impacts included patient and public contributors helping to choose the primary outcome for the trial and to increase recruitment through their contacts and networks. Diffuse impacts largely entailed interactions between researchers and patients and public contributors that helped to reassure the research team and increase or maintain their confidence and motivation for the trial.

Specific examples of the impact of patient and public involvement in researchers' professional lives are those where patients become co-authors in publications, co-applicants in projects call and trainers [4,54–56] or even becoming new editors in a prestigious journal, such as the BMJ Sexual and Reproductive Health [57], who wrote: 'Patient involvement is an important piece of cultural change in healthcare, and will require commitment, time and sensitivity, and a willingness to experiment and learn from mistakes'.

6. Conclusions

Without the active participation of citizens, sexual and reproductive health research risks losing relevance and validity. Health research funders and regulators are promoting patient and public involvement in research but there is a lack of quality tools for involving patients. International collaborations, and the cooperation between healthcare professionals, academic institutions and the community are essential to promote quality research in sexual and reproductive women's health. A significant development in women's health will be achieved when the involvement of citizens in the research process becomes standard.

Author Contributions: Conceptualization, K.K.S., G.-M.M., B.-C.A.; original draft preparation, A.-P.C.; writing—review and editing, B.-C.A., G.-M.M., H.A.W.B., J.J.S. and K.K.S. All authors have read and agreed to the published version of the manuscript.

Funding: This research received no external funding.

Acknowledgments: Professor Khan is Distinguished Investigator at University of Granada funded by the Beatriz Galindo (senior modality) program of the Spanish Ministry of Education. We would like to thank all women involved in health researching and those taking part in the development of the research process.

Conflicts of Interest: The authors declare no conflict of interest.

References

1. Handberg, C.; Beedholm, K.; Bregnballe, V.; Nellemann, A.N.; Seibaek, L. Reflections on patient involvement in research and clinical practice: A secondary analysis of women's perceptions and experiences of egg aspiration in fertility treatment. *Nurs. Inq.* **2018**, *25*, e12210. [CrossRef] [PubMed]
2. Dudley, L.; Gamble, C.; Preston, J.; Buck, D.; Hanley, B.; Williamson, P.; Young, B. What Difference Does Patient and Public Involvement Make and What Are Its Pathways to Impact? Qualitative Study of Patients and Researchers from a Cohort of Randomised Clinical Trials. *PLoS ONE* **2015**, *10*, e0128817. [CrossRef] [PubMed]
3. Factsheet. European Comision Health. Available online: https://ec.europa.eu/info/sites/info/files/research_and_innovation/knowledge_publications_tools_and_data/documents/ec_rtd_factsheet-health_2019.pdf (accessed on 22 October 2020).
4. Moss, N.; Daru, J.; Lanz, D.; Thangaratinam, S.; Khan, K.S. Involving pregnant women, mothers and members of the public to improve the quality of women's health research. *BJOG Int. J. Obstet. Gynaecol.* **2017**, *124*, 362–365. [CrossRef] [PubMed]
5. Khan, K.; O'Donovan, P. The CROWN Initiative: Journal editors invite researchers to develop core outcomes in women's health. *BMC Women's Health* **2014**, *14*, 75. [CrossRef]
6. Mockford, C.; Staniszewska, S.; Griffiths, F.; Herron-Marx, S. The impact of patient and public involvement on UK NHS health care: A systematic review. *Int. J. Qual. Health Care J. Int. Soc. Qual. Health Care* **2012**, *24*, 28–38. [CrossRef] [PubMed]
7. International Conference on Population and Development (ICPD). Cairo. 1994. Available online: http://www.ipci2014.org/en/node/64 (accessed on 9 September 2020).
8. WHO Geneva. Sexually Transmitted Infections (STIs). Available online: https://www.who.int/news-room/fact-sheets/detail/sexually-transmitted-infections-(stis) (accessed on 20 September 2020).
9. WHO Regional Office for Europe Committee. Action Plan for Sexual and Reproductive Health. Towards Achieving the 2030 Agenda for Sustainable Development in Europe—Leaving No One Behind. 2020. Available online: https://www.euro.who.int/__data/assets/pdf_file/0003/322275/Action-plan-sexual-reproductive-health.pdf?ua=1 (accessed on 22 October 2020).
10. WHO Regional Office for Europe. Copenhagen. Sexual and Reproductive Health. 2020. Available online: https://www.euro.who.int/en/health-topics/Life-stages/sexual-and-reproductive-health (accessed on 15 September 2020).
11. World Health Organization. A Decade of Tracking Progress for Maternal, Newborn and Child Survival: The 2015 Report. 2015. Available online: http://www.who.int/pmnch/media/events/2015/countdown/en/ (accessed on 10 September 2020).
12. United Nations. The Sustainable Development Goals Report; 20 September 2020. Available online: https://unstats.un.org/sdgs/report/2020/The-Sustainable-Development-Goals-Report-2020.pdf (accessed on 20 September 2020).
13. Prather, C.; Fuller, T.R.; Jeffries, W.L.T.; Marshall, K.J.; Howell, A.V.; Belyue-Umole, A.; King, W. Racism, African American Women, and Their Sexual and Reproductive Health: A Review of Historical and Contemporary Evidence and Implications for Health Equity. *Health Equity* **2018**, *2*, 249–259. [CrossRef]
14. European Commisions. Horizon 2020. Work Programme 2018–2020. Health, Demographic Change and Wellbeing. Available online: https://ec.europa.eu/research/participants/data/ref/h2020/wp/2018-2020/main/h2020-wp1820-health_en.pdf (accessed on 20 September 2020).

15. Frew, P.M.; Saint-Victor, D.S.; Isaacs, M.B.; Kim, S.; Swamy, G.K.; Sheffield, J.S.; Edwards, K.M.; Villafana, T.; Kamagate, O.; Ault, K. Recruitment and retention of pregnant women into clinical research trials: An overview of challenges, facilitators, and best practices. *Clin. Infect. Dis. Off. Publ. Infect. Dis. Soc. Am.* **2014**, *59* (Suppl. 7), S400–S407. [CrossRef]

16. The Faculty of Family Planning and Reproductive Health Care of the Royal College of the Obstetricians and Gynaecologists. Framework for Clinical Guideline Development. Available online: https://www.fsrh.org/site-search/?keywords=PPI (accessed on 15 October 2020).

17. The National Institute for Health Research. What Is Public Involvement in Research? 2019. Available online: https://www.invo.org.uk/find-out-more/what-is-public-involvement-in-research-2/ (accessed on 20 September 2020).

18. Staniszewska, S.; Jones, N.; Newburn, M.; Marshall, S. User involvement in the development of a research bid: Barriers, enablers and impacts. *Health Expect. Int. J. Public Particip. Health Care Health Policy* **2007**, *10*, 173–183. [CrossRef]

19. Pandya-Wood, R.; Barron, D.S.; Elliott, J. A framework for public involvement at the design stage of NHS health and social care research: Time to develop ethically conscious standards. *Res. Involv. Engagem.* **2017**, *3*, 6. [CrossRef]

20. Stevens, T.; Wilde, D.; Hunt, J.; Ahmedzai, S.H. Overcoming the challenges to consumer involvement in cancer research. *Health Expect. Int. J. Public Particip. Health Care Health Policy* **2003**, *6*, 81–88. [CrossRef]

21. Wright, D.; Corner, J.; Hopkinson, J.; Foster, C. Listening to the views of people affected by cancer about cancer research: An example of participatory research in setting the cancer research agenda. *Health Expect. Int. J. Public Particip. Health Care Health Policy* **2006**, *9*, 3–12. [CrossRef] [PubMed]

22. Walker, D.M.; Pandya-Wood, R. Can research development bursaries for patient and public involvement have a positive impact on grant applications? A UK-based, small-scale service evaluation. *Health Expect. Int. J. Public Particip. Health Care Health Policy* **2015**, *18*, 1474–1480. [CrossRef] [PubMed]

23. Boote, J.D.; Dalgleish, M.; Freeman, J.; Jones, Z.; Miles, M.; Rodgers, H. 'But is it a question worth asking?' A reflective case study describing how public involvement can lead to researchers' ideas being abandoned. *Health Expect. Int. J. Public Particip. Health Care Health Policy* **2014**, *17*, 440–451. [CrossRef]

24. Vale, C.L.; Thompson, L.C.; Murphy, C.; Forcat, S.; Hanley, B. Involvement of consumers in studies run by the Medical Research Council Clinical Trials Unit: Results of a survey. *Trials* **2012**, *13*, 9. [CrossRef]

25. Edwards, V.; Wyatt, K.; Logan, S.; Britten, N. Consulting parents about the design of a randomized controlled trial of osteopathy for children with cerebral palsy. *Health Expect. Int. J. Public Particip. Health Care Health Policy* **2011**, *14*, 429–438. [CrossRef] [PubMed]

26. Iliffe, S.; McGrath, T.; Mitchell, D. The impact of patient and public involvement in the work of the Dementias & Neurodegenerative Diseases Research Network (DeNDRoN): Case studies. *Health Expect. Int. J. Public Particip. Health Care Health Policy* **2013**, *16*, 351–361.

27. Goodman, C.; Mathie, E.; Cowe, M.; Mendoza, A.; Westwood, D.; Munday, D.; Wilson, P.M.; Crang, C.; Froggatt, K.; Iliffe, S. Talking about living and dying with the oldest old: Public involvement in a study on end of life care in care homes. *Bmc Palliat. Care* **2011**, *10*, 20. [CrossRef]

28. Hewlett, S.; Wit, M.; Richards, P.; Quest, E.; Hughes, R.; Heiberg, T.; Kirwan, J. Patients and professionals as research partners: Challenges, practicalities, and benefits. *Arthritis Rheum.* **2006**, *55*, 676–680. [CrossRef]

29. Boivin, A.; Currie, K.; Fervers, B.; Gracia, J.; James, M.; Marshall, C.; Sakala, C.; Sanger, S.; Strid, J.; Thomas, V. Patient and public involvement in clinical guidelines: International experiences and future perspectives. *Qual. Saf. Health Care* **2010**, *19*, e22. [CrossRef]

30. Delbanco, T.; Berwick, D.M.; Boufford, J.I.; Edgman-Levitan, S.; Ollenschläger, G.; Plamping, D.; Rockefeller, R.G. Healthcare in a land called PeoplePower: Nothing about me without me. *Health Expect. Int. J. Public Particip. Health Care Health Policy* **2001**, *4*, 144–150. [CrossRef]

31. National Institute for Health Research. INVOLVE Initiative. Available online: http://www.invo.org.uk (accessed on 9 September 2020).

32. International Association for Public Participation. Advancing the Practice of Public Participation. 2020. Available online: https://www.iap2.org/mpage/Home (accessed on 9 September 2020).

33. European Union. European Patients' Forum. EPF 2019 Work Programme. Available online: http://www.eu-patient.eu (accessed on 9 September 2020).

34. Nielsen, C.P.; Lauritsen, S.W.; Kristensen, F.B.; Bistrup, M.L.; Cecchetti, A.; Turk, E. Involving stakeholders and developing a policy for stakeholder involvement in the European network for health technology assessment, EUnetHTA. *Int. J. Technol. Assess. Health Care* **2009**, *25* (Suppl. 2), 84–91. [CrossRef]

35. Laverty, H.; Gunn, M.; Goldman, M. Improving R&D productivity of pharmaceutical companies through public-private partnership: Experiences from the Innovative Medicines Initiative. *Expert Rev. Pharm. Outcomes Res.* **2012**, *12*, 545–548.

36. Pushparajah, D.S.; Geissler, J.; Westergaard, N. Collaboration between patients, academia and industry to champion the informed patient in the research and development of medicines. *J. Med. Dev. Sci.* **2015**, *1*, 74–80. [CrossRef]

37. Price, A.; Schroter, S.; Snow, R.; Hicks, M.; Harmston, R.; Staniszewska, S.; Parker, S.; Richards, T. Frequency of reporting on patient and public involvement (PPI) in research studies published in a general medical journal: A descriptive study. *BMJ Open* **2018**, *8*, e020452. [CrossRef] [PubMed]

38. Price, A.; Albarqouni, L.; Kirkpatrick, J.; Clarke, M.; Liew, S.M.; Roberts, N.; Burls, A. Patient and public involvement in the design of clinical trials: An overview of systematic reviews. *J. Eval. Clin. Pract.* **2018**, *24*, 240–253. [CrossRef]

39. Mann, C.; Chilcott, S.; Plumb, K.; Brooks, E.; Man, M.S. Reporting and appraising the context, process and impact of PPI on contributors, researchers and the trial during a randomised controlled trial—The 3D study. *Res. Involv. Engagem.* **2018**, *4*, 15. [CrossRef]

40. Caron-Flinterman, J.F.; Broerse, J.E.; Bunders, J.F. The experiential knowledge of patients: A new resource for biomedical research? *Soc. Sci. Med.* **2005**, *60*, 2575–2584. [CrossRef]

41. Prior, L. Belief, knowledge and expertise: The emergence of the lay expert in medical sociology. *Sociol. Health Illn.* **2003**, *25*, 41–57. [CrossRef] [PubMed]

42. Brett, J.; Staniszewska, S.; Mockford, C.; Herron-Marx, S.; Hughes, J.; Tysall, C.; Suleman, R. Mapping the impact of patient and public involvement on health and social care research: A systematic review. *Health Expect. Int. J. Public Particip. Health Care Health Policy* **2014**, *17*, 637–650. [CrossRef]

43. Brett, J.; Staniszewska, S.; Mockford, C.; Herron-Marx, S.; Hughes, J.; Tysall, C.; Suleman, R. A systematic review of the impact of patient and public involvement on service users, researchers and communities. *Patient* **2014**, *7*, 387–395. [CrossRef] [PubMed]

44. Entwistle, V.A.; Renfrew, M.J.; Yearley, S.; Forrester, J.; Lamont, T. Lay perspectives: Advantages for health research. *BMJ (Clin. Res. Ed.)* **1998**, *316*, 463–466. [CrossRef] [PubMed]

45. Kirkham, J.J.; Gargon, E.; Clarke, M.; Williamson, P.R. Can a core outcome set improve the quality of systematic reviews?—A survey of the Co-ordinating Editors of Cochrane Review Groups. *Trials* **2013**, *14*, 21. [CrossRef] [PubMed]

46. Williamson, P.R.; Altman, D.G.; Bagley, H.; Barnes, K.L.; Blazeby, J.M.; Brookes, S.T.; Clarke, M.; Gargon, E.; Gorst, S.; Harman, N. The COMET Handbook: Version 1.0. *Trials* **2017**, *18* (Suppl. 3), 280. [CrossRef] [PubMed]

47. Chiarotto, A.; Ostelo, R.W.; Turk, D.C.; Buchbinder, R.; Boers, M. Core outcome sets for research and clinical practice. *Braz. J. Phys. Ther.* **2017**, *21*, 77–84. [CrossRef]

48. Evidence-Based Medicine. A new approach to teaching the practice of medicine. *JAMA* **1992**, *268*, 2420–2425. [CrossRef]

49. Haerry, D.; Landgraf, C.; Warner, K.; Hunter, A.; Klingmann, I.; May, M.; See, W. EUPATI and Patients in Medicines Research and Development: Guidance for Patient Involvement in Regulatory Processes. *Front. Med.* **2018**, *5*, 230. [CrossRef]

50. Lander, J.; Langhof, H.; Dierks, M.L. Involving patients and the public in medical and health care research studies: An exploratory survey on participant recruiting and representativeness from the perspective of study authors. *PLoS ONE* **2019**, *14*, e0204187. [CrossRef]

51. Staniszewska, S.; Brett, J.; Simera, I.; Seers, K.; Mockford, C.; Goodlad, S.; Altman, D.G.; Moher, D.; Barber, R.; Denegri, S. GRIPP2 reporting checklists: Tools to improve reporting of patient and public involvement in research. *Res. Involv. Engagem.* **2017**, *3*, 13. [CrossRef]

52. Meyrick, J.; Gray, D. Evidence-based patient/public voice: A patient and public involvement audit in the field of sexual health. *BMJ Sex. Reprod. Health* **2018**. [CrossRef]

53. Wilson, P.; Mathie, E.; Keenan, J.; McNeilly, E.; Goodman, C.; Howe, A.; Poland, F.; Staniszewska, S.; Kendall, S.; Munday, D. Health Services and Delivery Research. In *ReseArch with Patient and Public invOlvement: A RealisT Evaluation—The RAPPORT Study*; NIHR Journals Library: Southampton, UK, 2015.
54. Allotey, J.; Fernandez-Felix, B.M.; Zamora, J.; Moss, N.; Bagary, M.; Kelso, A.; Khan, R.; Van der Post, J.A.M.; Mol, B.W.; Pirie, A.M. Predicting seizures in pregnant women with epilepsy: Development and external validation of a prognostic model. *PLoS Med.* **2019**, *16*, e1002802. [CrossRef]
55. Moss, N. Meaningful consent can only really be established after an emergency. *BJOG Int. J. Obstet. Gynaecol.* **2018**, *125*, 1755. [CrossRef] [PubMed]
56. Thangaratinam, S.; Marlin, N.; Newton, S.; Weckesser, A.; Bagary, M.; Greenhill, L.; Rikunenko, R.; D'Amico, M.; Rogozinska, E.; Kelso, A. AntiEpileptic drug Monitoring in PREgnancy (EMPiRE): A double-blind randomised trial on effectiveness and acceptability of monitoring strategies. *Health Technol. Assess. (Winch. Engl.)* **2018**, *22*, 1–152. [CrossRef]
57. Pepper, L. Patient and public involvement in sexual and reproductive health: A new editor, and a new tool. *BMJ Sex. Reprod. Health* **2018**, *44*, 237–238. [CrossRef] [PubMed]

International Journal of
Environmental Research
and Public Health

Brief Report

Association between Sexual Habits and Sexually Transmitted Infections at a Specialised Centre in Granada (Spain)

Raquel Casado Santa-Bárbara [1], César Hueso-Montoro [2,*], Adelina Martín-Salvador [3,*], María Adelaida Álvarez-Serrano [4], María Gázquez-López [4] and María Ángeles Pérez-Morente [5]

[1] Gregorio Marañón Hospital, 28007 Madrid, Spain; raquelcasado60@hotmail.com
[2] Department of Nursing, Faculty of Health Sciences, University of Granada, 18016 Granada, Spain
[3] Department of Nursing, Faculty of Health Sciences, University of Granada, 52005 Melilla, Spain
[4] Department of Nursing, Faculty of Health Sciences, University of Granada, 51001 Ceuta, Spain;
     adealvarez@ugr.es (M.A.Á.-S.); mgazquez@ugr.es (M.G.-L.)
[5] Department of Nursing, Faculty of Health Sciences, University of Jaén, 23071 Jaén, Spain;
     mmorente@ujaen.es
*   Correspondence: cesarhueso@ugr.es (C.H.-M.); ademartin@ugr.es (A.M.-S.)

Received: 8 August 2020; Accepted: 19 September 2020; Published: 21 September 2020

Abstract: Sexually transmitted infections are an important public health issue. The purpose of this study is to analyse the association between different sexual habits and the prevalence of sexually transmitted infections in the population of Granada who consult with a specialised centre. An observational, cross-sectional study was conducted based on the medical records of 678 people from the Sexually Transmitted Diseases and Sexual Orientation Centre of Granada, who were diagnosed positively or negatively with a sexually transmitted infection, during the 2000–2014 period. Sociodemographic and clinical data, as well as data on frequency and type of sexual habits, frequency of condom use and sexually transmitted infection positive or negative diagnosis were collected. Univariate and bivariate analyses were conducted. The most popular sexual habits were vaginal intercourse, oral sex (mouth–vagina and mouth–penis) and the least popular were anus–mouth and anal sex. The use of condom is frequent in vaginal and anal sex and less frequent in oral sex. Sexually transmitted infection is associated with mouth–penis ($p = 0.004$) and mouth–vagina ($p = 0.023$) oral sex and anal sex ($p = 0.031$). It is observed that there is a relationship between the presence of STIs and oral sex practices, people having such practices being the ones who use condoms less frequently. There is also a relationship between anal sex and the prevalence of STIs, although in such sexual practice the use of condom does prevail.

Keywords: sexually transmitted infections; sexual behaviour; sexual health

---

1. Introduction

Risky sexual habits and behaviour strongly affect public health. According to the WHO, more than one million people acquire a sexually transmitted infection (STI) every day. The most recent averaged data indicate that one out of 25 people has at least one STI, maybe being infected with several of them at the same time [1].

Complications derived from an STI have a great impact on children's, teenagers', and adults' health and life on a worldwide basis, thus compromising their quality of life. Furthermore, they facilitate indirectly the transmission of HIV and cause cellular changes that lead to some types of cancer. They also impact on domestic economy and national health systems [2].

STIs are more common in people having high-risk sexual behaviour and attitude such as having sexual intercourse without using condoms, having contact without buccal–genital protection, having multiple sex partners, having a high-risk partner (someone having many sex partners or other risk factors), having anal sex or a partner who has an STI, having sexual intercourse with sex workers, or having sexual intercourse with someone who injects or has injected drugs before [3,4]. Moreover, an increase in the use of the Internet and mobile applications to easily meet sexual partners, thus increasing high-risk sexual practices, should be highlighted [5,6].

Hence, it is important to know the population's sexual habits in order to prioritise sexual health strategies that shall reduce the incidence of STIs, and therefore reduce the morbidity and mortality load worldwide derived from sexually transmitted pathogens.

Research has focused on studying the use of condom and its effectiveness in people with high-risk sexual contact [7–9], however, there are only a few studies on general population, which show that they are less aware of risks and motivation for condom use decreases [10,11]. In the United States, among 15–44-year-old people, prevalence of anal intercourse slightly increased from 2002 to 2015 both in women (30–33%) and men (34–38%), condoms being used in 32% in women and 46% in men, and only 7% during oral sex; such low percentages regarding condom use in high-risk sexual practices lead to an increase in the likelihood of getting an STI [12].

In Spain, a Blanc et al.'s study [13] revealed that the use of condom is higher in vaginal intercourse, and diminishes in anal sex and it is used in a lower percentage during oral sex (29.2%, 17.0%, and 2.0%, respectively). These results support the idea that young people use condoms more to avoid pregnancy than to prevent STIs.

Most published studies address sexual habits, condom use, and prevalence of STIs independently [14–16]; therefore, the purpose of this study is to analyse the association between different sexual practices and an STI diagnosis in people consulting with the Sexually Transmitted Diseases and Sexual Orientation Centre of Granada (Spain).

2. Materials and Methods

An observational, retrospective and cross-sectional study was conducted based on the medical records of people consulting with the Sexually Transmitted Diseases and Sexual Orientation Centre of Granada, attached to the Andalusian Health Service (Spain). It is the STI reference centre in the Granada province, it belongs to the public health system and it carries out preventive, diagnoses-related and therapeutic activities.

The sample was taken from a database composed of 1437 medical records of subjects attending for consultation associated with the presence or suspicion of an STI between 2000–2014. The sampling and data collection process may be checked in detail in a previous publication [17]. From said database, 678 medical records were taken for this research, which records complied with the criterion of having a test made to confirm or rule out the presence of an STI. Each history corresponds to one subject, so that there is no duplicate information in the final case selection. Every history was reviewed by a member of the research team, who took out data to include them in a data collection sheet created to that effect for this study. One of the disadvantages of this process was related to missing data, so the sample sizes defined for every analysis performed are specified in the data results.

The STI diagnosis was considered a dependent variable, collected as a binary qualitative variable (Yes, No). Sexual habits were established as an independent variable, identifying five practices: vaginal intercourse, oral (mouth–vagina), oral (mouth–penile), oral (anus–mouth), anal (penile–anus) intercourse. Subjects were asked about the frequency of such practices, so every habit was collected as a binary qualitative variable (Never or sporadic, Frequent or always). Additionally, data on condom use were collected for each and every described practice, following the same categorization in answers. To classify the sample, sociodemographic variables (age, sex, nationality, occupation, working status, education level, marital status, and sexual orientation), medical variables (reason for consultation) and risk indicators (age of first sexual intercourse, last time having sex without a condom, number of partners in the last month and in the past 12 months, use of drugs, and prior STI) were collected.

A univariate analysis was conducted by calculating the median (Me), interquartile range (IQR), frequency (*n*), and percentage (%), according to the variable type. The bivariate analysis was performed in order to analyse the association between STIs (dependent variable) and sexual habits (independent variable), via the chi-square test, by calculating the effect size through the Cramer's V (V), plus the Odds ration (OR) (Confidence Interval (CI) 95%).

In every case, a significant association was considered with $p < 0.05$. Univariate and bivariate analyses were conducted using the Statistical Package for the Social Sciences (SPSS) program, version 22, (IBM, New York, NY, USA).

The study protocol was approved by the Biomedical Research Ethics Committee of the province of Granada (research protocol approved on 12 February 2012, and 1 April 2015), as well as by the Management Directorate of the Granada-Metropolitan Health District, to which the centre where data were collected is attached.

3. Results

Table 1 shows results corresponding to sociodemographic variables, medical care, and risk indicators. The general profile of the sample was a young subject, Spanish, with higher-level education, employed, mostly single, and heterosexual. The sample included a similar proportion of men and women, although with a slightly higher percentage of men. They consulted with the STI centre for some reason related to STI contagion or suspicion, highlighting the presence of symptoms as the reason for consultation.

Table 1. Sociodemographic features, medical care and risk indicators.

Variables	Me (IQR)
Age (*n* = 678)	26 (23–33)
Age at first sexual intercourse (*n* = 322)	17 (16–18)
	n (%)
Sex (*n*= 678)	
Male	391 (57.7)
Female	287 (42.3)
Citizenship (*n* = 671)	
Spanish	511 (76.2)
Immigrant	160 (23.8)
Occupation (*n* = 630)	
Other occupation	316 (50.2)
Student	229 (36.3)
Sex workers/former workers	85 (13.5)
Employment (*n* = 615)	
Active	302 (49.1)
Unemployed	77 (12.5)
Retired	7 (1.1)
Student	229 (37.2)
Education level (*n* = 640)	
Without education	4 (0.6)
Primary	116 (18.1)
Secondary	150 (23.4)
Vocational training	70 (10.9)
Higher (University)	300 (46.9)

Table 1. *Cont.*

Variables	Me (IQR)
Marital status (*n* = 675)	
Single	544 (80.6)
Married/Domestic partner	92 (13.9)
Separated/Divorced	38 (5.6)
Widower	1 (0.1)
Sexual orientation (*n* = 657)	
Heterosexual	547 (83.3)
Homosexual	81 (12.3)
Bisexual	29 (4.4)
Reason for consultation * (*n* = 678)	
HIV	219 (32.3)
STI symptoms	429 (63.3)
Control	20 (2.9)
Follow-up of contacts	10 (1.5)
Last time having sex without condom (*n* = 385)	
Never used it	31 (8.1)
Last month	178 (46.2)
1–6 months	147 (38.2)
6–12 months	12 (3.1)
Over 1 year	17 (4.4)
Partners in the last month (*n* = 636)	
0–1	465 (73.1)
2	54 (8.5)
3–5	30 (4.7)
+5	9 (1.4)
Sex workers	78 (12.3)
Partners in the last year (*n* = 633)	
0–1	229 (36.2)
2	102 (16.1)
3–5	130 (20.5)
6–10	68 (10.7)
11–20	21 (3.3)
+20	10 (1.6)
Sex workers	73 (11.5)
Drug use (*n* = 288)	
No	185 (64.2)
Yes	103 (35.8)
Prior STIs (*n* = 542)	
No	412 (76.0)
Yes	130 (24.0)

* HIV: Human Immunodeficiency Virus; STI: Sexually Transmitted Infections; Control: people going to the centre for an STI control; Follow-up of contacts: people who go to the centre because they have had a risky contact.

Out of 678 analysed cases, 65.5% (*n* = 444) got an STI positive diagnosis, as opposed to 34.5% (*n* = 234) of negative cases. Table 2 shows the results of sexual habits and condom use variables in such sexual practices.

Table 2. Sexual habits and condom use.

Variables	Vaginal Intercourse		Oral (Mouth–Vagina)		Oral (Mouth–Penile)		Oral (Anus–Mouth)		Anal (Penile–Anus)	
Frequency	**H *** n (%)	**U.C. *** n (%)	**H *** n (%)	**U.C. *** n (%)	**H *** n (%)	**U.C. *** n (%)	**H *** n (%)	**U.C. *** n (%)	**H *** n (%)	**U.C. *** n (%)
Never-Sporadic	3 (0.8)	131 (36.8)	43 (22.5)	143 (98.6)	38 (15.3)	193 (86.5)	6 (100)	1 (50.0)	216 (72.7)	61 (43.0)
Frequent-Always	358 (99.2)	225 (63.2)	148 (77.5)	2 (1.4)	210 (84.7)	30 (13.5)	0 (0)	1 (50.0)	81 (27.3)	81 (57.0)
Total	361 (100)	356 (100)	191 (100)	145 (100)	248 (100)	223 (100)	6 (100)	2 (100)	297 (100)	142 (100)

* H = habit; U.C. = use of condom.

Table 3 shows the results corresponding to the association between STIs and sexual habits. STI diagnosis was significantly associated with the practice (frequent-always) of mouth–penile oral sex ($p = 0.004$), mouth–vagina oral sex ($p = 0.023$), and anal sex ($p = 0.031$). In the three cases, the effect size was low.

Table 3. Sexual habits and STI diagnosis.

Variables	STI Yes n	STI Yes %	STI No n	STI No %	p	V	OR (CI 95%)
Vaginal intercourse ($n = 361$)							
Frequent-always	224	62.6	134	37.4	n.s *	n.a *	n.a *
Never-sporadic	2	66.7	1	33.3			
Oral (mouth–penile) ($n = 248$)							
Frequent-always	145	69.0	65	31.0	0.004	0.184	2.756 (1.364–5.567)
Never-sporadic	17	44.7	21	55.3			
Oral (anus–mouth) ($n = 6$)							
Frequent-always	0	0.0	0	0.0	n.s *	n.a *	n.a *
Never-sporadic	3	50.0	3	50.0			
Oral (mouth–vagina) ($n = 191$)							
Frequent-always	94	63.5	54	36.5	0.023	0.164	2.199 (1.104–4.378)
Never-sporadic	19	44.2	24	55.8			
Anal (penile–anus) intercourse ($n = 297$)							
Frequent-always	60	74.1	21	25.9	0.031	0.125	1.854 (1.052–3.268)
Never-sporadic	131	60.6	85	39.4			

* n.s.: not significant; n.a.: not applicable.

4. Discussion

From among the results of this research, it is worth mentioning the association observed between the STI diagnosis and the frequency of oral sex (mouth–penile and mouth–vagina) and anal sex. Before describing such finding more deeply, there are aspects of the sample features that should be pointed out.

The age of the first sexual intercourse was around 17 years old, which approximates to other studies which establish the commencement of sexual intercourse between 17 and 18 years old [6,18], unlike other studies which point out slightly lower figures, reporting median ages between 15 and

16 years old [19,20]. It has been shown that a reduction in the age of the first sexual intercourse is a factor contributing to an increase in STIs [21,22].

As regards other risk indicators, subjects stated they had one sexual partner or no sexual partner during the last month and approximately 50% stated they had two or fewer partners during the last year, which results are similar to the ones observed in another study [4]. Although our study does not observe an elevated number of sexual partners, an increase in the number of sexual partners is known to be related to an increase in STIs [21,23]. With respect to the use of drugs, most of them stated they did not use or used them sporadically, which findings also coincide with another research [6]. As regards such finding, it should be noted that factors leading to a higher risk of acquiring HIV and STIs include not only early sexual intercourse and a higher number of sexual partners, but also drug use during said intercourse [22,23]. In Europe, an earlier initiation of sexual intercourse is observed, as well as an increase in the number of sexual partners, which would contribute to an increase in the incidence of STIs [21,23].

Regarding the sexual habits analysed, vaginal intercourse stands out as to frequency, followed by mouth–penis and mouth–vagina oral sex. In contrast, there were very few reported cases of mouth–anus oral sex as well as anal sex. Such data are similar to data from another study where higher percentages were found for vaginal intercourse [24], or where oral sex was more frequent than anal sex [18]. Our findings with respect to a higher frequency of vaginal sex as opposed to anal sex may be conditioned by mostly heterosexual subjects in our sample. The importance of anal intercourse as opposed to vaginal intercourse resides in the fact that the former constitutes a higher-risk sexual practice regarding transmission of HIV both for men and women [25].

In relation to condom use, results show an inconsistent use when asking subjects about the time elapsed from the last time they had sex without using a condom, which could increase contagion and transmission of STIs. It is worth mentioning that evidence is firm when relating the low use of condom to a higher risk of getting an STI [13,24,26,27]. In general, the epidemiological studies available has shown that, when condoms are used constantly and correctly, they are highly effective to prevent HIV infection and they reduce the risk of other STIs [28,29].

Upon analysing the use of condom in different sexual practices in detail, it is observed that its use in vaginal intercourse occurs always or frequently, closely followed by its use in anal sex. However, in mouth–vagina and mouth–penis oral sex, the use of condom is non-existent or sporadic. Previous studies consulted have varied results; some of them agree on the use of condom in vaginal intercourse and its non-frequent use in oral sex [13,30], and others report a lower use than the one observed in our study regarding anal sex [13]. Specifically, in the "Encuesta nacional sobre sexualidad y anticoncepción entre los jóvenes españoles" [19] ("National survey on sexuality and contraception among Spanish young people"), the main reason for not always using contraception methods, like condom, resides in the number of occasions in which oral sex is practised (59.1%). A possible explanation for the low use of condom in oral sex could be the fact that people practising it are less aware of the risk of acquiring an STI during oral sex or because they see condoms as a barrier that adversely affects sexual pleasure in the couple.

Our study shows a significant association between mouth–penis and mouth–vagina oral sex and an STI positive diagnosis. As observed, in such sexual practices, the frequency of condom use is low, which could be one of the causes explaining such association. Coincidences with another study were found, which study suggests that many young people are still unaware of the ways in which STIs may be transmitted in oral–genital contact [13] and, although some studies have demonstrated that the risk of getting HIV in oral sex with an infected partner (whether by giving or by receiving it) is much lower than the risk of getting this virus in anal or vaginal intercourse with an infected partner, it is possible that this may not be the case for other STIs [31]; hence, the importance of condom use in oral sex as well.

Furthermore, a significant association between anal sex and an STI positive diagnosis was found; although, it is worth recalling that descriptive data showed this practice is not frequent in the sample studied. Anal intercourse is the riskiest sexual practice as regards transmission of STIs such as HIV, chlamydia, or gonorrhea [32], the risk being higher when a passive role is adopted [33,34], with a probability of infection 13 times higher than for the sexually active partner, due to the fact that the lining of the rectum is thin and may allow HIV to enter the body during anal sex [25].

This study has some limitations, one of them being related to the high percentage of values lost in some variables, since not in all of them data on the same number of subjects were registered; when working with medical records, not all variables are completed. In addition, another limitation is the fact that it is a single-centre and single-province study, which, when extrapolating the results obtained, would limit their external validity. Nevertheless, the WHO has emphasized, in its latest reports, the availability of local data to improve the approach to STIs [2]. Finally, the methodological design used does not allow for the establishment of causal relations, so the associations observed must be considered as hypotheses to be supported by more complex designs and analyses.

5. Conclusions

In conclusion, the most frequent sexual practices reported in this research were vaginal and oral (mouth–vagina and mouth–penis) sex, unlike the low frequency of anal sex. Likewise, the use of condom was frequent in vaginal and anal sex, as opposed to the figures observed in oral sex. A statistically significant relationship is established between the presence of STIs and oral and anal sex.

The results obtained in this study may contribute to the design of sex education policies aimed at reducing the risk of certain sexual practices through strategies oriented to an improvement in sexual health and a minimization of exposure to and contagion of STIs. As previously mentioned, the findings of this study are in line with the WHO's proposal [2] of worldwide strategies to prevent sexually transmitted infections, where disparity in reports among different regions and countries, as well as within every region and every country, is worth mentioning because of the difficulty derived from the lack of local data to globally address STIs, in order to measure the advance towards their control.

Author Contributions: Conceptualization, R.C.S.-B. and A.M.-S.; methodology R.C.S.-B. and C.H.-M.; software, C.H.-M; investigation, R.C.S.-B., C.H.-M., and M.Á.P.-M.; data curation, R.C.S.-B., C.H.-M., and M.Á.P.-M.; writing—original draft preparation, R.C.S.-B., C.H.-M., A.M.-S., M.A.Á.-S., M.G.-L., and M.Á.P.-M.; writing—review and editing, R.C.S.-B., C.H.-M., A.M.-S., M.A.Á.-S., M.G.-L., and M.Á.P.-M.; and supervision, C.H.-M. and M.Á.P.-M. All authors have read and agreed to the published version of the manuscript.

Funding: This research received no external funding.

References

1. World Health Organization. Sexually Transmitted Infections. Available online: https://www.who.int/es/news-room/fact-sheets/detail/sexually-transmitted-infections-(stis) (accessed on 13 May 2020).
2. Organización Mundial de la Salud. Estrategia Mundial del Sector de la Salud Contra Infecciones de Transmisión Sexual Para 2016–2021. Hacia el fin de las ITS. OMS. Available online: https://www.who.int/reproductivehealth/publications/rtis/ghss-stis/es/ (accessed on 6 September 2020).
3. An, Q.; Wejnert, C.; Bernstein, K.; Paz-Bailey, G. Syphilis Screening and Diagnosis Among Men Who Have Sex With Men, 2008–2014, 20 U.S. Cities. *J. Acquir. Immune Defic. Syndr.* **2017**, *75*, S363–S369. [CrossRef] [PubMed]
4. De Munain, J.L. Epidemiología y control actual de las infecciones de transmisión sexual. Papel de las unidades de ITS. *Enferm Infecc. Microbiol. Clínica* **2019**, *37*, 45–49. [CrossRef] [PubMed]
5. Walsh-Buhi, E.; Klinkenberger, N.; McFarlane, M.; Kachur, R.; Daley, E.; Baldwin, J.; Blunt, H.D.; Hughes, S.; Wheldon, C.W.; Rietmeijer, C. Evaluating the Internet as a sexually transmitted disease risk environment for teens: Findings from the communication, health, and teens study. *Sex. Transm. Dis.* **2013**, *40*, 528–533. [CrossRef]
6. Folch, C.; Alvarez, J.L.; Casabona, J.; Brotons, M.; Castellsagué, X. Determinantes de las conductas sexuales de riesgo en jovenes de Cataluña. *Rev. Española Salud Pública* **2015**, *89*, 471–485. [CrossRef] [PubMed]

7. Yamamoto, N.; Ejima, K.; Nishiura, H. Modelling the impact of correlations between condom use and sexual contact pattern on the dynamics of sexually transmitted infections. *Theor. Biol. Med. Model.* **2018**, *15*, 6. [CrossRef] [PubMed]

8. Medina-Perucha, L.; Family, H.; Scott, J.; Chapman, S.; Dack, C. Factors Associated with Sexual Risks and Risk of STIs, HIV and Other Blood-Borne Viruses Among Women Using Heroin and Other Drugs: A Systematic Literature Review. *AIDS Behav.* **2018**, *23*, 222–251. [CrossRef]

9. Thoma, B.C.; Huebner, D.M. Parent-Adolescent Communication about Sex and Condom Use among Young Men who Have Sex with Men: An Examination of the Theory of Planned Behavior. *Ann. Behav. Med.* **2018**, *52*, 973–987. [CrossRef]

10. Buttmann-Schweiger, N.; Nielsen, A.; Munk, C.; Liaw, K.-L.; Kjær, S.K. Sexual risk taking behaviour: Prevalence and associated factors. A population-based study of 22,000 Danish men. *BMC Public Health* **2011**, *11*, 764. [CrossRef]

11. Vranic, S.M.; Aljicevic, M.; Segalo, S.; Joguncic, A. Knowledge and Attitudes of Sexually Transmitted Infections among High School Students in Sarajevo. *Acta Med. Acad.* **2019**, *48*, 147–158.

12. Habel, M.A.; Leichliter, J.S.; Dittus, P.J.; Spicknall, I.H.; Aral, S.O. Heterosexual Anal and Oral Sex in Adolescents and Adults in the United States, 2011–2015. *Sex. Transm. Dis.* **2018**, *45*, 775–782. [CrossRef]

13. Molina, A.B.; Rojas-Tejada, A.J. Uso del preservativo, número de parejas y debut sexual en jóvenes en coito vaginal, sexo oral y sexo anal. *Rev. Int.* **2018**, *16*, 8–14. [CrossRef]

14. Giménez-García, C.; Nebot-García, J.; Bisquert-Bover, M.; Elipe-Miravet, M.; Gil-Llario, M.D. Infecciones de transmisión sexual en población joven ¿qué mantiene su exposición al riesgo? *Int. J. Dev. Educ. Psychol. Rev. INFAD Psicol.* **2019**, *5*, 547–554. [CrossRef]

15. Sviben, M.; Ljubin-Sternak, S.; Meštrović, T.; Vraneš, J. Sociodemographic, Sexual Behavior, and Microbiological Profiles of Men Attending Public Health Laboratories for Testing for Sexually Transmitted Diseases. *Acta Derm. Croat. ADC* **2017**, *25*, 125–132.

16. Provenzano, S.; Santangelo, O.E.; Alagna, E.; Giordano, D.; Firenze, A. Sexual and reproductive health risk behaviours among Palermo university students: Results from an online survey. *La Clin. Ther.* **2018**, *169*, e242–e248.

17. Pérez-Morente, M.Á.; Sánchez-Ocón, M.T.; Martínez-García, E.; Martín-Salvador, A.; Hueso-Montoro, C.; García-García, I. Differences in Sexually Transmitted Infections between the Precrisis Period (2000–2007) and the Crisis Period (2008–2014) in Granada, Spain. *J. Clin. Med.* **2019**, *8*, 277. [CrossRef] [PubMed]

18. Bravo, B.N.; Segura, L.R.; Latorre, J.M.; Villafruela, J.C.E.; Honrubia, V.L.; Marchante, M.R. Hábitos, Preferencias y Satisfacción Sexual en Estudiantes Universitarios. *Revista Clínica de Medicina de Familia* **2010**, *3*, 150–157. [CrossRef]

19. Sociedad Española de Contracepción. Encuesta Nacional Sobre Sexualidad y Anticoncepción Entre los Jóvenes Españoles (16-25 Años). Available online: http://sec.es/encuesta-nacional-sobre-sexualidad-y-anticoncepcion-entre-los-jovenes-espanoles-16-25-anos (accessed on 6 September 2020).

20. Moreno, C.; Ramos, P.; Rivera, F.; Sánchez-Queija, I.; Jiménez-Iglesias, A.; García-Moya, I.; Moreno-Maldonado, C.; Paniagua, C.; Villafuerte-Díaz, A.; Ciria-Barreiro, E.; et al. La Adolescencia en España: Salud, Bienestar, Familia, Vida Académica y Social. Resultados del Estudio HBSC 2018. Ministerio de Sanidad, Consumo y Bienestar Social. Available online: https://www.mscbs.gob.es/profesionales/saludPublica/prevPromocion/promocion/saludJovenes/estudioHBSC/docs/HBSC2018/HBSC2018_ResultadosEstudio.pdf (accessed on 6 September 2020).

21. Calatrava, M.; Burgo, C.L.-D.; De Irala, J. Factores de riesgo relacionados con la salud sexual en los jóvenes europeos. *Med. Clínica* **2012**, *138*, 534–540. [CrossRef]

22. Teva, I.; Bermúdez, M.P.; Buela-Casal, G. Variables sociodemográficas y conductas de riesgo en la infección por el VIH y las enfermedades de transmisión sexual en adolescentes: España, 2007. *Rev. Española Salud Pública* **2009**, *83*, 309–320. [CrossRef]

23. Morán Arribas, M.; Rivero, A.; Fernández, E.; Poveda, T.; Caylá, J.A. Burden of HIV infection, vulnerable populations and accessbarriers to healthcare. *Enferm. Infecc. Microbiol. Clin.* **2018**, *36*, 3–9. [CrossRef]

24. Ward, J.; Wand, H.; Bryant, J.; Delaney-Thiele, D.; Worth, H.; Pitts, M.; Byron, K.; Moore, E.; Donovan, B.; Kaldor, J.M. Prevalence and Correlates of a Diagnosis of Sexually Transmitted Infection Among Young Aboriginal and Torres Strait Islander People. *Sex. Transm. Dis.* **2016**, *43*, 177–184. [CrossRef]

25. Centers for Disease Control and Prevention (CDC). Relaciones Sexuales Anales y el Riesgo de VIH. Available online: https://www.cdc.gov/hiv/spanish/risk/substanceuse.html (accessed on 5 September 2020).

26. Velo, C.; Cuéllar-Flores, I.; Sainz-Costa, T.; Navarro, M.L.; García-Navarro, C.; Fernández-McPhee, C.; Ramírez, A.; Bisbal, O.; Blázquez-Gamero, D.; Ramos-Amador, J.T.; et al. Jóvenes y VIH. Conocimiento y Conductas de Riesgo de un Grupo Residente en España. *Enferm. Infecc. Microbiol. Clínica* **2019**, *37*, 176–182. [CrossRef] [PubMed]
27. Pérez Morente, M.A.; Campos Escudero, A.; Sánchez-Ocon, M.T.; Hueso-Montoro, C. Características sociodemográficas, indicadores de riesgo y atención sanitaria en relación a infecciones de transmisión sexual en población inmigrante de Granada. *Rev. Esp. Salud. Pública* **2019**, *93*, 1–13.
28. Centers for Disease Control and Prevention (CDC). Los Condones y las ETS: Hoja Informativa Para el Personal de Salud Pública. Available online: https://www.cdc.gov/condomeffectiveness/docs/Condoms_and_STDS_spanish.pdf (accessed on 5 September 2020).
29. Teva, I.; Bermúdez, M.P.; Ramiro, M.T. Sexual satisfaction and attitudes towards the use of condoms in adolescents: Evaluation and analysis on their relationship with condom use. *Rev. Latinoam. Psicol.* **2014**, *46*, 127–136.
30. Grupo de Trabajo Sobre Tratamientos del VIH (gTt-VIH). Transmisión sexual del VIH. Guía para Entender las Pruebas de Detección y el Riesgo en las Prácticas Sexuales (2ª Edición). Available online: http://gtt-vih.org/files/active/0/GUIA_transmision_sexual_web_2Edi.pdf (accessed on 13 May 2020).
31. Centers for Disease Control and Prevention (CDC). Available online: https://www.cdc.gov/std/healthcomm/stdfact-stdriskandoralsex.htm (accessed on 13 May 2020).
32. Centers for Disease Control and Prevention (CDC). HIV Risk Reduction Tool. Available online: https://hivrisk.cdc.gov/can-i-get-or-transmit-hiv-from/?mid=anal-sex-1 (accessed on 13 May 2020).
33. Zeng, X.; Zhong, X.; Peng, B.; Zhang, Y.; Kong, C.; Liu, X.; Huang, A. Prevalence and associated risk characteristics of HIV infection based on anal sexual role among men who have sex with men: A multi-city cross-sectional study in Western China. *Int. J. Infect. Dis.* **2016**, *49*, 111–118. [CrossRef]
34. Mustanski, B.; Feinstein, B.A.; Madkins, K.; Sullivan, P.; Swann, G. Prevalence and Risk Factors for Rectal and Urethral Sexually Transmitted Infections from Self-Collected Samples among Young Men who Have Sex with Men Participating in the Keep it Up! 2.0 Randomized Controlled Trial. *Sex. Transm. Dis.* **2017**, *44*, 483–488. [CrossRef] [PubMed]

MDPI
St. Alban-Anlage 66
4052 Basel
Switzerland
Tel. +41 61 683 77 34
Fax +41 61 302 89 18
www.mdpi.com

International Journal of Environmental Research and Public Health Editorial Office
E-mail: ijerph@mdpi.com
www.mdpi.com/journal/ijerph